Tools for Critical Thinking in Biology

Tools for Critical Thinking in Biology

Stephen H. Jenkins

OXFORD

UNIVERSITY PRESS

OXFORD
UNIVERSITY PRESS

Oxford University Press is a department of the University of
Oxford. It furthers the University's objective of excellence in research,
scholarship, and education by publishing worldwide.

Oxford   New York
Auckland   Cape Town   Dar es Salaam   Hong Kong   Karachi
Kuala Lumpur   Madrid   Melbourne   Mexico City   Nairobi
New Delhi   Shanghai   Taipei   Toronto

With offices in
Argentina   Austria   Brazil   Chile   Czech Republic   France   Greece
Guatemala   Hungary   Italy   Japan   Poland   Portugal   Singapore
South Korea   Switzerland   Thailand   Turkey   Ukraine   Vietnam

Oxford is a registered trademark of Oxford University Press
in the UK and certain other countries.

Published in the United States of America by
Oxford University Press
198 Madison Avenue, New York, NY 10016

Cataloging-in-Publication data is on file at the Library of Congress
ISBN 978–0–19–998104–5

9 8 7 6 5 4 3 2
Printed in the United States of America
on acid-free paper

Contents

Preface

In 2004 I published *How Science Works: Evaluating Evidence in Biology and Medicine.* I asked a series of questions: "Can police dogs identify criminal suspects by smell?" "Why are frogs in trouble?" "Why do we age?" "How does coffee affect health?" I explained how scientists have tried to answer these questions using various kinds of evidence ranging from comparative observations and correlations to experiments to statistical analyses. My goal was to show nonscientists some of the different ways that science works, which I hoped would encourage them to engage with new stories about scientific discoveries in the media and help them to interpret these new stories.

There's been an explosion of research on these questions and many others since 2004. For example, the Web of Science lists 2,215 studies about coffee and cancer since 1970, with more than half published since 2004. Defining information broadly to include opinions as well as facts and knowledge, the amount of readily available information is much greater in 2014 than in 2004, thanks in part to the Internet.

I was employed as a teacher for 42 years, mostly at the college level. My experience in the classroom and in individual instruction taught me that the primary goal of teachers must be to help students develop their critical thinking skills. This is even more true in 2014 than it was in 1970 because of the information explosion. Do a Web search on any topic. How do you determine what search results give credible facts and what results give bogus facts? How do you determine what results give information that genuinely contributes to understanding and knowledge? How do you determine what results lead to well-reasoned opinions consistent with well-established knowledge? For any topic that might interest you, there's virtually unlimited information available at your fingertips, but the only way to judge the value of that information is by thinking about it critically.

This poses a problem, however. Critical thinking isn't easy. It's not an intuitive human skill. It takes lots of practice to overcome some natural impediments to critical thinking. We jump to conclusions before considering all the evidence, much less weighing the strengths and weaknesses of evidence for and against our conclusion. We "make evidence subservient to belief, rather than the other way around" (van Gelder 2005:46). These are just a few examples of several challenges to critical thinking.

These challenges and my retirement from active teaching in 2011 started me thinking about a sequel to my 2004 book. I wanted to make the same argument that I made in 2004, but more persuasively because I've become even more convinced of the importance of critical thinking in both individual lives and the public sphere. I wanted to make the argument more directly than I did in 2004 and I wanted to make it more accessible to nonscientists by avoiding technical language and discussion. I didn't want to avoid quantitative issues altogether, because these are fundamental in modern science, but I wanted to explain these in as friendly a way as possible. I wanted to use different examples and write about current research.

Critical thinking takes practice, so just reading this book won't make you a better critical thinker. Instead, I hope the book will introduce you to the joys of critical thinking and be a guidebook to some key tools for critical thinking. Each chapter includes a few questions that involve critical thinking and a list of resources for further practice. I urge you to use these resources and look for additional opportunities to develop your critical thinking skills. The quality of your life may depend on it, if you face a difficult decision about medical care that requires you to evaluate different treatments offered by your physician. The future of human society and the natural world certainly depends on more and better critical thinking by all of us.

Organization of the Book

This book is called *Tools for Critical Thinking in Biology* because I use mostly biological examples. Another title might have been *Tools for Critical Thinking in Life Illustrated by Biological Examples*. I use biological examples because I'm a biologist and know biological stories better than ones from other sciences. Of course critical thinking is used in all areas of scholarship—there might be parallel books called *Tools for Critical Thinking in History*, or *Tools for Critical Thinking in English Literature*. In fact, if you are a student of either of these fields, you might try applying some of the general methods illustrated here to questions in history or English literature. In some ways, critical thinking seems closer to the surface in the sciences—physical, biological, and social—than in the humanities. Therefore I hope you find this book useful whatever your primary academic (and other) interests.

Chapter 1 introduces the kinds of questions biologists ask. This may seem pretty elementary, but it actually has important implications. First, these questions are the same as those asked by journalists investigating a story—who, what, when, where, how, and why. Indeed, they are the same questions that we all ask in daily life. Second, and of more importance, these questions fall into two categories—questions about basic facts (who, what, when, and where) and questions about causation (how and why). Both types of questions are important and answering questions about causation depends on getting the basic facts right. Ultimately, however, answering questions about causation gives deeper understanding than answering questions about basic facts. As illustrated in Chapter 1 and several later chapters, understanding causation can be deeply satisfying and have great practical importance. One prerequisite for critical thinking is asking the right questions and knowing how different kinds of questions are related to each other; the main example in Chapter 1 of migration by monarch butterflies illustrates these ideas.

The basic raw material for all discovery in science, as in life itself, is observation. Critical thinking depends on learning how to assess the validity of observations. Is the ivory-billed woodpecker really extinct, or does it still exist in bottomland swamp forests of the southeastern United States? How about wolverines in California? As a more practical example, how should we evaluate eyewitness identifications of alleged perpetrators of crime? Chapter 2 uses these examples to illustrate the challenges of interpreting basic observational evidence, especially in light of our tendency to follow a form of belief preservation called confirmation bias.

All sciences, including biology, are quantitative. While Chapter 2 focuses on using individual observations as evidence, Chapter 3 builds on the fact that much science depends on evaluating multiple observations, that is, data. Therefore I introduce some basic tools for organizing and analyzing data in Chapter 3. These tools include descriptive statistics and graphs. Both of these tools are important for nonscientists to appreciate, partly because they are used commonly in the media, often without explanation. Just as one picture can be worth 1,000 words, graphs are especially important in science and for critical thinking in general, so I use lots of graphs in the book.

I've mentioned the importance of understanding causation, and Chapter 4 explains how scientists use experiments to answer questions about causation. I illustrate experimentation with one example from medicine and a very different example from evolutionary biology to reinforce the idea that a full understanding of causation involves answering both how and why questions. If you are a nonscientist, you probably won't have much opportunity to include experiments in your personal toolkit for critical thinking. However, you need to evaluate evidence from experiments in making important practical decisions. For example, the pros and cons of alternative medical treatments often depend on experimental evidence, and the same is increasingly true in comparing educational programs that might be used in your child's classroom.

Another major source of evidence in science and as a basis for public policy is comparative and correlational data, as discussed in Chapter 5. Does exposure to violence in movies, TV, and video games increase violent behavior in children? Many people reach this conclusion from reports of correlations between exposure to violence in media and subsequent violent behavior, while others are quick to point out that correlation doesn't imply causation. Correlational and comparative evidence often isn't as convincing as results of a well-designed experiment, but some questions don't lend themselves to experimental research so we have to rely on correlations and comparisons. Chapter 5 introduces some tools for evaluating the credibility of this kind of evidence, by being critical where criticism is warranted but not rejecting all such evidence out of hand.

Chapter 6 introduces modeling as another fundamental tool in science. Most models are quantitative, but Chapter 6 includes two non-quantitative examples that are important for very different reasons. The other two examples in Chapter 6 are quantitative, in keeping with the general theme that biology depends on math, but I've tried to develop them in a transparent way. I hope that the topics of these examples are interesting enough that when you finish reading about them you'll think back and say to yourself that you just read some math, almost without realizing it. One of the examples involves choices about childhood vaccination; if or when you become a parent, you'll want to apply your critical thinking skills to these choices, and a brief introduction to some math related to this issue will be helpful.

Chapters 1 to 6 introduce several of the fundamental tools of critical thinking in science. If you aren't a scientist, you probably won't have an opportunity to apply these tools directly—by doing an experiment or making a model. However, you'll make decisions throughout life that depend on scientific knowledge, so it's important to know how to evaluate scientific evidence. These decisions may be strictly personal or may have consequences for society at large, as illustrated in Chapter 6 by decisions about childhood vaccination. Chapters 7 and 8 broaden the stage to consider the complexity of causation. How can we understand phenomena that have multiple interacting causes? How should we evaluate multiple lines of evidence that may be in partial conflict? Many decisions, both personal and social, raise these issues. One mark of a critical thinker is accepting the complexity of our world and making nuanced decisions in light of incomplete and uncertain evidence.

Chapter 8 and the associated Appendix 6 also introduce a specific tool to aid critical thinking that you can use in your own process of making decisions. This tool is argument mapping, which is a way to visualize alternative arguments and the strengths and weaknesses of different lines of evidence bearing on these arguments. Argument mapping helps identify hidden assumptions and avoid confirmation bias and other breakdowns in logic.

Chapter 9 explains the social process of science, including peer review, which is one basis for the effectiveness of science. I use two recent examples of scientific controversies to illustrate the strengths and pitfalls of peer review. As a nonscientist who thinks critically, you could emulate this social process of science by exposing your ideas to those who might disagree with you rather than simply sharing them with like-minded friends.

Finally, Chapter 10 outlines six key principles of critical thinking that can be used in science and life in general. James Lett explained these principles in "A Field Guide to Critical Thinking" using an evaluation of claims for paranormal phenomena such as extrasensory perception to illustrate the principles. I apply Lett's principles to global climate change because this is the most important challenge of our time. I hope you will find ways to use these principles in personal decisions and that you will embrace the challenge of thinking critically and acting responsibly about climate change.

Acknowledgments

I thank several anonymous reviewers of my proposal for this book for their encouragement and their advice about content and style. One of these reviewers introduced me to Lett's principles of critical thinking represented by the acronym FiLCHeRS, which I apply to climate change in Chapter 10.

I thank Tom Albright, Amy Altick, J. H. Atkinson, Marilyn Banta, John Basey, Kathleen Boardman, Elissa Cameron, Sandra Carroll, Deborah Davis, Ned Dochtermann, Cynthia Downs, Richard Duncan, Chris Elphick, Jim Estes, Fred Grinnell, Ian Hardy, Martha Hildreth, Mary Hinton, Chris Jenkins, Wally Jenkins, Nancy Krieger, Bill Longland, Grant Mastick, Howard Mielke, Katie Moriarty, Tom Newman, Jon Nickles, David Nieman, Stefanie Scoppettone, Mel Sunquist, Eric Turkheimer, Joe Veech, and Michael Weisberg for comments on drafts of one or more chapters.

I thank Deborah Davis, Chris Elphick, Martha Hildreth, Anne Leonard, Tom Nickles, Eric Turkheimer, and David Zeh for key references.

I thank Lee Dyer, Matt Forister, Matt Jenkins, and Tom Nickles for valuable discussion.

The following individuals provided photographs and artwork for which I am most appreciative: Jim Gagnon (cover and Plate 2d), Katie Moriarty (Figure 2.1), Greg Hodges (Figure 6.8a), Christopher Wilmers (Box Figure 8.1), Andrew Davidhazy (Figure A3.1), Richard Crossley (Plate 1c), Jakob Vinther (Plate 3), Medhi Yokubol (Plate 6a), Cathrine Macort (Plate 7), Howard Mielke (Plate 9), Michael Weisberg (Plate 11), Dave Cowles (Plate 13a), Derek Lohuis (Plate 14a), Chuck Kopczak (Plate 14b), Ross Haley (Plate 16a), Matt Lavin (Plate 16c), and Ned Dochtermann (Plate 16d). I also thank Tanya Wolfson and Elissa Cameron for providing the data plotted in Figures 4.4 and 4.8, respectively; Judy Silberg for providing the twin data used in Chapter 7; and David Duffy and Eric Turkheimer for additional data. Jim Estes provided data used in Figures 8.4 and 8.8

and an unpublished manuscript that he encouraged me to discuss in Chapter 8 and that formed the basis for the argument map illustrated in Appendix 6. Georgia Grundy, Maggie Ressel, and Amy Shannon, librarians at the University of Nevada, Reno, helped me obtain important materials and navigate the process of getting permission to use those under copyright.

I thank my editor at Oxford University Press, Jeremy Lewis, for his encouragement and support.

My wife Kathie Jenkins was the first reader of every paragraph of this book. She is not a scientist, so told me when my writing was too technical or too murky for general readers. She combines good critical thinking with wonderful artistic sensibility. I thank her for her patience while I was absorbed in this project and for her inspiration through the 45 years of our partnership in life, and I dedicate this book to her.

Tools for Critical Thinking in Biology

Discovery and Causation

The Great Pyramid of Giza (Plate 1a) is one of the original Seven Wonders of the World, a list of magnificent engineering projects built around the eastern Mediterranean in ancient times. Indeed, this structure is the only original wonder that survives today. Modern lists include the Taj Mahal and the Great Wall of China, with the Great Pyramid of Giza retained as an eighth, honorary member of a list proposed by the New7Wonders Foundation in 2007. There are now lists of seven natural wonders, seven underwater wonders, seven industrial wonders, seven wonders of Portugal, and more. What about biological wonders? The BBC asked various scientists to discuss their selections, and the evolutionary biologist Richard Dawkins included the spider's web, the bat's ear, and the pianist's fingers. In truth, it would be difficult to limit such a list to seven phenomena, but animal migration would certainly be one of my choices.

Several examples illustrate why migration belongs on a short list of biological wonders. Arctic terns (Plate 1b) breed during summer along arctic coasts of North America and Eurasia. When fall comes in the northern hemisphere, these small birds of about 100 grams migrate as far south as the coast of Antarctica. This migration of about 35,000 kilometers was considered the longest migration of any animal, but a group of researchers using new technology recently showed that total distance traveled was even greater. Carsten Egevang and colleagues attached miniature GPS units to 11 birds nesting in Greenland or Iceland and tracked them to the Weddell Sea off Antarctica and back. Average roundtrip distance was about 71,000 kilometers. Since these birds can live for at least 30 years, an individual might make the equivalent of three roundtrips to the moon in its lifetime (see Appendix 1 for a guide to translating from the metric units used here to English units that may be more familiar).

Snowbirds are retired people who migrate from their homes in cities like Chicago or New York to warmer places like Florida or Arizona in fall, returning home the following spring. Many migrating birds show similar site fidelity when they return to their breeding grounds. Harris's sparrows (Plate 1c) breed near the boundary between boreal forest and tundra in far northern Canada and spend the winter in the central Great Plains of the United States. Because of the remote location of the breeding grounds, very little was known about their breeding behavior until a graduate student at the University of Kansas named Christopher Norment began studying them in the 1980s. By attaching numbered tags to the legs of birds, Norment could discover which ones returned to his study site in the Northwest Territories in subsequent years. About 34% of all birds did so, but 86% of those that were successful in one year returned the next year. More remarkably, males often returned to the same territories they had held the previous year, although females never did so. Instead, they bred with a different male on a different territory at the study site. If snowbirds were like Harris's sparrows, men would return to the houses that they owned in Chicago or New York each spring, while their wives would return to different houses in the same neighborhood. It's easy to understand how snowbirds find their way home in the spring, but how do sparrows do it without any of the tools that we use, from maps and road signs to smartphones? The remarkable site fidelity of migrating animals is another reason why migration is a biological wonder.

Birds and retirees aren't the only organisms that migrate seasonally. One of the most dramatic examples of mammalian migration occurs in the Serengeti of East Africa where wildebeest, zebras, and gazelles (Plate 1d) move about 3,000 kilometers in a large circle around the landscape each year. The seasons in this tropical environment aren't hot and cold, but wet and dry, and these hoofed mammals migrate in response to rainfall patterns that determine availability of the grass that they eat. They don't travel nearly as far as arctic terns or even Harris's sparrows, but they do travel in huge herds of up to 1.5 million animals. Such mass movement adds another reason to consider migration as one of the biological wonders of the world.

I've introduced you to arctic terns that travel from one end of the Earth to the other and back each year, to male Harris's sparrows that reestablish the same breeding territory this year as last year after spending the intervening months thousands of kilometers away, and to wildebeest that follow the rains around the Serengeti with more than a million of their fellows. But the most astonishing story of migration is that of monarch butterflies (Plate 2a). While monarchs are larger than most other butterflies, they are tiny compared to migrating birds. Yet some monarchs migrate 4,000 kilometers from southern Canada to the mountains of central Mexico. One reason for the fascination of the monarch story is the decades-long search for the wintering area in Mexico. Researchers couldn't attach GPS units to monarchs as they did with arctic terns; they couldn't even use numbered leg bands since monarchs weigh only about 1% of what Harris's sparrows weigh. After much trial and error, Fred Urquhart devised a marking system that enabled him to gradually track monarchs from his home near Toronto to points farther and farther south. The search for the wintering area illustrates the discovery process that is the foundation of science. This process is fundamentally the same whether it involves miniature GPS units transmitting their signals to satellites or small paper disks with Urquhart's phone number attached to the wings of butterflies. By telling the monarch story, I hope to show you that discovery in science doesn't depend on modern, high-tech tools, although these certainly can be very helpful in some cases. But the monarch story will reveal the core features of discovery because of the simplicity of the methods involved. Besides, it's quite an adventure.

Journalism students learn to build stories around six questions: who, what, where, when, how, and why? Doing science is not that different, and finding the wintering area of monarch butterflies meant answering the *where* question. Questions about *who* migrates, *what* happens, and *when* migration occurs are similar to questions about *where* butterflies end up—all of these involve discovery of basic facts. In biology, the *how* and *why* questions are quite different, however. These are questions about what causes migration. *How* do animals know when to begin migrating in the fall? *How* do they find their way from eastern Canada to central Mexico? *Why* do they migrate, since other species of butterflies spend their whole lives in Canada, overwintering as eggs or even adults in a state of suspended animation? Said a bit differently, *why* is migration an advantage to monarch butterflies but apparently not to other species that remain in the north all year?

While discovery of basic facts is the foundation of science, all scientists are ultimately interested in going beyond discovery to understand causation. Migration by monarch butterflies is on my short list of biological wonders of the world not just because it's difficult to imagine these tiny creatures traveling such long distances from all over eastern North America to a tiny target in the highlands of central Mexico, but because researchers have used clever observations and experiments to understand the causes of migration. In short, this story of migration by monarch butterflies encapsulates the complete scientific process, from discovery to understanding, with knowledge gained at each step of the process raising new questions for future research.

Discovering the Winter Home of Monarch Butterflies

Like many professional biologists, Fred Urquhart became interested in natural history as a child. Some children are fascinated by dinosaurs, others by snakes or birds. Starting at about age 5, Urquhart focused on insects. He eventually became a professor of zoology at the University of Toronto and studied monarch butterflies for most of his long life (1911–2002).

Monarch butterflies are familiar to many people because of their large size and bright coloration, their broad distribution in North America, and their abundance in diverse habitats including cities, towns, and agricultural areas. Adult females lay tiny eggs on milkweed plants. These eggs hatch into caterpillars (Plate 2b) that live for about two weeks. They feed voraciously on milkweed leaves and grow rapidly to about 5 centimeters in length and three times the weight of an adult butterfly. Then each caterpillar attaches to some object in the environment and forms a pupa or chrysalis around itself

(Plate 2c). Within the pupa, the caterpillar undergoes an amazing transformation into an adult butterfly in about 10 days. In spring and summer, successive generations of adults live for two to six weeks, but the last generation to emerge from pupae in the fall doesn't reproduce and die within a few weeks but rather lives several months until the following spring, when females finally lay eggs to initiate the first generation of the next year. These long-lived monarchs migrate to traditional wintering sites in fall and fly partway back to where they originated the next spring. Unlike birds and mammals, each monarch that migrates makes the trip only once. This means that the locations of wintering sites aren't passed down from generation to generation by learning, even though each migrating generation of monarchs returns to the same sites used in previous years. How do they do it? Before tackling this question, we need to answer a more basic one—*where* do they go?

Two things have been known about monarch butterflies for a long time:

- they disappear from most of North America in fall, flying south in huge numbers, and
- large aggregations form in winter in wooded areas along the Pacific coast of California.

These winter aggregations along the Pacific coast imply that some butterflies migrate to California, but do all monarchs do this or only those from nearby? The only way to find out was to tag butterflies in fall and try to find them at the wintering sites. Fred Urquhart started experimenting with tagging methods in 1935 but had no success for many years. Tags needed to be light enough to be attached to the wings of the butterflies but large enough to hold individual identification numbers and instructions for reporting a find to Urquhart at the University of Toronto. They also had to resist falling off in rainy or windy weather. After serving as a weatherman for the Royal Canadian Air Force in World War II, Urquhart returned to his tagging experiments in the 1950s. He devised a method of attaching tags that gave some success, with tagged specimens sent to him from as far away as 1,300 kilometers. But he was still using water-soluble glue, which he tested further on a trip to California in 1954. He and his wife Norah tagged more than 1,000 monarchs in a park near Monterey. During the following night it rained steadily, and there was thick fog. By morning, "the monarch butterflies were still in large clusters on the branches of the pines, but our tags, like scattered confetti, littered the ground" (Urquhart 1987:124).

Urquhart was nothing if not persistent. Back in Toronto, he asked chemists at the university and local merchants for suggestions, one of which was to use a label like that used to mark prices on glass merchandise. These adhere to smooth surfaces, so Urquhart thought they might work after scales were removed from part of a butterfly wing (Plate 2d). Indeed they did, adding only about 2.5% to the weight of the monarch. They could be mass produced, and an experienced field worker could tag one butterfly in about 8 seconds. In the following years, thousands of volunteers joined the Urquharts in tagging monarchs, and hundreds of thousands of butterflies were tagged.

Urquhart initially wondered if all monarch butterflies migrated to wintering sites along the Pacific coast south of San Francisco, where winter aggregations had been observed for many years. When he got reports of sightings of tagged individuals, he plotted the straight-line distances between the tagging locations and the sightings of these specimens on a map of North America. As more and more of these "release-recapture lines" filled his map, Urquhart realized that the overwintering populations in California originated west of the Rocky Mountains but that butterflies from eastern North America had to be going elsewhere. Release-recapture lines of monarchs tagged in eastern Canada and the northeastern United States were directed toward the southwest; those from the upper Midwest pointed more directly to the south (Figure 1.1). Although many of these lines headed toward the Gulf of Mexico, no wintering aggregations could be found along the Gulf Coast or in Texas.

In 1972, the Urquharts started looking for volunteers farther south in Mexico. Ken Brugger from Mexico City started searching in early 1973. Finally, on January 9, 1975,

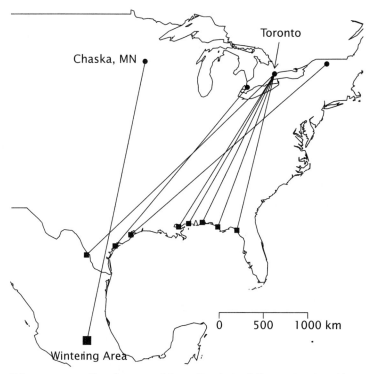

FIGURE 1.1 Release-recapture lines for monarch butterflies during fall migration plotted by Fred Urquhart. Circles at north ends of lines show release locations of marked butterflies, squares at south ends show recapture locations. Most butterflies from the northeastern United States and Canada fly along the Gulf Coast into Mexico, but some fly across the Gulf of Mexico; these movements aren't shown. The larger square in Mexico shows the location of the wintering area for monarchs east of the Rocky Mountains, part of which is now preserved as the Monarch Butterfly Biosphere Reserve. The line ending at the wintering area represents a butterfly tagged in Chaska, Minnesota, and recovered by Fred Urquhart on his first visit to the wintering area.

Brugger called the Urquharts to say that he and his wife Cathy had found millions of monarchs resting in a grove of evergreen trees at almost 10,000 feet elevation in central Mexico. This was 40 years after Fred Urquhart had started tagging monarchs and 20 years after he developed a reliable method. The Urquharts visited this site with the Bruggers in early 1976. Here is how Urquhart describes the thrill of discovery:

> Then we saw them. Masses of butterflies—everywhere! In the quietness of semidormancy, they festooned the tree branches, they enveloped the oyamel trunks, they carpeted the ground in their tremulous legions. Other multitudes . . . filled the air with their sun-shot wings. . . .
>
> One of our guides, Juan Sanchez, added up the tall trees. He estimated more than 1,000, every one garbed in monarchs!
>
> While we stared in wonder, a pine branch three inches thick broke under its burden of languid butterflies and crashed to earth, spilling its living cargo. I stooped to examine the mass of dislodged monarchs. There, to my amazement, was one bearing a white tag!
>
> By incredible chance I had stumbled on a butterfly tagged by one Jim Gilbert, far away in Chaska, Minnesota. Later Mr. Gilbert sent me a photograph of the very field of goldenrod where he had marked this frail but tireless migrant.
>
> (URQUHART 1976:166, 173)

How did the Bruggers and Urquharts discover the winter home of millions of monarch butterflies in central Mexico? The methods couldn't have been more basic—tagging and releasing specimens, recording their recapture locations, and plotting the data on large maps. Making this discovery took time, a willingness to follow the trail wherever it led, and the help of thousands of volunteers who tagged butterflies and, especially, reported captures of tagged specimens to Urquhart's lab in Toronto. Most important, making the discovery required long-term dedication to an elusive goal. Fred Urquhart spent more than half of his professional career on the search, even though other entomologists questioned whether an overwintering site for the eastern population even existed.

What Did Dinosaurs Look Like?

Other basic discoveries in science involve more sophisticated methods. For example, how can we learn what dinosaurs looked like? There are plenty of fossils of dinosaurs, but these are mostly fossilized bones, and the last dinosaurs lived about 65 million years ago. Natural history museums exhibit reconstructions of dinosaurs, and the presumed accuracy of these reconstructions has increased as paleontologists have collected more samples (articulated skeletons instead of scattered bones) and learned how to better interpret their samples. You might imagine that it would be relatively easy to determine the size, shape, and even posture of a long-extinct dinosaur from fossilized bones but impossible to say what it *really* looked like—its color, for example. Notwithstanding the 1993 movie *Jurassic Park*, you would have been right—until 2010, when a group of researchers from the United States and China reported on their studies of a chicken-sized dinosaur called *Anchiornis huxleyi*. They found a very well-preserved specimen of this species in a 150-million-year-old rock formation in northeastern China. The specimen included feathers from all parts of the body except the tail, adding to evidence that dinosaurs didn't really go extinct 65 million years ago but persist to this day as birds.

Anchiornis huxleyi was not the first discovery of a fossil dinosaur with feathers, but Quanguo Li and colleagues were able to take the next step and figure out the color patterns of these feathers. Coloration of feathers is determined by structures inside the cells of the feathers called melanosomes, which contain the pigment melanin. Li's group observed the fossilized remains of melanosomes in their specimen using scanning electron microscopy. Although most of the melanin is degraded after 150 million years, the sizes, shapes, and relative positions of the melanosomes determine the colors of feathers. Using statistical comparisons of these features of melanosomes in their fossil to sizes, shapes, and positions of melanosomes in feathers of living birds, the researchers were able to make a colored picture of *Anchiornis* (Plate 3). No one knows for sure what the color pattern of this handsome animal meant for its daily life, but, like birds, it may have helped in attracting mates or deterring predators.

These are two examples of discovery in biology. One is several decades old, the other is more recent. I was able to outline the methods used to discover the winter home of monarch butterflies in a few paragraphs, but it would take many pages to describe the scanning electron microscopy and advanced statistical methods used to learn what a small dinosaur looked like. However, both of these examples illustrate three common features of all scientific research, whether the goal is to discover basic facts or to understand causation. In all cases, the fundamental motivation of a scientist is *curiosity*. Did you ever wonder whether dinosaurs had bright colors or any color patterns at all? Maybe not, but a few scientists got intensely curious about this question at a time when enough background knowledge had accumulated that they thought they might be able to answer it. Jakob Vinther was a graduate student at Yale and a member of the team that reported on *Anchiornis*. He told the science journalist Carl Zimmer "We had a dream: to put colors on a dinosaur" (Zimmer 2010).

A second feature of scientific research that is illustrated by both the monarch butterfly and feathered dinosaur examples is *creativity*. Fred Urquhart tried many different tagging methods for monarchs before he got one to work. The researchers who studied *Anchiornis* developed a convincing story by combining field work in China with detailed microscope work in the lab at Yale and sophisticated statistical analyses. To do this successfully, they had to imagine that these very different methods could be woven together to realize their dream.

Finally, like all discoveries, finding the winter home of monarchs and putting colors on a dinosaur led to more questions. In these cases and others, further questions are often about causation. What causes monarchs to migrate in fall? What causes them to fly toward central Mexico regardless of whether they start in Maine or Kansas? What caused the evolution of a reddish plume of feathers on the head of a 150-million-year-old dinosaur? Biologists ask two kinds of questions about causation, reflecting different perspectives on explaining nature. The first type of question is about how things work. Think about what causes monarchs to migrate in fall. How can we be more specific in stating this question? We might ask, what environmental changes happen in fall to trigger migration? Or what physiological changes happen to monarchs that cause them to stop mating and laying eggs and start flying south instead? Or, even better, we might ask both of these questions, since changes in the external environment may be linked to changes in the internal physiological state of monarchs. Questions like these are about external and internal mechanisms that directly affect individual organisms. This is the realm of *proximate—"nearby"—causation.*

Evolution provides the framework for another set of questions that biologists ask about causation. These are questions about longer-term processes that occur over many generations. For example, why did the specimen of *Anchiornis* studied by Li and colleagues have a reddish plume of feathers on its head? In many species of modern birds, males are more colorful than females; the peacock is an extreme example of such *sexual dimorphism.* One reason this can occur is because females choose more colorful males as mates. Suppose the *Anchiornis* specimen was a male and the social behavior of this dinosaur was like that of modern birds. If females preferred to mate with more colorful males and if genetic differences contributed to color differences, then a reddish plume could evolve in males over many generations because males with bigger, brighter plumes would have more offspring. This process is called *sexual selection* because different traits are favored in males and females. It's difficult to imagine how this hypothesis could be directly tested, although it might be possible to discover if male and female *Anchiornis* differed in appearance. In other cases, it may be easier to get direct evidence of evolutionary causation. This is the realm of *ultimate causation*—not because evolutionary aspects of causation are more important than proximate aspects but because ultimate causation acts at a "long distance," that is, over many generations.

Proximate Causes of Migration by Monarch Butterflies

During summer, monarch butterflies live a few weeks as eggs, caterpillars, pupae, and then adults. Males and females mate, females lay the next generation of eggs, and then both sexes of adults die. As summer gives way to fall, the adults that will make the long trip to Mexico differ from their parents in several ways. They don't mate right away but instead eat and store fat. They live much longer than their predecessors. And they start flying south or southwest toward the wintering area hundreds or thousands of kilometers away. They finally mate and lay eggs the following spring, when they return north. What causes these changes and what enables monarchs to find the traditional wintering sites, which are very small targets a long way away that the butterflies have never seen?

The proximate factors that induce migration in monarchs include both external environmental factors and internal physiological factors. Several things happen in northern environments in late summer—days get shorter, temperatures drop, and plants become senescent and eventually die. Liz Goehring and Karen Oberhauser at the University of Minnesota tested the effects of these factors on the delay in reproduction from late summer until after migration the following spring. They used observations of wild butterflies and experiments in captivity to study this process of *reproductive diapause*. For example, they used light timers in the lab to expose monarch larvae to long day length (16 hours of light and 8 hours of darkness, 16:8) or decreasing day lengths (starting at 15:9 and gradually changing to 13:11 over about one month). The latter condition represented how day length (also called photoperiod) changes in Minnesota from late July until late August, when butterflies in natural populations start showing signs of reproductive diapause—females no longer produce mature eggs and males have smaller ejaculatory ducts.

No larvae exposed to long days entered reproductive diapause as adults, but about 50% exposed to shorter and shorter days during larval development were in reproductive diapause as adults. When larvae were given old milkweed plants to eat, they were more likely to be in diapause as adults than when they were given young plants. Finally, when the temperatures of treatment chambers were reduced at night while larvae grew, they were more likely to develop into diapausing adults than if temperatures were the same day and night.

It seems that all three of the most obvious environmental factors that change from summer into fall—photoperiod, temperature, and food quality—influence reproductive diapause of monarch butterflies. In order to migrate, the fall generation of monarchs must have a longer lifespan than summer generations, and reproductive diapause is a prerequisite for extended life. But important questions remain about the proximate environmental cause of migration. Although changing photoperiod had the clearest effects on reproductive diapause in these experiments in Minnesota, no more than 56% of monarchs entered diapause in any of the treatments used by Goehring and Oberhauser. By varying treatments further, researchers might discover conditions that lead to nearly 100% diapause. By repeating experiments farther south, researchers could see if cues triggering migration are the same in different environments. For instance, day length doesn't change as much from summer to fall in Alabama as in Minnesota, so changing day length might not be as reliable a signal to prepare for migration in Alabama. This example illustrates a common theme that will reappear throughout the book—answers to questions about causation are often incomplete and usually lead to further questions.

Proximate causes of behavior such as migration and breeding involve both external environmental factors and internal physiological factors. For example, songbirds in temperate environments breed in the spring. Increasing day length causes increasing levels of testosterone in males, which in turn triggers sperm production and behaviors such as territoriality that are necessary to attract a mate. Parallel changes occur in females.

For monarchs, the physiological cause of increased lifespan and delayed reproduction in the generation that migrates to central Mexico is similar to that for birds. Insects, like vertebrates, use hormones to regulate physiology and behavior, although the hormones are different than those such as estrogen and testosterone that are familiar to you as a vertebrate. One of the most widespread and important hormones in insects is called *juvenile hormone*, produced by a small gland behind the brain. Levels of juvenile hormone in the migrating generation of monarchs are only 1% of levels in the summer generations that live fast, breed profusely, and die young. Juvenile hormone influences development of reproductive structures in summer monarchs so that adult butterflies can mate and lay eggs. Summer generations invest much of their energy in reproduction rather than growth and accumulation of fat reserves, so these generations don't live very long. In fall, with reduced juvenile hormone, monarchs don't develop reproductive structures or devote energy to reproduction, so lifespan is much longer.

William Herman and Marc Tatar tested this scenario with two kinds of experiments. By removing the glands that produce juvenile hormone from summer monarchs, they caused these butterflies to stop reproducing and live longer than control butterflies that were anesthetized and operated on without having glands removed. By applying a small concentration of juvenile hormone dissolved in acetone to the abdomens of the migrating generation of monarchs, they caused these butterflies to have shorter lives than controls that just got the acetone treatment. Butterflies that migrate eventually mate and lay eggs after returning north the next spring. This may be triggered by increasing day length causing juvenile hormone levels to increase, much like the role of photoperiod and sex hormones in the reproduction of songbirds in spring. In both cases, an environmental factor that changes with the seasons has profound influences on the biology of animals, although the nature of this factor isn't as certain for butterflies as for songbirds. Both cases also involve hormonal changes that influence the physiology and behavior of the animals.

Proximate Causes and Navigation

Once monarchs start migrating in fall, how do they find their way to traditional wintering sites in central Mexico or along the coast of California? The mechanism of navigation is another fascinating aspect of the proximate cause of migration, especially because migrating monarchs make the trip only once and the wintering area is hundreds to thousands of kilometers away.

Researchers have studied navigation by migrating birds for many years, but research on monarchs has a shorter history. Many birds use the sun as a compass, just as you might if you were lost in the wilderness. The Owens Valley is a long, broad valley in eastern California between the White Mountains to the east and the Sierra Nevada to the west.

The Owens Valley is famous for providing much of the water used in Los Angeles and for containing the trailhead of the main climbing route to Mount Whitney in the Sierra, the tallest mountain in North America outside Alaska. Suppose you want to explore the less well-known White Mountains east of Owens Valley. You start hiking into the mountains but get lost. There are no streams nearby, so you can't follow a stream downhill and out of the mountains. In fact, you are in a small bowl that you have to climb out of before starting your descent to Owens Valley. Let's further stipulate that you don't have a cell phone (reception may be spotty in this remote area anyway) and the battery is dead in your GPS unit. As long as you know what time it is, you can use the sun as a compass—hiking away from the direction of the sun in the morning and toward it in the afternoon should take you west toward Owens Valley, where you can reward yourself with a bearclaw at Schat's Bakkerÿ in Bishop.

Not surprisingly, using the sun for directional cues is called sun-compass navigation. Birds do it in migration, and bees do it to lead colony mates to new food sources, although, with apologies to Cole Porter, I'm not so sure that educated fleas do it. How about monarch butterflies? They only fly during the day, so they could use the sun as a compass to navigate to wintering areas. But their brains are minuscule compared to those of birds and their navigation task involves much greater distances than that of foraging honeybees. How could we test the ability of monarch butterflies to use sun-compass navigation?

Since using the sun as an accurate compass depends on knowing what time it is, bird researchers designed experiments to test for this ability in their subjects by tricking them about the time. Suppose you don't know where you are but you do know that you have to travel south to get to where you are going. You plan to start at dawn. Since the sun rises in the east, you'll begin traveling 90 degrees clockwise from the position of the sun. However, what if you think it's noon when you start, although it's really shortly after dawn? If it *were* noon, the sun would be in the south (assuming you are in the northern hemisphere), so you'll travel directly toward the sun. But it's actually dawn, so you'll travel east instead.

Tests of sun-compass navigation by messing with the internal clocks of animals are called clock-shifting experiments. Researchers keep birds or butterflies in the lab with artificial lights and use timers to gradually change the on and off times. After a week or two, "dawn" in the lab might happen six hours earlier or six hours later than dawn outdoors. Subjects are then tested outdoors to see if their flight directions are shifted. One nice feature of this design is that researchers can make specific predictions about the direction and magnitude of the shift in flight direction, just like in the human example in the last paragraph.

Sandra Perez and colleagues did the first test of sun-compass navigation by monarch butterflies in September 1996 in Kansas. Kansas is due north of the wintering area in Mexico, and most wild monarchs were flying south at this time (Figure 1.2). The researchers kept two groups of butterflies in the lab. For the control group, the light-dark cycle was the same as outdoors. For the clock-shifted group, simulated sunrise and sunset were delayed by six hours. The researchers released butterflies outdoors and recorded their flight paths for up to five minutes. Most butterflies in the control group flew south, like wild monarchs. But clock-shifted butterflies acted as if the real sun was on the same cycle as lights in the lab. For example, think about one of these butterflies released at noon. True dawn was six hours ago, but for this butterfly lights just came on in the lab before it was released outdoors. Since the sun rises in the east and the direction of migration is south, if this butterfly assumes that it's dawn, it will head off 90° clockwise from the position of the sun. But it's actually noon outdoors, so the clock-shifted butterfly heads west instead of south. Indeed, the majority of the clock-shifted butterflies followed this prediction and flew west (Figure 1.2).

Perez and her colleagues tracked 272 butterflies in this experiment by chasing them on foot and recording their flight paths with a hand-held compass. Mouritsen and Frost used considerably more sophisticated methods five years later. They built a large, outdoor flight simulator in which they tethered individual butterflies in such a way that they could fly in any direction, which they did for several hours. The researchers used a miniature wire attached to a computer to record continuously the flight paths of the butterflies. Mouritsen and Frost did their experiments at Queen's University in Kingston, Ontario, where wild monarchs fly southwest during migration. Their control subjects, kept in the lab under the same light-dark cycle as outdoors until being tested in the flight simula-tor, flew southwest, as expected. Besides using this automated apparatus, Mouritsen and Frost extended the studies of Perez and colleagues in two ways. Like the Kansas research-ers, they delayed the light-dark cycle of one group of monarchs by six hours, but they

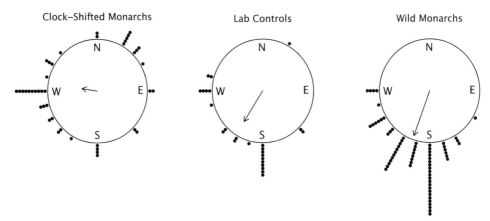

FIGURE 1.2 Initial flight directions of monarch butterflies studied by Perez, Taylor, and Jander in Kansas. Each point shows the direction of an individual butterfly, while arrows show the average direction of but-terflies in each of three groups: monarchs held in the lab with the light-dark cycle delayed by six hours compared to outdoors (clock-shifted monarchs), monarchs held in the lab on the same light-dark cycle as outdoors (lab controls), and wild monarchs. The lengths of the arrows are proportional to the consistency of flight directions of the monarchs in each group; clock-shifted monarchs were more variable than lab controls and wild monarchs. Each point represents one butterfly except for wild monarchs, where each point represents three butterflies.

advanced the cycle by six hours for another group. They also used a translucent plastic cover on the flight simulator to simulate overcast conditions for another set of monarchs. This allowed diffuse light into the simulator but prevented the monarchs from seeing the sun. How do you interpret their results (Figure 1.3)? Do the clock-shifted monarchs behave as expected if they use sun-compass navigation? Are these results consistent with those of Perez and her colleagues? Do monarchs have a backup method of navigation when they can't use the sun?

The evidence seems pretty strong that monarch butterflies, like birds, use sun-compass navigation in directed migration. However, as you'll see repeatedly in this book, the story is incomplete. Some birds are able to migrate at night by sensing the Earth's magnetic field and using that information to orient themselves in the proper direction. Monarchs fly on overcast days during migration, when a simple sun compass doesn't work because they can't see the sun. They might use magnetic information under these conditions or be able to detect the planes of polarization of sunlight through cloud cover, but evidence about these possibilities is inconclusive. It's also unclear how monarchs navigate the last part of their journey to locate roosting sites in high elevation forests in Mexico that are only a few square kilometers in area. Do they home in on these trees by smell? Do social factors play a role, with late-arriving monarchs attracted to sites where early arrivals have already settled in for the winter?

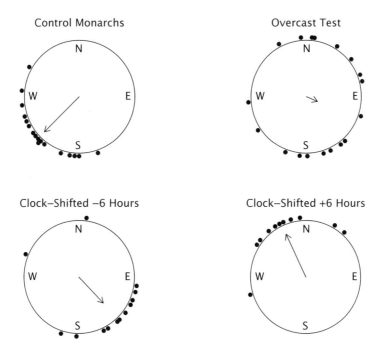

FIGURE 1.3 Average flight directions of monarch butterflies studied by Mouritsen and Frost. Each point represents the direction of an individual butterfly, while arrows show the average direction of butterflies in four groups described in the text. The lengths of the arrows are proportional to the consistency of flight directions of monarchs in each group.

How Do Monarchs Tell Time?

Studies of sun-compass navigation raise another question—what is the nature and location of the clock that animals use to tell time? In fact, monarch butterflies have become a model system for physiological and molecular studies of biological clocks. From mammals to insects, biological clocks involve a feedback process of synthesis and degradation of proteins in key cells, usually in the brain. This process occurs in a cycle of "about a day," so is called a *circadian rhythm*. In one animal, the cycle might happen in 23.5 hours; in another, it might take 26 hours. If this was all that was happening, it wouldn't take many days before these intrinsic cycles of physiology and behavior of animals were mismatched to cycles of light and dark in nature caused by sunrise and sunset, since these last exactly 24 hours. So the intrinsic circadian rhythms of animals are reset daily by detection of sunlight. The activity patterns of beavers in summer and winter illustrate this process. In summer, they are active mostly at night—swimming in their ponds, coming ashore to harvest plants for food, and building and maintaining their dams. They live in lodges in the middle of these ponds. In fall, they cache a pile of tree branches in the water near their lodge. Ponds freeze over in winter, and beavers swim out under the ice to cut branches from the cache. They bring these branches back into the lodge where they have a platform above water level that they use for sleeping and eating (see Chapter 9 for more about beavers). But they can't see the sun rise or set from inside their lodges, and researchers have used cameras mounted in lodges to show that activity cycles gradually shift through the winter until the beavers may be active half the day and half the night instead of just at night. By spring, their circadian rhythms again become matched to the daily light-dark cycle, and the beavers are active at night and rest in their lodges during the day.

Amazingly enough, monarch butterflies have three molecular clocks. One is in the brain and responds to light detected by the eyes. The other clocks are in the antennae, which respond directly to light. When the antennae are painted black, monarchs are unable to orient in an appropriate direction for migration, implying that the antennal clocks are involved in sun-compass navigation. The brain cells that contain the molecular machinery for the other clock are responsible for regulating the timing of other processes in the lives of monarchs, such as egg laying. In 2011, researchers reported the complete genetic sequence of monarch butterflies, providing the basis for even more detailed understanding of the proximate causes of migration at the molecular and cellular level.

Ultimate Causes of Migration by Monarch Butterflies

The genome of monarchs was sequenced by a group at the University of Massachusetts Medical School headed by Steven Reppert. Reppert and two colleagues summarized the status of research on the mechanisms of navigation in 2010. They concluded their report by listing 11 "outstanding questions in the field of monarch butterfly migration."

I mentioned a few of these questions above; for example, can they use a magnetic sense to navigate on overcast days? But Reppert's list doesn't include any questions about the evolution of migration, even though this is another major aspect of causation that interests biologists. How can we begin to understand the evolutionary causes of migration by monarch butterflies, considering that there are few fossils of any butterflies and none of monarchs and that molecular data suggest that the group of species containing monarchs originated almost 5 million years ago?

A few species of butterflies migrate short distances, but monarchs are the only ones to migrate hundreds to thousands of kilometers between breeding areas in the United States and Canada and wintering areas in Mexico. One clue to the origin of migration comes from thinking about the closest relatives of monarchs. If these are species that spend their whole lives at temperate latitudes and overwinter as eggs or dormant adults in soil or leaf litter, then we may ask how monarchs might benefit from flying south to spend the winter. If instead the relatives of monarchs are tropical species, then we may ask how monarchs might benefit from flying north to breed. In fact, the closest relatives of monarchs live their entire lives in the tropics. An earlier generation of entomologists reached this conclusion based on morphological comparisons of butterflies, and the classification of monarchs with other tropical species was confirmed by more recent DNA comparisons.

Since monarchs are related to other tropical butterflies, we can infer that monarchs themselves or their immediate ancestors once spent their entire lives in the tropics. Migration probably evolved gradually, with the first generations traveling short distances north to breed and returning home for the winter and subsequent generations traveling farther and farther north. The process might be something like prehistoric South Pacific islanders gradually extending their range by rowing across the ocean to more and more distant islands, except that the human example involved cultural change while the monarch example involved genetic change.

The evidence bearing on this argument about the origin of migration in monarch butterflies differs from the evidence that they use the sun as a compass to navigate to the wintering area. Most of the latter evidence came from experiments, but it's difficult to imagine experiments that could support the hypothesis that migration originated deep in evolutionary time by tropical butterflies gradually flying farther and farther north to breed. Instead, this hypothesis is a logical consequence of comparative data showing that monarchs are related to other tropical butterflies rather than species that live in temperate regions. Yet this hypothesis leads to a specific question about the evolution of migration that can be tested experimentally, as we will soon see.

If the first step in the evolution of migration was a search for new breeding areas, then we can ask how monarchs might benefit from this. It seems pretty certain that migration has major costs, so it could only evolve if there were significant benefits. Recall that the migrating generation of monarchs doesn't breed until it makes the long and arduous trip to Mexico, spends several months roosting in high elevation forests where it is cool and damp, and then returns to the southern United States. Individuals face many risks during this process—they may get caught in a severe storm while migrating and perish, they may get eaten by a predator at the wintering site, or they may roost with hundreds of their fellow butterflies on a rotten branch that falls to the ground, killing many of them. If any of these things happen, they don't breed at all so have no opportunity to pass their genes to the next generation. Therefore migration would only be worthwhile if the potential benefits of breeding in the north compensate for these risks.

Animals might benefit in several ways from traveling to a new area to breed. There might be more food resources to support reproduction, there might be fewer predators, or there might be less risk of parasitism or disease. These potential benefits suggest three general hypotheses for the ultimate cause of migration in monarch butterflies—that migrants gain access to more and better food, avoid predators, or reduce risk of parasitism or disease. Researchers have paid most attention to the parasitism hypothesis in recent years, reporting various kinds of evidence consistent with the hypothesis that minimizing parasitism might account for the evolution of migration in monarchs. We can't yet reject the alternative hypotheses about food and predation, but at least the studies of parasitism suggest a plausible scenario for the evolution of migration. Of course these are studies of contemporary populations of monarchs, so we must assume that conditions were similar when migration originated. Despite these reservations, the story illustrates how biologists can bring together different kinds of evidence to start answering questions about ultimate causation.

Monarch butterflies harbor a parasite of the same type that causes malaria in humans, although the monarch form doesn't infect humans. Both the parasite that causes malaria and the parasite that infects monarchs are large, single-celled organisms with complex life cycles called protozoans. One type of evidence that migration by monarchs is related to parasitism comes from comparing prevalence of parasites in different monarch populations. As you know, migratory populations east of the Rocky Mountains are separate from those west of the Rockies, and the eastern populations migrate longer distances to their overwintering sites (Figure 1.1). About 30% of western monarchs are heavily infected with parasites, but fewer than 8% of eastern monarchs are heavily infected. Even more telling, western migrants that breed farther from the overwintering area along the Pacific coast have lower parasite loads than those that breed closer to the coast (Figure 1.4). Perhaps migration allows animals to escape areas where parasites are abundant, and the farther animals migrate, the better chance they have of escaping parasites.

Sonia Altizer asked another question motivated by the idea that parasitism might be related to the evolution of migration—what impacts do parasites have on monarch butterflies? She found that monarchs with very high levels of parasites had lower survival as larvae and adults, but there were no direct effects of lower parasite loads on survival. Altizer wondered whether lower parasite loads had other effects on butterflies, so she teamed up with Catherine Bradley to test the effects of low to moderate doses of parasites on flying ability. They infected one group of monarch larvae in the lab with these doses and raised another uninfected group as controls. When these larvae became adults, the researchers compared their performance in "flight mills," that is, treadmills for butterflies. The infected butterflies flew more slowly, had less endurance, and lost

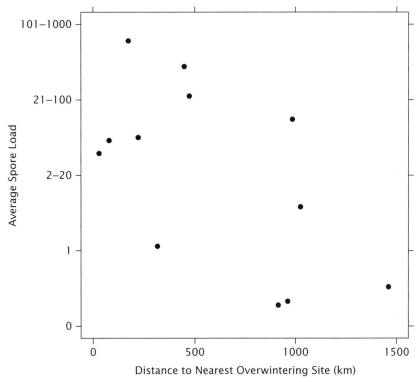

FIGURE 1.4 Parasitism of monarch butterflies at various locations in western North America as described by Sonia Altizer, Karen Oberhauser, and Lincoln Brower. Each dot shows the average spore load for a sample of monarchs from one site.

more weight during flight than the controls. Bradley and Altizer estimated that an uninfected monarch would take nine weeks flying nine hours each day to travel 2,500 kilometers from eastern North America to central Mexico, but an infected monarch would need an extra week for the trip. This is an extra week during which it might encounter a killing storm or another source of mortality. Thus it appears that a common parasite of monarchs has effects on their fitness even at low to moderate doses and that butterflies may escape parasites by migrating longer distances. This combination of experimental and observational evidence is consistent with the parasite hypothesis for the evolution of migration, although certainly not the end of the story of attempts to understand the ultimate cause of migration by monarch butterflies.

Broadening Our Perspective on Proximate and Ultimate Causation

Biologists ask the same kinds of questions as journalists—who, what, where, when, how, and why? Once we understand some basic facts about the natural history of a species or phenomenon, we devote most of our attention to answering *how* and *why* questions about proximate and ultimate causation. These two kinds of questions about causation can be asked about almost anything in biology, so let's consider another example to broaden our perspective beyond migration by monarch butterflies, as interesting as I hope you found that story.

I started this chapter by arguing that migration is one of the biological wonders of the world, and my first example was the incredibly long migration undertaken by arctic terns. Since many arctic terns live for 30 years or more, they may travel as far as three Apollo missions to the moon. Were you surprised that arctic terns live so long? Larger animals generally have longer lives than smaller ones, but arctic terns aren't that big. In fact, they're about the same size as eastern chipmunks, but chipmunks live at most to age 8 and more typically only about two years. Here's an even more dramatic example of a small animal with a long lifespan—Brandt's bats weigh about 7 grams, less than one tenth the size of arctic terns and chipmunks, but the record longevity of a free-living Brandt's bat carrying a numbered tag for identification is an astonishing 41 years!

Why do we die? There are lots of specific reasons, but one general reason all of us eventually die—arctic terns, eastern chipmunks, Brandt's bats, and humans—is that we grow old and our bodies wear out. This process is called aging, or senescence. We may die before this happens from predation (not so likely with humans these days, but certainly a risk for other species), disease, or an accident, but our potential lifespan is determined by our own physiological limitations. You won't be surprised that medical scientists have devoted *lots* of attention to understanding the causes of senescence. The general picture that has emerged from decades of research is that senescence is caused by the inevitable accumulation of toxic byproducts of metabolism in cells. Metabolism is the set of chemical reactions necessary to sustain life, but these reactions produce byproducts that damage DNA, proteins, and cell membranes, eventually causing cancer and other deadly diseases.

This is a well-studied and well-supported hypothesis about the proximate cause of aging—changes in the bodies of individual organisms that cause their physiological deterioration and eventual death. But the long lives of arctic terns and Brandt's bats, as well as many other birds and bats, suggest that there is more to the picture. If all we need to know is that metabolism produces toxic compounds that damage cells, then how can we explain the long lives of birds and bats compared to terrestrial animals of similar size? If anything, birds have higher metabolic rates than similar sized mammals, so should accumulate toxins faster and die sooner.

Perhaps we can gain some insight by thinking about the ultimate, or evolutionary, cause of senescence. Within a species, some individuals will have more offspring than others. If genetic differences between individuals contribute to these differences in reproduction, then the genes of those that reproduce more will be more common in the next generation. This is a concise description of *natural selection*, the major cause of adaptive evolution. Since one way to have more offspring is to live longer, it would seem that natural selection would favor longer and longer lifespan in most species. Yet some organisms live fast and die young while others live much longer, and no organisms are immortal. This basic look at how natural selection works doesn't seem to hold much promise for explaining the long lives of birds and bats compared to terrestrial mammals. Perhaps my claim that genetic differences between individuals contribute to these differences in reproduction is too simplistic. In fact, some genes do seem to have simple effects on organisms that carry them. For example, sickle cell disease is caused by a genetic variant that influences the shape of hemoglobin molecules. Red blood cells are packed with hemoglobin, and people with two copies of the typical gene for making hemoglobin have disk-shaped red blood cells that travel smoothly through capillaries in the body. However, people with two copies of the sickle cell variant have altered hemoglobin that causes red blood cells to be shaped like sickles and get caught in capillaries, causing severe anemia (see Appendix 2 for a more complete discussion of the role of natural selection in evolution).

Other genes have complex effects on multiple physiological processes. Some genes that influence growth and development may enable individuals to reproduce early and abundantly but also shorten their lives by speeding up the production of toxic byproducts of metabolism. This connection may be unavoidable because of the way such genes exert their effects. If it's beneficial to reproduce early and abundantly, these genes may be favored by natural selection even though they shorten lifespan.

What kinds of species would benefit from reproducing early and abundantly? Suppose a species doesn't have good defenses against predation. Individuals may get taken by a predator at any time; unless they reproduce before this happens, none of their genes will be represented in the next generation. This means that genes that speed up growth and development, thus accelerating reproduction, will be favored by natural selection, even if these genes also accelerate senescence. By contrast, slower growth and development and a longer reproductive period may be beneficial in species that aren't so vulnerable to predation, favoring alternative forms of these genes that cause slower growth and development and delayed senescence.

This is one of several evolutionary hypotheses that have been proposed to account for differences in senescence between species. Ecologists have shown that flying organisms, birds and bats among the vertebrates, are less vulnerable to predation than similar sized terrestrial vertebrates. Thus the ultimate cause of the remarkably long lives of birds and bats may be their lower risk of accidental mortality at the claws or talons of a predator. Despite all the attention of medical researchers to proximate causes of senescence, ultimate causes are interesting too. This comparative perspective on senescence may even have some practical importance. Many birds and bats live much longer than small mammals that are typically used in medical research. Do birds and bats have mechanisms to reduce or repair the molecular damage that results in senescence? If so, a species of bird or bat that can be studied in the lab might be a new model system for learning more about proximate causes of senescence.

In this chapter I've described the discovery process that is the foundation of science. I've also introduced two kinds of causation that inspire biological research—proximate and ultimate causation. Researchers who focus on proximate causation ask questions about how things work, for example, how environmental factors influence physiological processes of individual organisms to determine their behavior. Researchers who focus on ultimate causation ask questions about why organisms have certain traits, that is, how these traits evolved over multiple generations. As illustrated for senescence, these kinds of questions are complementary because answers to questions about ultimate causation can lead to new research directions in studying proximate causation, and vice versa.

Causation in biology is a complex process, and the distinction between proximate and ultimate causation is only one dimension of this complexity. You've seen several examples in this chapter of how scientists answer questions using observations, comparisons, and experiments. We'll consider these methods in more detail in the next several chapters, before returning to other aspects of the complexity of causation in Chapters 7 and 8.

Questions to Ponder

1. Fred Urquhart's account of seeing the Mexican wintering area of monarch butterflies for the first time appeared in *National Geographic* in 1976. It's pretty clear from Urquhart's writing style that seeing millions of butterflies roosting in the trees was a tremendously exciting experience. How important to the story was the sight of one tagged butterfly on a branch that fell from a tree? Would your answer differ if you read this story as a scientist, rather than someone just interested in a good adventure?

2. Plate 3 shows how Quanguo Li's team interpreted their evidence on the color of one specimen of *Anchiornis huxleyi* from a 150-million-year-old fossil bed in China. It's not possible to determine the sex of an individual dinosaur fossil, but what if you found a group of well-preserved *Anchiornis* fossils that could all be tested for color patterns of their feathers. Suppose that about half of them had reddish plumes and half didn't. How would you interpret this evidence?

3. One hypothesis for the evolution of migration by monarch butterflies is that it allows animals to escape areas where parasites are abundant. Figure 1.4 shows that the further monarchs in western North America migrate from their wintering areas along the coast, the fewer parasites that they carry. If escape from parasitism is associated with migration, what parasite loads would you predict for non-migratory monarchs that occur in Florida and Hawaii, in comparison to migratory populations in eastern and western North America?

4. Plate 1d shows the Serengeti grassland in Tanzania with a herd of wildebeest as far as the eye can see, plus two zebras in the foreground. The migration of ungulates across the Serengeti includes about 1.3 million wildebeest and 200,000 zebras. Besides participating in this massive movement of animals, wildebeest mothers give birth during a very brief period of time. In a typical year, about 500,000 calves are born, most within two to three weeks near the beginning of the rainy season. This highly synchronized calving of wildebeest is another biological wonder worth pondering.

In addition to its huge herds of grazing ungulates, the Serengeti is home to several iconic predators—lions, hyenas, cheetahs, and wild dogs. Unlike some other ungulates like pronghorn antelope of the American west, there is no place for mother wildebeest to hide their newborns from predators. Instead, the calves can stand and run within about seven minutes of birth, although they are still highly vulnerable to predators for the first few days of life.

Consider a pregnant wildebeest that gives birth at the same time as many other mothers in the herd. Predators have many calves to choose from, so the calf of our focal mother has a decent chance of survival, even through its most vulnerable first few days of life. However, a pregnant wildebeest that gives birth earlier or later than most of the population will produce a calf that is less likely to survive, because predators have fewer alternatives at these times. In short, the mother's chance of passing her genes to the next generation may depend on when she gives birth compared to other mothers. This describes the *predator-satiation hypothesis* for birth synchrony in wildebeest—it's beneficial for an individual female to give birth at the same time as many other females because predators become satiated with this huge supply of newborns, so each individual mother has a greater chance of her own calf surviving.

Thinking about reproductive physiology offers another perspective on birth synchrony in wildebeest. These medium-sized ungulates have a gestation period of about eight months, not too different from that of humans. Wildebeest have estrous cycles, which are similar to the menstrual cycles of humans and some other primates. Eggs are released from the ovaries at a particular time during estrous or menstrual cycles, and if an egg is fertilized within a short time after this, gestation begins and birth can be expected eight months later for a wildebeest mother or nine months later for a human mother.

One way in which births might be synchronized in wildebeest would be if estrous cycles were synchronized, as long as there wasn't much variation in length of the gestation period. The four-day estrous cycles of female rats kept in the lab become synchronized if the rats are exposed to each other's odors. In fact, Martha McClintock reported in 1971 that roommates in a dormitory at Wellesley College were more likely to have synchronized menstrual cycles than women living in the same dorm but not in the same room, although this human example has been discredited by more recent research. Perhaps wildebeest have some mechanism to synchronize their estrous cycles, which causes highly synchronized calving eight months later. If this mechanism is shared odors, as in laboratory rats, we can call this the *pheromone hypothesis* for birth synchrony, since a pheromone is a chemical released by one animal that influences the behavior or physiology of another animal.

We now have two hypotheses for birth synchrony in wildebeest, the predator-satiation hypothesis and the pheromone hypothesis. Are these hypotheses attempts to explain proximate causation or ultimate causation? Are they competing or complementary hypotheses?

Figure 1.5 compares birth synchrony of wildebeest and zebras in the Serengeti. It also shows how births in the two species are related to an important measure of diet quality for herbivores, the protein content of their food. One difference between wildebeest and zebras that may help you interpret this graph is that zebras live in small bands containing one adult male, several females, and young. When attacked by a predator, the adults attempt to defend the young by making a circle and keeping the young inside it. Is the evidence shown in Figure 1.5 consistent with the predator satiation or pheromone hypothesis for birth synchrony in wildebeest? Why or why not?

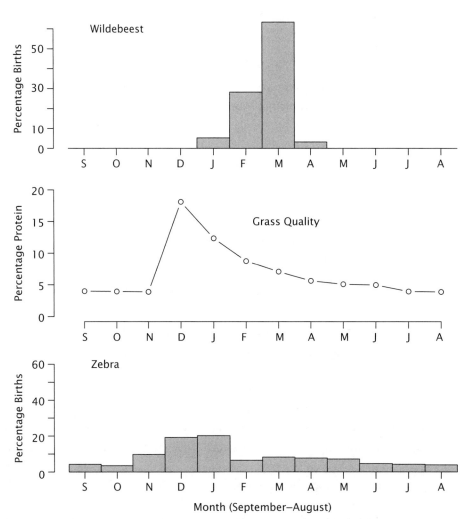

FIGURE 1.5 Birth synchrony and nutritional quality of food for wildebeest and zebras in the Serengeti as reported by A. R. E. Sinclair, Simon Mduma, and Peter Arcese. The bars in the top and bottom panels show percentages of births in each month; for each species, these percentages sum to 100%. The middle panel shows corresponding monthly changes in the protein content of the grass that both species eat.

Katerina Thompson studied a group of sable antelope at the Conservation and Research Center of the National Zoo in Virginia. Many ungulates, including wildebeest and sable antelope, have a characteristic behavior called flehmen in which one animal retracts its upper lip after sniffing the rear end or fresh urine of another animal. Researchers most often see males showing flehmen after sniffing females, but Thompson observed flehmen by females in her captive group of sable antelope. She found that pairs of females that tested each other's urine most often in one year were most synchronized in giving birth the following year. Is this evidence consistent with the predator satiation or pheromone hypothesis for birth synchrony in wildebeest? Why or why not? What are the strengths and limitations of these two kinds of evidence for testing the predator satiation and pheromone hypotheses?

Resources for Further Exploration

I hope this chapter has whetted your curiosity to learn more about biology. If so, here are some resources that may be helpful.

1. The National Center for Case Study Teaching in Science promotes and supports the use of case studies for teaching and learning at all levels in all the sciences. Here are three cases related to major themes of this chapter:
   a. *A Deadly Passion: Sexual Cannibalism in the Australian Redback Spider*, by Erin Barley and Joan Sharp. This case "teaches students about the distinction between proximate and ultimate causes of behavior using the fascinating courtship and mating rituals of the Australian redback spider." Available at http://sciencecases. lib.buffalo.edu/cs/collection/detail.asp?case_id=548&id=548.
   b. *I'm Looking over a White-Striped Clover: A Natural Selection Case*, by Susan Evarts, Alison Krufka, and Chester Wilson. Why do some clovers have toxic cyanide in their white-striped leaves, while others lack both cyanide and stripes? Available at http://sciencecases.lib.buffalo.edu/cs/collection/detail.asp?case_id=272&id=272.
   c. *Seven Skeletons and a Mystery*, by Clyde F. Herreid. This case uses an analysis of *Archaeopteryx* to illustrate the interpretation of fossil evidence about the relationship of dinosaurs and birds and the origin of flight. Available at http://science-cases.lib.buffalo.edu/cs/collection/detail.asp?case_id=440&id=440.
2. SimBio produces software to teach concepts in biology through simulation. Several packages in their EvoBeaker collection deal with evolution within populations as well as across long periods of geological time. All of these simulations emphasize interpreting various kinds of evidence of evolutionary change. Packages are usually purchased by instructors for their classes at a modest cost per student, typically $3.00. *Darwinian Snails* in the EvoBeaker collection is one of the best ways that I know to learn how natural selection works. Another simulation, *Evolutionary Evidence*, uses lizards to illustrate how biologists infer adaptive radiation of species. There are other

simulations that can be used to learn aspects of human evolution. Information available at http://simbio.com/.

3. Understanding Evolution is a website developed and maintained by the University of California Museum of Paleontology and the National Center for Science Education. It has a wealth of resources for learning about evolution, from *Evolution 101* to *Evo in the News*. It is available at http://evolution.berkeley.edu/.

Observations as Evidence

Just what *is* science? Here are two recent definitions. According to Understanding Science, a website of the University of California Museum of Paleontology, "Science is both a body of knowledge and a process" that is "exciting, . . . useful, . . . ongoing, . . . [and] a global human endeavor." As a scientist, I appreciate this complimentary definition. But I wonder if these features of science describe other human activities too. How about art or music?

In 2008, the US National Academy of Sciences gave a more specific definition of science: "The use of evidence to construct testable explanations and predictions of natural phenomena, as well as the knowledge generated through this process." This reinforces the fact that science is both knowledge and a process for gaining knowledge but emphasizes the importance of evidence in doing science. We'll focus on this use of evidence in science because there are many different kinds of evidence, and evaluating evidence can be challenging. If you learn how to evaluate evidence, however, you'll gain a valuable skill that will help you interpret new ideas in nutrition, health, environmental policy, and other areas. This, in turn, may help you answer questions like these: What are the chances that a new diet will improve my health? My child has a disease for which there are two alternative treatments—which is safest and has the greatest chance of success? Should I vote for candidate P who supports environmental policy X or candidate Q who supports policy Y?

Chapter 1 introduced various kinds of evidence about migration by monarch butterflies. As early as 1857, observations of butterflies disappearing from the northeastern and midwestern United States in fall and reappearing the next spring suggested that they might migrate, although the wintering area in central Mexico remained unknown until 1975. With the help of thousands of volunteers, Fred and Norah Urquhart accumulated tens of thousands of observations of tagged butterflies that culminated in the discovery of a tagged butterfly from Minnesota in a huge winter aggregation in central Mexico.

Evidence from comparative studies was also important in the monarch story. For example, monarchs that migrated the greatest distances to overwintering sites had the fewest parasites (Figure 1.4), consistent with the hypothesis that avoiding parasites may have been important in the evolution of migration. You also saw evidence from experiments in Chapter 1. These experiments included studies of both proximate and ultimate causes of migration. Later chapters will include more examples of comparative studies and experiments, but this chapter focuses on "simple" observations because these are the foundation of comparative and experimental studies. This means that all empirical evidence is rooted in simple observations, so a good understanding of the nature of simple observations and their strengths and limitations is necessary for appreciating the more complex kinds of evidence that you'll see in later chapters.

Two Case Studies of Observations as Evidence

Simple observations in biology sometimes earn headlines, as in *The Washington Post* on April 29, 2005: "Extinct? After 60 Years, Woodpecker Begs to Differ." This front-page story reported the rediscovery of the ivory-billed woodpecker, which hadn't been sighted in the United States since 1944 and was thought to be extinct. The ivory-billed woodpecker (Plate 4a) is a large, elegant bird of virgin forests of the southeastern United States. In the last century, these birds were most often found in bottomland swamp forests, but Jerome Jackson suggests that before humans logged much of their habitat "the ivory-billed woodpecker was associated with extensive old-growth forests, the solitude of wilderness, and the availability of immense beetle larvae that were its principal food." Locals knew this majestic bird as the "ghost bird" or the "Lord God bird," as in "Lord God, what's that giant thing flying through the forest?"

The *Washington Post* story was based on a press conference attended by two members of President George W. Bush's cabinet, the president of the Nature Conservancy, and John Fitzpatrick, head of the Cornell Laboratory of Ornithology. Fitzpatrick and 16 other scientists and birders described the search for ivory-billed woodpeckers in the Big Woods area of eastern Arkansas in a report in *Science* magazine in June 2005. A local birder named Gene Sparling had been kayaking in the Cache River National Wildlife Refuge a year earlier when he saw a large, red-crested woodpecker that he thought had the distinguishing marks of an ivory-billed woodpecker rather than a pileated woodpecker (Plate 4b), a species that is commonly seen and the only species that could be visually confused with an ivory-billed woodpecker. Sparling's sighting led to an intensive search of the Big Woods over the next several months, resulting in six more brief sightings, four seconds of videotape of a bird thought to be an ivory-billed woodpecker, and seven reports of hearing double-knock drumming on tree trunks by ivory-billed woodpeckers. Since news about endangered species is often discouraging, this report that ivory-billed woodpeckers might not be extinct after all was widely heralded as a conservation success story, validating efforts to preserve habitat for the species. Between 2004 and 2010, the US Fish and Wildlife Service spent about $14 million for further habitat preservation as well as research on ivory-billed woodpeckers.

What do you think about this evidence that ivory-billed woodpeckers may still exist? Does it seem a little thin? Although ivory-billed woodpeckers are somewhat larger than pileated woodpeckers and have lighter bills and more white coloration on the tops of their wings, the two species occur in the same habitats and these habitats are dark and densely vegetated, making identification of these woodpeckers uncertain. Four seconds of videotape isn't much, and the image is blurred and pixilated (http://www.sciencemag. org/content/suppl/2005/06/02/1114103.DC1/1114103S1.mov). In addition, Fitzpatrick's group admitted that calls by blue jays could have been mistaken for double-knock displays of ivory-billed woodpeckers.

The excitement of researchers at Cornell who apparently rediscovered the ghost bird, birders anxious to add a new species to their life lists, and politicians eager for good news about an environmental issue wasn't shared by all ornithologists. In early 2006, a group of researchers published a detailed critique of the June 2005 *Science* report about the rediscovery. The authors of the original report refuted these critiques and continued searching for more evidence of ivory-billed woodpeckers in the southeastern United States for five more years but suspended their search in 2010 after thousands of hours in the field yielded some additional sightings but no irrefutable physical evidence of this elusive species.

What counts as an observation in this story and more generally? Once we have some observations, how do we assess their validity as evidence to test an idea, such as the hypothesis that ivory-billed woodpeckers still exist?

The ordinary definition of observing something is to see it, so we certainly should consider each sighting of an ivory-billed woodpecker as an observation. In science, however, *observation* has a broader meaning than in everyday usage. Observations include not only what we see but also what we hear, smell, touch, and taste, that is, input from all of our senses. Observations also include things that we sense indirectly. A videotape or audio recording is an observation as is a photograph through a microscope or telescope. Measurements count as observations too. In the case of the alleged rediscovery of ivory-billed woodpeckers, the sightings, the four-second videotape, the reports of hearing double-knock displays, and the subsequent audio recordings of noises attributed to woodpeckers were all observations. The pictures drawn and measurements recorded in field notebooks also became observations that could be interpreted by people who read those notebooks.

These various kinds of observations differ along several dimensions, with the existence of a physical record being one of the most important. There is a long tradition in natural history of collecting specimens as evidence of finding a species at a particular location. John James Audubon shot many birds that he used to make his famous paintings in *Birds of America*; natural history museums around the world document the diversity of life with drawer after drawer filled with specimens. Photographs, video and audio recordings, tissue samples, and feces all count as physical evidence of organisms in contrast to reports of seeing a plant or animal in the field. Physical records are important because anyone can examine them and make an independent interpretation of their relevance as evidence. By contrast, when Gene Sparling reported his 2004 sighting of an ivory-billed woodpecker on the web, this was second-hand evidence for me—it meant nothing unless I assumed his honesty, reliability, and accuracy.

Scientific controversy about the rediscovery of ivory-billed woodpeckers illustrates the importance of physical evidence that can be independently evaluated by different researchers. *Science* is one of the premier journals in the world for publishing new findings in all of the sciences—mathematical, physical, biological, and social. Space in the pages of *Science* is precious, so research reports are very concise—three printed pages for "Ivory-Billed Woodpecker (*Campephilus principalis*) Persists in Continental North America" by John Fitzpatrick and his coauthors. These authors used a little less than half of their text to explain and interpret the four-second videotape that was the main physical evidence of an ivory-billed woodpecker in eastern Arkansas. David Sibley and his colleagues devoted all three pages of their technical comment published in *Science* in 2006 to refuting the claim that the video showed an ivory-billed woodpecker. Sibley's group argued that several aspects of the video were more consistent with the appearance of a pileated woodpecker than an ivory-billed woodpecker. In their response to Sibley's group, Fitzpatrick and colleagues spent another three pages defending their original interpretation. (Although papers published in *Science* are very concise, each of these printed papers was supported by online supplementary material, for a total of 35 additional pages.) A year later, Martin Collinson published a new study of videotape of pileated woodpeckers at a bird feeder in Ohio in a different journal, *BMC Biology*. This study supported Sibley's interpretation of the original videotape by Fitzpatrick's group, who didn't contest Collinson's conclusions, at least in print. Finally, Jerome Jackson wrote a broader review of the controversy in an ornithology journal called *The Auk* in 2006, which led to an exchange between Fitzpatrick and Jackson in later issues of this journal. Jackson's subtitle was "Hope, and the Interfaces of Science, Conservation, and Politics." Although Jackson discussed all of the evidence for persistence of ivory-billed woodpeckers, including sightings by Gene Sparling and others, most of the controversy about validity of evidence has focused on the four-second videotape. This was something tangible that any researcher could examine in great detail and evaluate independently, unlike a reported sighting, which the researcher would have to take on trust based on the report of the observer.

Another biological observation that was headline news in 2008 involved a wolverine in California. In this case, the observation was a photograph of an animal that wasn't thought to be extinct but was found in an unexpected place. Like ivory-billed woodpeckers, wolverines have an aura of mystery because they are rarely seen. Like the story of ivory-billed woodpeckers discussed in the last few paragraphs, this story about wolverines involves both anecdotal and physical evidence, but the wolverine story has a different resolution.

Wolverines are fierce predators related to weasels. They weigh up to 32 kilograms, about the size of a medium to large dog, and they take a wide range of prey species including beavers and deer. They occur in alpine and forested habitat in Canada, Alaska,

Siberia, and northern Europe as well as north-central Washington and Idaho, western Montana, and northwestern Wyoming. Individuals are solitary and have very large home ranges, and population densities are very low. These traits make wolverines difficult to study in nature.

Wolverines used to be found in California, but no specimens have been collected since 1922, and most researchers consider the species extinct in the state. However, many people have reported sightings of animals, tracks, or other signs like feces. As for the apparent sightings of ivory-billed woodpeckers, these sightings engender some doubt about the hypothesis that wolverines are extinct in California, although the sightings aren't conclusive because it's difficult to judge the reliability of the witnesses.

In winter 2008, a graduate student named Katie Moriarty was studying American martens north of Lake Tahoe in the Sierra Nevada of California. Martens are much smaller relatives of wolverines that are widely distributed in North America, although reclusive like wolverines. Because of the low population density and cryptic behavior of martens, Moriarty used a set of 30 remotely activated cameras to get evidence of their activity. These cameras used in the field by wildlife biologists are connected to sensors for heat and motion. When a sensor is triggered, ideally by an animal passing in front of it rather than a pine cone falling from a nearby tree, the camera takes a picture.

On February 28, 2008, one of Moriarty's cameras took a picture of an animal that was much larger than a marten (Figure 2.1). Based on size, shape, and the distinctive band

FIGURE 2.1 First photograph of a wolverine taken by Katie Moriarty and colleagues using a remotely activated camera in the Sierra Nevada of California on February 28, 2008.

of light fur across the haunches and base of the tail, this was clearly a wolverine. Moriarty and her crew quickly deployed additional cameras and got 17 more photographs at three sites. They also found hair and fecal samples at one of these sites. This gave them multiple observations, in the form of different kinds of physical evidence, which allowed them to ask important follow-up questions: Who was this wolverine that showed up in California in 2008? Was it a member of a small population that had existed in the area despite no documented physical evidence for 100 years? Was it an immigrant from elsewhere? Or was it a captive animal that was released accidentally or on purpose?

Moriarty and her colleagues addressed these questions by extracting DNA from these hair and fecal samples. Other researchers had extracted DNA from museum specimens of wolverines collected in California before 1922 and compared the DNA of these wolverines to DNA from animals living today in other parts of their range. Different populations have unique DNA signatures, so Moriarty's group was able to see if their wolverine was most likely a relic of the original population in California or a member of another population. The researchers also used DNA analysis to determine the sex of the animal that they captured on film.

The researchers discovered that all six of the hair and fecal samples that they collected came from the same male. The most likely source of this male was the Sawtooth Mountains of Idaho, about 650 kilometers away. They couldn't definitely reject the possibility that it was a captive animal that was released at their study site, although no known captive facilities use wolverines from Idaho, and the last open trapping season in Idaho was about 50 years ago. The DNA analysis did show clearly that their wolverine was not closely related to animals that used to live in California.

These stories about ivory-billed woodpeckers and wolverines illustrate several kinds of observational evidence in field biology. Both stories involved eyewitness reports as well as various kinds of physical evidence—a video of a woodpecker, still photos of a wolverine, audio recordings that might have been woodpeckers, and DNA from hair and fecal samples of the wolverine. Researchers were unable to corroborate the eyewitness reports of ivory-billed woodpeckers with physical evidence because the identity of the bird in the brief video made in 2004 was uncertain and no one was able to collect further photographic evidence. For wolverines, scattered eyewitness reports suggested that a population persisted in the mountains of California, and photographs of a wolverine north of Lake Tahoe in 2008 were consistent with this hypothesis. However, DNA analysis of hair and feces of this animal implied that it was not part of a native California population but instead an immigrant from Idaho.

Each of these pieces of evidence was a simple observation in the sense that it was not part of a set of data collected in a systematic way to test an hypothesis. There were a handful of sightings of birds that might have been ivory-billed woodpeckers, one video of a bird flying in the Big Woods of Arkansas, several photos of a wolverine taken by remotely activated cameras, and six samples of wolverine hair and feces from which Moriarty extracted DNA. The DNA profiles of these samples were also simple observations of what turned out to be a single individual, even though the technology used to determine the profiles was not simple. However, these DNA profiles were more valuable than purely anecdotal observations because they could be compared with existing profiles for different wolverine populations, including the one that existed in California into the early 1900s. These published data provided a context for interpreting the profile of the wolverine photographed in California in 2008, specifically for testing three alternative hypotheses about the origin of that wolverine—that it was a California native, an immigrant from elsewhere, or a captive animal that was released in the Sierra.

Observations in Court

Crimes sometimes have witnesses, and eyewitness identification of suspects can be persuasive evidence for judges and juries. Yet reliance on eyewitness identification has some of the same pitfalls as reliance on sightings of rare species—both depend on the reliability of witnesses, and there is no way to verify this anecdotal evidence without additional physical evidence. How frequently are eyewitness identifications mistaken?

One approach to answering this question is to study cases of people convicted of crimes who were later cleared. To this end, researchers at the University of Michigan Law School and Northwestern University School of Law established the National Registry of Exonerations with detailed data on 873 exonerations in the United States between 1989 and 2012. Most of the defendants had been convicted of murder or sexual assault; 101 received death sentences. The defendants served an average of about 11 years in prison before their convictions were overturned and they were released. Exoneration was based on DNA evidence in many of the cases but not all.

Perjury, false accusations, and mistaken eyewitness identification of suspects were the major factors leading to wrongful conviction. Errors in eyewitness identification were involved in 375 of the 873 cases (43%). These were false positive errors, a type of error that undermines the basic premise of our legal system that individuals are "innocent until proven guilty."

The National Registry of Exonerations documented a large number of wrongful convictions for serious crimes but wasn't able to estimate the frequency of false positives in eyewitness identification because it had no data on *correct* identification by eyewitnesses. Some prisoners convicted of crimes based on eyewitness identification must actually be guilty, but there's no way to guess how many of those still imprisoned for serious crimes are truly guilty and how many are innocent. Even if DNA evidence could definitively resolve guilt or innocence in all cases (which it can't), perpetrators of crimes don't always leave their DNA at the crime scene.

How can we resolve this problem of estimating the likelihood of false positive identification by eyewitnesses? The Exoneration Registry shows that eyewitness identifications of suspects are sometimes inconsistent with physical evidence such as DNA collected at the scene of a crime, but we don't know whether errors by eyewitnesses are common or rare. There may be no good way to figure this out in the messy world of real people committing real crimes seen briefly under stressful conditions by real witnesses. But psychologists have done experiments that provide some insight even though they have the artificial elements of all experiments. In fact, Nancy Steblay and two colleagues summarized results of 72 experiments performed by various researchers between 1985 and 2010 that used a total of 13,143 volunteer eyewitnesses.

A typical procedure in these experiments was to stage a crime in front of a volunteer witness and then present the witness with a set of photos that might or might not include the perpetrator. Atul Gawande described an early experiment by Gary Wells, one of the pioneers of eyewitness research, as follows:

> He [Wells] asked people in a waiting room to watch a bag while he left the room. After he went out, a confederate got up and grabbed the bag. Then he dropped it and picked it up again, giving everyone a good look at him, and bolted. (One problem emerged in the initial experiment: some people gave chase. Wells had to provide his shill with a hiding place just outside the room.)

Police departments sometimes use lineups of real people, including one suspect and several foils who are known to be innocent, but often use photo lineups with a similar composition. All of the experiments discussed by Steblay's group involved photo lineups, and 89% used a total of six photos, corresponding to typical police practice. Some of the researchers who did these experiments attempted to match police procedures as closely as possible, but one fundamental difference was that all of the psychology studies were experiments while the police lineups were simple observations. These experiments on eyewitness identification illustrate three of the common features of all experiments—researchers compare two or more conditions or circumstances, they assign multiple subjects to each condition, and they make these assignments randomly. In this case, the subjects were people who volunteered to participate in the experiment. The conditions, or *experimental treatments*, are described in the next two paragraphs.

When police use lineups, either with real people or photos, they include a suspect and five foils who often are superficially similar to the suspect in appearance. For example, if the suspect was a short white male, it wouldn't make sense to include a tall black female as a foil. However, the police don't know if the suspect is the true culprit. This means that police use two types of lineups, culprit-present and culprit-absent, but don't know which type they are using in any particular case. In the psychology lab, by contrast, researchers can purposely create these two types of lineups and compare responses of witnesses to them. In other words, each type of lineup is an experimental treatment. Many of the studies reviewed by Steblay's group used these two treatments. This enabled the researchers to distinguish several types of choices that eyewitnesses might make. If the culprit was present in the lineup, the witness might (i) correctly identify him, (ii) mistakenly identify one of the foils as the culprit, or (iii) mistakenly make no choice. If the culprit was absent, the witness might correctly make no choice or mistakenly select one of the foils.

All of the experiments discussed by Steblay and her colleagues included another treatment comparison, between simultaneous and sequential lineups. In a simultaneous lineup, the witness views all photos at the same time; in a sequential lineup, she views photos one at a time. Psychologists were interested in this comparison because of the possibility that witnesses viewing a culprit-absent lineup might have a tendency to pick the closest match to the culprit who they saw committing a crime, even if they were told that the culprit might not be present and they didn't have to pick someone. Psychologists thought this error might be less likely if witnesses saw photos in sequence rather than simultaneously. Keep in mind that police officers don't know if their lineups contain the culprit or not, so they can't assess whether witnesses make this error. Psychologists, on the other hand, design their experiments to know whether lineups do or don't contain culprits, so they can study tendencies of witnesses to select close matches even when culprits are absent from lineups.

Twenty-seven of the 72 experiments reviewed by Steblay's group used all combinations of both of these experimental treatments—simultaneous culprit-present lineups, sequential culprit-present lineups, simultaneous culprit-absent lineups, and sequential culprit-absent lineups. These complete experiments allowed researchers to disentangle effects of experimental treatments on eyewitness accuracy and test the strengths and weaknesses of simultaneous and sequential lineups with and without culprits.

In lineups with culprits present, Steblay and her colleagues found that eyewitnesses correctly identified culprits 52% of the time with simultaneous lineups and 44% of the time with sequential lineups. This isn't very impressive, although if witnesses made choices completely by chance, they would be correct only about 14% of the time with one correct choice out of seven possible choices (the culprit, one of the five foils, or none of the six members of the lineup).

These results show that eyewitnesses are slightly better at identifying culprits if they see photos of the culprit plus the foils together than if they see these photos one at a time. What if the culprit isn't present in the lineup? In this case, the correct choice is no selection because all the photos are of foils, and none shows the culprit who the witness actually saw doing the staged crime. In lineups with culprits absent, Steblay's group found that eyewitnesses correctly rejected all choices 68% of the time with sequential lineups and 46% of the time with simultaneous lineups. In other words, accuracy of witnesses with culprits absent was substantially better for sequential lineups than for simultaneous lineups.

In summary, eyewitnesses do slightly better with simultaneous lineups if they contain culprits but much better with sequential lineups if culprits are absent. Unfortunately, unlike psychologists doing experiments, police investigating real crimes don't know whether culprits are present in their lineups or not. For each type of lineup, there is a relatively high probability of errors by eyewitnesses—missing the culprit if he is present in the lineup (a false negative) or identifying an innocent person as the culprit if the true culprit is absent (a false positive, which may lead to conviction of an innocent person as described above). Psychologists have studied many different features of lineups, and some police departments have modified their conduct of lineups in response to this research, although few have instituted sequential lineups.

Why does this matter? Eyewitness identification of criminal suspects, like sightings of ivory-billed woodpeckers or wolverines, are observations that may be used as evidence. In each of these cases, the observations may be supported by little or no physical evidence or lots of physical evidence. The accuracy of eyewitness identification hasn't been tested in the case of naturalists sighting woodpeckers or wolverines but has been extensively tested for witnesses of crimes, where it seems to be better than chance though not especially impressive. The most serious problem, however, is not so much the modest accuracy of eyewitnesses but the fact that members of juries consider eyewitness identification highly credible. The most dramatic moment in a high-stakes criminal case may be when the prosecuting attorney asks a witness if the person who he saw committing a crime is in the courtroom and the witness points to the defendant. This is not only a moment of high drama but often of great persuasiveness for the jury, despite the experimental evidence that witnesses often identify suspects as culprits even when lineups don't contain culprits.

Imagine that you are a juror in a criminal case. The prosecutor introduces as evidence results of a lineup in which a witness identified the defendant as the perpetrator of a crime and then puts the witness on the stand to verify this identification. Later in the trial, the defense attorney calls a psychologist as an expert witness. The psychologist testifies that eyewitnesses make false positive identifications about half the time in controlled experiments in the lab. Which evidence will you find most convincing—the concrete, even visceral experience of seeing the witness point to the defendant in court or the more technical and abstract report of the scientific expert questioned by the defense attorney? For some of you, this scenario may be imaginary now but real later in life when you are called for jury duty.

Overreliance on simple observations like eyewitness identifications in court matters not just because it may lead to convictions of innocent people, as important as this may be. It also matters because it illustrates a very common way that humans evaluate evidence that may lead us astray in making all sorts of decisions. Our tendency is to credit evidence supporting our ideas more strongly than evidence opposing our ideas, to seek out positive evidence, even to twist ambiguous evidence so it seems more supportive than it really is. This process is called *confirmation bias*, meaning "seeking or interpreting of evidence in ways that are partial to existing beliefs, expectations, or a hypothesis in hand" according to Raymond Nickerson in a review of this phenomenon that he published in 1998.

Confirmation bias can take many forms. One more example from the legal arena comes from another type of experiment done by psychologists who stage mock trials to evaluate how juries make decisions. One common result of such experiments contradicts how our system of trial by a jury of peers is supposed to work. At the beginning of a trial, the judge typically instructs the jurors to reserve judgment until all the evidence has been presented. When attorneys for the prosecution and defense have completed their cases, the jurors are instructed to meet in the jury room and weigh all the evidence before forming their individual opinions and reaching their collective judgment. In fact, it often happens that jurors form opinions about guilt or innocence of defendants early on and then interpret subsequent evidence presented by the two sides to bolster their initial opinions. This doesn't necessarily indicate a problem with how the jury system works. Instead, if citizens serving on juries show confirmation bias by forming opinions early and then giving more weight to evidence consistent with those opinions, this pattern of thinking is likely due to human nature rather than the special circumstances of the courtroom because it is seen in many other human activities as well.

As an example of confirmation bias in another realm, consider fortune tellers, who rely on confirmation bias to make a living. A fortune teller reads your palm or uses tarot cards to make a vague prediction about something in your future—"you will soon be successful in love." Two weeks later you attend a party. Early in the evening, you have an argument with an acquaintance that you've dated once or twice in the past few months. Later in the evening, you have a pleasant conversation with someone new. The second event brings to mind the fortune teller's prediction; you think—"Wow, maybe she was right!" This is confirmation bias in action.

Confirmation Bias in Normal Science

Science is as different from fortune telling as a human activity can be, so you might think that confirmation bias would be less powerful in science than in activities like fortune telling. If so, you'd be in agreement with the mathematician George Polya who wrote:

> The mental procedures of the trained naturalist are not essentially different
> from those of the common man, but they are more thorough. Both the common
> man and the scientist are led to conjectures by a few observations and they are
> both paying attention to later cases which could be in agreement or not with the
> conjecture. A case in agreement makes the conjecture more likely, a conflicting
> case disproves it, and here the difference begins: Ordinary people are usually more
> apt to look for the first kind of cases, but the scientist looks for the second kind.
>
> (POLYA 1954:40)

As you'll see, this is an idealization of science that doesn't describe the actual practice of scientists very well. Scientists, like ordinary mortals, have favorite hypotheses, look hard for supporting evidence for those hypotheses, and sometimes discount or rationalize inconsistent evidence. Later we'll discuss cases in which confirmation bias has led scientists astray, but first let's consider an example that illustrates how it contributes to normal progress in science.

Louis Pasteur was a giant of nineteenth-century biology who was revered by his French countrymen and not just because he improved the making of wine and beer by explaining how fermentation works. He made fundamental advances in chemistry, microbiology, and immunology, including important contributions to public health such as pasteurization of milk and vaccination against rabies. His practical work was grounded in research in basic biology; for example, he was the first to show that fermentation was not a purely chemical process but a result of the metabolism of yeast and bacteria. One of his most important contributions to basic science was his demonstration in the 1860s that microbial life comes only from preexisting microbial cells, not from nonliving organic matter. This finally refuted the concept of *spontaneous generation* that had been articulated by Aristotle in about 350 BC and generally believed for two millennia.

Pasteur's work was motivated in part by a prize offered by the French Academy of Sciences in 1859 for resolution of the question of spontaneous generation once and for all. He competed with another French scientist, Felix Pouchet, who was a strong advocate of spontaneous generation. Pasteur won the prize in 1862, although it took another 10 to 20 years before all scientists were fully convinced of Pasteur's germ hypothesis—that all life, including single-celled microorganisms, comes only from preexisting life.

Pasteur was a giant among scientists not only because of the breadth and practical importance of his work but also because he was very clever at designing experiments to test hypotheses. By the middle of the nineteenth century, standard procedure to test the spontaneous generation and germ hypotheses was to put sterilized growth media in a flask and cover the top. If spontaneous generation was true, microbes should have developed from the nutrients in the growth medium. If the germ hypothesis was true, there was no source of microbes and thus no way for a microbial population to develop.

Up to this time, most researchers had used hay infusions as growth media for testing these alternative hypotheses. A hay infusion was made by putting dead grass or hay in water. But experiments with sterilized hay infusion in covered flasks gave conflicting results—sometimes microbes grew on the media; sometimes they didn't. The germ hypothesis predicted that microbes should never grow on sterilized hay infusion in a closed flask, so Pasteur might have accepted the evidence that microbes sometimes did grow in this environment as evidence against the germ hypothesis. Instead, in an example of confirmation bias, he believed strongly enough in the germ hypothesis that he focused on the inconsistency of results of experiments with hay infusion and decided that something must be wrong with this technique. So he developed a different medium containing sugar and yeast for his own experiments. Later, researchers found that hay infusions sometimes contained bacterial spores resistant to heat, so these infusions could only be sterilized by heating them for a longer time at higher temperatures than were typically used.

One premise of the germ hypothesis was that microbes existed in air that could colonize sterile culture media in open flasks, but no one had ever tested this premise. Pasteur tested it by filtering air through cotton, then dissolving the cotton fibers in a solution of alcohol and ether. When he looked at the resulting liquid through a microscope, he saw particles that looked like microbes, and when he treated sterile culture media with this liquid, populations of microbes developed.

Unlike the inconsistent results with hay infusions obtained by previous researchers, Pasteur always got microbial growth in open flasks containing sterilized yeast plus sugar and didn't get growth in closed flasks with this medium. But supporters of spontaneous generation didn't abandon their hypothesis based on these results. Instead, they argued that air contained some unspecified chemical needed for microbes to develop from nonliving organic matter, and the reason microbes didn't grow in closed flasks was because of the absence of this chemical. This is another example of confirmation bias in science: rather than giving up their hypothesis of spontaneous generation in the face of negative evidence, advocates of spontaneous generation attempted to rescue their hypothesis by adding another requirement for spontaneous generation—not just the organic nutrients present in sterile media but an unknown chemical present in air.

Pouchet designed a different experiment to test this idea. He prepared sterile media in flasks and covered the media with mercury rather than closing the flasks. He reasoned that the mercury would prevent microbes in the external air from colonizing the media but not block the unknown chemical in air from triggering spontaneous generation in the media. Indeed, microbial populations did develop under these conditions, leading us back to Pasteur, who repeated the experiment with a variety of modifications and also found microbial growth. As another example of confirmation bias, Pasteur set these results aside and continued with other experiments to try to support the germ hypothesis that he preferred. After completing these other experiments and winning the prize of the French Academy, he eventually returned to the work with mercury and found that it contained dust particles that might have harbored microbial cells that contaminated the media underneath.

Pasteur ultimately designed an experiment that clearly distinguished between the germ and spontaneous generation hypotheses. He made flasks with curved necks called swan-neck flasks (Figure 2.2) and added sterile media to them. The flasks were open at the end, so if there was a chemical in the air that stimulated spontaneous generation of microbes in the media, this chemical should have had access to the media and spontaneous generation should have occurred. On the contrary, if growth of microbes in the sterile media depends on the media being seeded by microbial cells from the external air, the swan-neck flasks should have prevented this growth because the cells would have settled on the bends in the necks of the flasks. Pasteur found that he could keep these swan-neck flasks for several months with no growth of microbes in the media but that microbial populations developed in a few days once the necks of the flasks were broken. This was the critical experiment finally refuting spontaneous generation and supporting the hypothesis that all life comes from preexisting life.

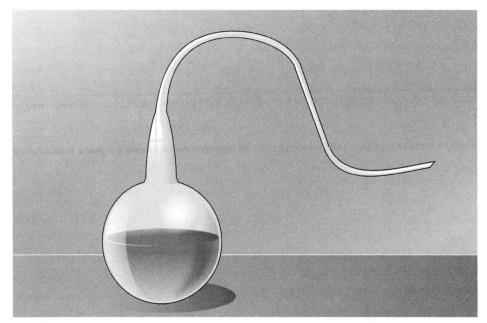

FIGURE 2.2 Depiction of swan-neck flask used by Louis Pasteur to discriminate between spontaneous generation and the germ hypothesis for the development of microbes on culture media.

The strong faith in their hypothesis that led supporters of spontaneous generation to invent a mysterious airborne chemical to account for evidence contradictory to the earlier version of their hypothesis seems silly in retrospect. Yet this kind of confirmation bias was as much a part of Pasteur's approach to science as of Pouchet's. We may look more kindly on it in Pasteur than in Pouchet because Pasteur eventually got the right answer. More fundamentally, however, we should realize that confirmation bias can make a positive contribution to science by keeping scientists from giving up on interesting ideas too easily.

Observations as Evidence in Medicine

All of us get sick at various times in our lives, whether with a simple cold or a life-threatening disease. It is at these times that science most impacts our lives because medical practice is grounded in science. Therefore, as a patient someday, you will benefit from understanding about different kinds of medical evidence.

As in other areas of science, experiments are a key part of medical research. Experiments may actually be more important in medicine than in some other research areas. There are two main reasons for this. First, experiments are more feasible in medicine than in some other areas. Recall the difficulties of imagining an experiment that could explain the evolutionary origin of migration by monarch butterflies that we discussed in Chapter 1. Second, experiments can give more definitive answers to questions about what causes particular diseases or what treatments work best than can other kinds of evidence. So medical researchers do many experiments, sometimes with enormous sample sizes of thousands of volunteer subjects recruited in national or international campaigns.

We'll consider medical experiments in more detail in Chapter 4 but focus here on basic observations as the foundation for experiments. The history of medicine offers classic examples of observations as inspiration for new hypotheses that were later tested by experiments and other systematic research. This important role of basic observations is illustrated by the discovery of penicillin and the first description of Alzheimer's disease.

Alexander Fleming was a Scottish biologist who won the Nobel Prize in Physiology or Medicine in 1945 for his discovery of penicillin, which became the first antibiotic. This discovery is famous as an example of serendipity, or taking advantage of an accidental observation. In the 1920s, Fleming was studying a type of bacteria called staphylococci. In August 1928, he went on vacation after preparing some cultures of staph in petri dishes that he left in his laboratory. When he returned to the lab on September 3, he saw that one dish had fungus growing in it and a bare area with no bacteria around the fungus, while the other dishes had no fungus and luxuriant growth of bacteria. Rather than simply discarding the contaminated dish, Fleming identified the fungus as a *Penicillium* and did a series of experiments testing the effects of the fungus on various disease-causing bacteria. After a long period of trial and error, other researchers figured out how to mass produce penicillin and use it in treating bacterial diseases of Allied troops near the end of World War II.

In 1901, a German psychiatrist named Aloysius Alzheimer was working at the "asylum for lunatics and epileptics" in Frankfurt where he began studying a patient named Auguste Deter. Although only 51, she showed memory loss, delusions, and other symptoms of dementia that got progressively worse until she died in 1906. Alzheimer then studied her brain using new methods developed by a colleague at the asylum. Based on Alzheimer's case report for his patient, including both his detailed notes on her behavior and his description of abnormalities in her brain, the disease later named for Alzheimer was identified. Today, of course, researchers throughout the world study Alzheimer's disease from almost every imaginable perspective using case studies, comparative research, experiments, and other methods.

Earlier in this chapter, I used the term *simple observations* to denote observations that were not collected as part of an experiment or other systematic study. Simple observations are usually observations of a single event or individual, whereas experiments and other comparative studies usually require observations of multiple events or individuals in different situations and often require statistical analysis of the results. Medicine has a special type of simple observation called a *case report*, which is a description of the condition of a single patient. This is what all doctors do in their daily practice—they examine patients, get lab results, make a diagnosis, and write a case report justifying their diagnosis and proposed treatment. If the case is sufficiently unusual, they may publish the case report in a medical journal.

As illustrated by the stories about penicillin and Alzheimer's disease, case reports have a long and distinguished history in medicine. So let me give another example that shows the limitations of this kind of evidence. This story was told by Thomas Newman, a pediatrician at the University of California, San Francisco, in an essay called "The Power of Stories over Statistics," published in the *British Medical Journal* in 2003. Newman is an expert on treating jaundice in newborns, but he got involved in a very different issue as a result of a policy statement made by the American Academy of Pediatrics (AAP). The policy statement was a response to airplane crashes in 1989 and 1994 in which infants sitting on the laps of adults died. Passengers survived both of these crashes, and people thought that the infants might have survived too if they had been restrained in safety seats. In 2001, the AAP supported proposed legislation to require parents traveling with children younger than 2 to use child-restraint seats on airplanes.

Newman wondered about the evidence bearing on this issue, especially since the AAP cited little evidence in their policy statement. He found that the most dramatic evidence was testimony at a congressional hearing by the head flight attendant on the plane that crashed in 1989. Jan Brown Lohr told the story of the crash, ending with what happened after she left the wrecked plane:

> The first person I encountered was a mother of a 22 month old boy—the same mother I had comforted and reassured right after the engine exploded. She was trying to return to the burning wreckage to find him, and I blocked her path, telling her she could not return. And when she insisted, I told her that helpers would find him. Sylvia Tsao then looked up at me and said, "You told me to put my baby on the floor, and I did, and he's gone."

This powerful story is a case report, or anecdote, or basic observation. Newman reports that he was "close to tears" when he read it, and it's hard to imagine a different reaction. Then why isn't this sufficient evidence to require that infants be in safety seats on airplanes? The reason is because the extra cost of a ticket for an infant would cause some families to drive instead of fly and automobile accidents are much more common than airplane accidents. The Federal Aviation Administration estimated that a regulation for infants to be in safety seats on planes might save five lives of infants in airplane crashes in 10 years but cause 92 additional deaths in car accidents. Despite "the power of stories over statistics," in Newman's words, the law wasn't changed, and infants can still ride in their parents' laps on airplanes.

As humans, we make decisions all the time based on incomplete evidence. Simple observations, also known as anecdotes or stories, are often prominent in the evidence we use to decide our activities. But this reliance on simple observations carries the risk of confirmation bias, which can cause us to do things that seem sensible but are harmful to health. A recent study of people who participate in extreme marathons illustrates this problem.

The Western States 100-Mile Endurance Run in California is one of the most challenging races in the world. It starts at 6,200-feet elevation at Squaw Valley in the Sierra Nevada, climbs 2,550 feet to Emigrant Pass in the first 4.5 miles, then climbs a total of 15,540 more feet and descends 22,970 feet, mostly on mountain trails, before ending in the Sierra foothills. Many runners who compete in this and other extreme marathons take the anti-inflammatory drug ibuprofen before the race, thinking that it should reduce inflammation and therefore make the experience less painful. David Nieman, director of the Human Performance Lab at Appalachian State University, decided to test this idea. He and his colleagues went to the Western States Run and recruited 63 volunteers for an experiment in which about half took typical amounts of ibuprofen before and during the race and the other half served as controls. Both males and females volunteered for the experiment, and they were randomly assigned to the ibuprofen and control treatments. The researchers collected blood and urine samples before and immediately after the race and measured several metabolites in these samples that allowed them to assess the effects of ibuprofen under the stressful conditions of the race.

Nieman and his colleagues found that ibuprofen users actually experienced greater inflammation than nonusers. Ibuprofen users also had mild kidney dysfunction and indications of impairment of their immune systems compared to nonusers. The most interesting result, however, came when Nieman presented his findings at annual medical conferences for runners in the Western States 500. Most of them said they would continue to take ibuprofen despite the evidence from Nieman's experiment that it did more harm than good. Their expectation that ibuprofen would reduce their pain while running was so strong that they discounted Nieman's contrary evidence from a rigorous experiment.

Conclusions

This chapter focused on observations because observations are the raw material of all kinds of empirical evidence in science. New hypotheses grow from observations, as in Fleming's discovery of penicillin and Alzheimer's discovery of the disease that bears his name. Simple observations provide evidence for testing hypotheses, as in the videotape of a bird in the Big Woods of Arkansas that might have been an ivory-billed woodpecker, or photos and tissue samples of a wolverine in California, or eyewitness identifications of suspects who may have committed crimes. Some observations are supported by physical records, so can be independently judged by anyone. Other observations, like eyewitness reports, have to be taken on faith, although these kinds of observations sometimes have more credibility than they deserve.

This chapter also showed that observations in science don't exist in a vacuum but are connected with ideas or hypotheses. As Darwin wrote, "How odd it is that anyone should not see that all observation must be for or against some view if it is to be of any service!" In our examples, this view might be that ivory-billed woodpeckers still exist in the southeastern United States, that wolverines from Idaho occasionally disperse to

California, that microbial life can arise spontaneously from nonliving organic matter, that infants should ride in child-restraint seats on airplanes, or that ibuprofen decreases inflammation in people who run ultra-marathons. We also saw that people have a tendency to interpret observations to support their preexisting views. This isn't necessarily a bad thing, as illustrated by Pasteur's research on spontaneous generation, but can lead us astray when personal experience or tradition trumps experimental evidence, as in the use of anti-inflammatory drugs by long-distance runners. In later chapters, we'll see how repeated observations become data; how observations are used in comparisons, correlations, and experiments to test hypotheses more rigorously; and how we can strengthen our inferences about the world by evaluating evidence both for and against alternative views.

Questions to Ponder

1. How do art and music compare to science as bodies of knowledge and processes that are exciting, useful, ongoing, and global human endeavors?

2. Katie Moriarty and her coworkers collected six samples of hair and feces from a wolverine in northern California (page 30). By comparing DNA from these samples, they found that all six came from one individual male that likely traveled from central Idaho, a distance of about 650 kilometers. In general terms, how do you think the researchers determined that the six samples came from the same individual and that this individual was a male? Spin a story about the mating behavior of wolverines based on the fact that this animal that was a long way from home was a male and not a female. What are the limitations of this story?

3. On December 28, 2011, a juvenile male gray wolf wearing a radio collar crossed into California from Oregon. This wolf was born to a pack in northeastern Oregon in spring 2009 and was captured and fitted with a radio collar along with other members of his pack. He was the first wild wolf known to be present in California since 1924. As of April 6, 2012, OR7 had traveled more than 2,000 miles in Oregon and California. By summer 2013, OR7 appeared to have settled down in south-central Oregon. Reconsider your answer to Question 2 in light of this new observation of a different species.

4. In experiments testing eyewitness identification, witnesses make fewer mistakes with simultaneous lineups when culprits are absent but make fewer mistakes with sequential lineups when culprits are present. What do you think accounts for this difference?

    With culprits absent in these experiments, eyewitnesses viewing sequential lineups correctly made no choice 68% of the time while eyewitnesses viewing sequential lineups were correct 46% of the time, for a difference of 22% in favor of sequential lineups. With culprits present, eyewitnesses viewing simultaneous lineups correctly identified the culprit 52% of the time while eyewitnesses viewing sequential lineups were correct 44% of the time, for a difference of 8% in favor of simultaneous lineups. Why do you think most police departments in the United States still use exclusively simultaneous lineups, considering this evidence?

5. On July 19, 2012, the Supreme Court of New Jersey ruled that judges in charge of cases in that state that include eyewitness testimony must tell the jury about problems with this kind of evidence, specifically, that human memory can be flawed. Will this cause fewer errors in trial outcomes? Why or why not?

Resources for Further Exploration

1. A British philosopher named Jeremy Stangroom maintains a website called Philosophy Experiments, available at http://www.philosophyexperiments.com/. Several of these experiments are interesting, but one that is particularly relevant for this chapter is called *Elementary, My Dear Wason?* Try it, and then think about how it relates to one theme of this chapter.

2. Understanding Science is a website developed and maintained by the University of California Museum of Paleontology. It has lots of examples of how science works. A good place to start is the flowchart linking exploration and discovery to testing ideas, community analysis and feedback, and benefits and outcomes. Available at http://undsci.berkeley.edu/index.php.

3. The National Center for Case Study Teaching in Science has cases related to the major themes of this chapter:

   a. *Tragic Choices: Autism, Measles, and the MMR Vaccine*, by Matthew P. Rowe. In the introduction to this chapter, I suggested that learning how to evaluate evidence would help you make choices that would benefit the health and nutrition of you and your children. This case involves a sobering, real-world example (see also Chapter 6). Available at http://sciencecases.lib.buffalo.edu/cs/collection/detail. asp?case_id=576&id=576.

   b. *Salem's Secrets: A Case Study on Hypothesis Testing and Data Analysis*, by Susan Nava-Whitehead and Joan-Beth Gow. More than 200 people were accused of witchcraft in Salem, now Massachusetts, in the late 1600s, and 20 of these people were executed. This case shows how different kinds of evidence can be brought to bear to interpret an historical event. Available at http://sciencecases.lib.buffalo. edu/cs/collection/detail.asp?case_id=307&id=307.

   c. *Extrasensory Perception—Pseudoscience? A Battle at the Edge of Science*, by Sarah G. Stonefoot and Clyde Freeman Herreid. This chapter began with the question, what is science? One way to answer this question is to think about theories that aren't scientific. Is extrasensory perception an example (see also Chapter 10)? Available at http://sciencecases.lib.buffalo.edu/cs/collection/detail. asp?case_id=229&id=229.

From Observations to Data

Guinness World Records (http://www.guinnessworldrecords.com/) is one source of well documented observations. Some of these observations are of no more than passing interest, like the world's fastest toilet, which is mounted on a motorcycle that can travel 42 miles per hour. Other entries in Guinness World Records have more potential significance. For example, the size of an animal is one of its most important traits because it influences basic aspects of its life such as lifespan, reproduction, food supply, predators and parasites, and responses to heat and cold. Because of the importance of body size for animals, I got to wondering about record sizes of humans. Who is the shortest person? Who is the tallest? Guinness World Records is a natural place to look for answers to these questions. As of August 21, 2012, the shortest living adult was Chandra Bahadur Dangi of Nepal at 21.5 inches, or just under 2 feet. The tallest was Sultan Kösen of Turkey at 8 feet, 3 inches.

Like many of the examples in Chapter 2, these record heights are simple observations. They have been validated by direct measurements made by judges employed by Guinness World Records. Of course they aren't absolute records because there may be very tall or very short people living in remote areas who haven't come to the attention of Guinness World Records. However, unlike some of the observations discussed in Chapter 2, we can safely assume that these observations are accurate. Still, they are just miscellaneous, isolated observations that raise questions that can't be answered without more observations collected in a more systematic way, that is, without *data*.

In particular, Guinness World Records can't answer two basic questions about unusually tall or short people—what are the causes and consequences of extreme size? Dr. Jaime Guevara-Aguirre asked these questions about a group of unusually short people living in remote mountain villages in Ecuador (Plate 5). Most of the 99 people studied by Guevara-Aguirre were less than 3.5 feet tall and all had Laron syndrome, one of several causes of dwarfism. Laron syndrome is caused by a genetic mutation that modifies a molecule of cell membranes that responds to growth hormone. In most people, growth hormone stimulates this molecule to release a signal inside liver cells, causing release of another hormone, IGF-1, that causes children to grow. Because of the mutation in people with Laron syndrome, the message carried to the liver by growth hormone in circulating blood isn't transferred into the interior of liver cells, so IGF-1 isn't produced and growth is inhibited.

People who are unusually short or tall or light or heavy often have health problems. For example, Sultan Kösen walks with crutches because of his extreme height. By contrast, the most remarkable thing about Ecuadoreans with Laron syndrome was that none of them had diabetes and only one had cancer. In comparison, 5% of relatives without Laron syndrome were diabetic and 17% had cancer.

Guevara-Aguirre and his colleagues studied how Laron syndrome might protect against cancer by adding serum of Laron individuals to laboratory cell cultures and then adding a chemical that damages DNA. They found less DNA damage in these cells than in control cultures without Laron serum. When DNA damage did occur in the presence of serum, these cells died rather than becoming cancerous. The researchers were able to reverse the protective effect of serum from Laron individuals by adding the hormone IGF-1 to these cultures.

Cancer and diabetes are major causes of death, especially of older people. If Laron syndrome protects against cancer and diabetes, does it prolong life for those who carry the mutation that causes this condition? Unfortunately not. Although Dr. Harry Ostrer of New York University described these individuals as "remarkably youthful in appearance," they didn't live any longer on average than relatives without the syndrome. Instead, they were more likely to die from accidents or alcoholism.

Ecuadoreans with Laron syndrome are descended from Sephardic Jews who left Spain and Portugal in the fifteenth century to escape the Inquisition. The mutation that causes this condition also occurs in a few people in Israel, Brazil, and Chile, suggesting that this mutation arose before the emigration of some Sephardic Jews to separate populations in South America. Dr. A. L. Rosenbloom of the University of Florida College of Medicine estimated that the 99 Ecuadorians studied by Guevara-Aguirre were about one-third of the total number of people with Laron syndrome worldwide. Although some medical research uses much larger sample sizes, this is an impressive set of data for a rare but very interesting human condition, especially since Guevara-Aguirre studied his subjects for 24 years, long enough for some of them to die, though not of cancer or diabetes.

Describing and Analyzing Human Heights and Weights

To see how observations become data, let's consider another example involving height—this time a set of people with more typical human heights. The US Army made detailed body measurements of some of their personnel in 1987–1988. They measured more than 100 features of 2,208 women and 1,774 men for the purpose of designing better uniforms, protective equipment, and vehicles. We'll eventually consider height and weight of both sexes but begin with heights of women.

We might start thinking about these data by listing them: 167, 153, 157, 184, 155, 152, 184, 163, 155, 163 for the first 10 women. But this isn't very helpful for two reasons. First, it would take several pages to list all 2,208 heights, and I'm pretty sure you'd stop reading way before you got to the end of the list. Second, I haven't yet told you the units of these measurements, so even if there were only 10 women in the sample, you couldn't make much sense of the numbers. In fact, I've listed these heights in centimeters (cm) in keeping with the standard use of the metric system for measurement in science (see Appendix 1). For a benchmark, 100 cm equals 1 meter, which is a little more than one yard. So these 10 women range in height from about 1.5 to 1.8 yards. More specifically, the shortest at 152 cm is just under 5 feet tall and the two tallest at 184 cm are just over 6 feet tall.

Finding the smallest and largest values in a list of observations like this is one way to begin understanding the data. If we expand our scope from the first 10 women to all 2,208, we find that the shortest is 143 cm and the tallest is 187 cm. By comparison, the range of heights for men is 150 to 204 cm. But these two numbers for each sex don't tell us much—a more complete picture would be nice. Indeed, one of the best ways to understand data like these is to draw pictures; in this case, *histograms* showing the full distributions of the data for each sex (Figure 3.1).

Figure 3.1 shows that the distributions of heights for women and men measured by the US Army are roughly symmetrical. Not surprisingly, the shortest women are shorter than the shortest men and the tallest men are taller than the tallest women, but there is quite a bit of overlap in heights of men and women between 160 and 180 cm. We can summarize the difference between men and women by calculating average heights. The arithmetic mean is the most familiar way of estimating the average value of a set of data and is defined as the sum of all the individual values divided by the total number of values. Symbolically, this is

$$\overline{X} = \frac{\sum_{i=1}^{n} X_i}{n}.$$

In this formula, n stands for the total number of observations, and i is an index for an individual observation, for example, $i = 1$ for the first woman, $i = 2$ for the second, and so on, up to $i = 2{,}208$ for the last. The Greek Sigma, $\sum_{i=1}^{n}$, represents the sum of all observations, from $i = 1$ to $i = n = 2{,}208$. Finally, $\overline{X}$ is the arithmetic mean. For these data, women have a mean height of 162.9 cm and men have a mean height of 175.6 cm. This is a difference between men and women of 12.7 cm, or 5 inches.

The median, or middle value, is another common way of representing the average of some data. In any data set, half of the individual values are below the median and half are above the median. For these data, the median is 162.8 cm for women and 175.6 cm

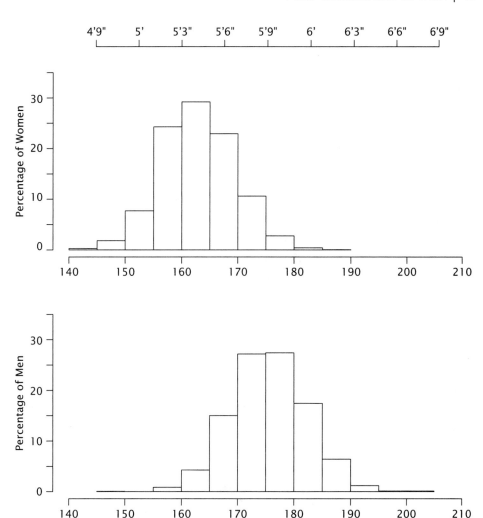

FIGURE 3.1 Heights of 2,208 women and 1,774 men in the US Army in 1987–1988. Horizontal axes below the histograms are in centimeters, with corresponding heights in feet and inches shown above the histogram for women.

for men. In this example, the mean and median are the same for men and differ by only 0.1 cm for women, so it doesn't make much difference which measure of the average that we use. But we'll soon see examples in which the mean and median for a data set differ dramatically, so the choice of mean versus median does matter.

The *minimum, maximum, range* (maximum—minimum), *mean*, and *median* are examples of *summary statistics*. Using them is not a substitute for looking at histograms of all the data, but graphs and summary statistics provide complementary windows into a set of data. Two more summary statistics are especially important for our purposes, the *variance* and the *standard deviation*. These are closely related to each other because the standard deviation is the square root of the variance, so I'll describe the variance first and then say a word about why we might want to use the standard deviation in some applications.

Means and medians tell us something about the middles of frequency distributions like those shown in Figure 3.1. It is also useful to have a summary statistic to represent the breadth, or *dispersion*, of a distribution. The range from minimum to maximum is one candidate for representing dispersion. This range is 44 cm for height of women, from 143 cm for the shortest to 187 cm for the tallest. But imagine a set of 2,200 women with 50 being 143 cm tall, 50 being 144 cm, 50 being 143 cm, , 50 being 186 cm, and 50 being 187 cm. In other words, there are 50 women with each of the 44 heights between 143 and 187 cm, for a total of $50 \times 44 = 2{,}200$ women. The histogram of this frequency distribution would be perfectly flat, quite different from the shape of the actual distribution, but the range would be the same. Although this particular example isn't realistic, it illustrates a problem with using the range as a measure of dispersion. The variance is more complicated to calculate, but gives a better sense of the dispersion of a distribution.

The basic elements that we need for calculating variance are deviations of individual values from the mean. These deviations are positive for values greater than the mean and negative for values less than the mean. We might simply calculate all these individual deviations and use the average of the deviations as a measure of dispersion, except that this would always be zero because the definition of the mean implies that deviations for values above the mean exactly cancel deviations for values below the mean. To get around this problem (and for deeper mathematical reasons that needn't concern us here), the variance is defined as the sum of *squared* deviations from the mean divided by 1 less than the total number of observations. Writing this formula in symbols rather than words,

$$Variance = \frac{\sum_{i=1}^{n}(X - \bar{X})^2}{(n-1)}.$$

You've seen all the individual components of this equation in the equation for the mean on page 46. The part in parentheses in the numerator of the equation for variance is the deviation of each individual observation (X_i), from the mean of all observations, ($\bar{X}$); note that each of these deviations is squared before it is added to the total. The denominator of the whole expression is 1 less than the sample size of n.

For our height data, the variances are 40.4 cm for women and 44.6 cm for men. This difference is consistent with the slightly greater breadth of the height distribution for men than for women (Figure 3.1). Finally, the standard deviations are the square roots of the variances, 6.36 cm for women and 6.68 cm for men. Because we squared values to get variances and took square roots to get standard deviations, the standard deviations are expressed in the same units as the raw values. This means that we can concisely summarize data like these by writing "Heights of women and men in the US Army are 162.9 ± 6.4 cm and 175.6 ± 6.7 cm, respectively." Alternatively, we could write 64.1 ± 2.5 inches for women and 69.1 ± 2.6 inches for men (and in either case we should specify that we are giving the mean plus or minus the standard deviation).

We can do a similar comparison of weights of these army personnel. Examination of graphs is often the most revealing part of data analysis, so Figure 3.2 is set up just like Figure 3.1 to help you compare the frequency distributions of weights of women and men measured by the Army in 1987–1988. Weights are shown in kilograms (kg), with a translation to pounds (lbs) at the top of the graph. The range of weights including both women and men is 41 to 128 kg; since 1 kg equals 2.2 lbs, this is 90.2 lbs for the lightest woman to 281.6 lbs for the heaviest man. Although we can't read the average weights directly from Figure 3.2, we can tell that the average weight of men is greater than that of women because the distribution of data for men is further to the right than that of women. If we use means and standard deviations to represent the averages and dispersions of the

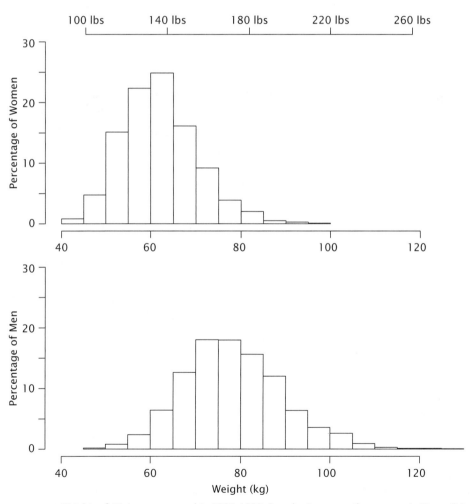

FIGURE 3.2 Weights of US Army personnel in 1987–1988. Sample sizes were the same as in Figure 3.1. Horizontal axes below the histograms are in kilograms, with corresponding weights in pounds shown above the histogram for women.

weight data for the two sexes, we would say that women in this sample weigh 62.0 ± 8.4 kg while men weigh 78.5 ± 11.1 kg. Men are not only heavier on average than women (78.5 kg vs. 62.0 kg), but weights of men are also more variable (standard deviations of 11.1 vs. 8.4). Visually (Figure 3.2), this difference in dispersion of the data is seen in the flatter histogram for weights of men than women. By contrast, the histograms for heights of men and women in Figure 3.1 are quite similar in shape.

We're looking at numbers and graphs in this chapter because all science is quantitative, including biology. More fundamentally, critical thinking sometimes depends on understanding or making quantitative arguments, as you'll see throughout this book. For example, suppose you wanted to compare the physical characteristics of people living in different countries. You might be interested in how typical diets in different parts of the world relate to prevalence of obesity. A simple index of risk of obesity is BMI, or *body mass index*, defined as weight in kilograms divided by height in meters squared:

$$BMI = \frac{Weight\ (kg)}{\left[Height(m)\right]^2}.$$

We could easily calculate BMI from these data for heights and weights of 3,982 US Army personnel. Would this be an appropriate sample for your study of BMI and diet in different countries? Why or why not? What would be a good sample to represent the United States or another country? How would you obtain a good sample for your study?

Sizes of Animals

The human species is just one of about 5,000 species of mammals. Some species are much smaller than us, a few are about the same size, and some are much larger. Size influences all aspects of an animal's life, so let's expand our scope from humans to mammals in general to learn more about understanding data as well as a little about physiology, behavior, and ecology.

In 2009, a group of researchers published a database called PanTHERIA that included as much data as they could find on body size, basic physiology, reproduction, lifespan, ecology, and geographic distribution of mammals. *Pan* means all and *therion* means "wild beast" in Greek, indicating the goal of this project. PanTHERIA includes mean body weights of 3,542 species of mammals, which is about 65% of all species alive today.

Figure 3.3 shows the distribution of average body weights of 265 species of primates in the PanTHERIA database. This picture couldn't be more different from the distribution of individual body weights of male and female humans in Figure 3.2. Sixty percent of primate species have weights of less than 5 kg, represented in the leftmost bar of Figure 3.3. The tiny bar at the far right is the rare eastern gorilla, which weighs about 149 kg, while the next bar at 112 kg is the more common western gorilla. The bars clustered around 50 kg are humans, chimpanzees, and orangutans.

FIGURE 3.3 Mean weights of 265 species of primates from the PanTHERIA database (http://esapubs.org/archive/ecol/E090/184/metadata.htm). This represents about 70% of the total number of species of primates included in the 2005 edition of *Mammal Species of the World.*

When we calculate averages for this distribution, we find that the mean is 5.9 kg while the median is only 3.0 kg. For these data, the mean is almost twice as large as the median. Medians differ from means for distributions of data that are not symmetrical, like the body weights of primates shown in Figure 3.3. These are called *skewed* distributions, and another classic illustration is household incomes. In the United States in 2004, for example, median household income was $44,389 while mean household income was $60,528. In skewed distributions, a small number of very large values—gorillas and other higher primates with body weights above 50 kg and households with incomes above $1,000,000 per year—can pull means above medians.

Histograms of skewed distributions like Figure 3.3 are difficult to interpret because many values are tightly clustered so we can't see details at one end of the scale. In this case, all we can tell about the smallest primates (those included in the leftmost bar) is that 60% of species weigh less than 5 kg, or 11 pounds. Are these small primates fairly evenly distributed in weight between 1 and 5 kg, or are most of them smaller than 2 kg or close to 5 kg? We can answer questions like these by calculating logarithms of body weight and plotting the data on a logarithmic scale instead of the arithmetic scale used in

Figure 3.3. The easiest way to understand logarithms is to think about powers of ten. For example, 10 squared = $10 \times 10 = 10^2 = 100$ and 10 cubed = $10 \times 10 \times 10 = 10^3 = 1,000$. We can continue powers of 10 in both directions: $10^4 = 10,000$, $10^5 = 100,000$ and $10^1 = 10$, $10^0 = 1$, $10^{-1} = 0.1$, and so on. Each step in this sequence is called an *order of magnitude*, so 1,000,000 is an order of magnitude greater than 100,000 and 0.01 is an order of magnitude less than 0.1. The exponents in these expressions are common logarithms of numbers on the right side of the equals signs: 2 is the logarithm of 100, 4 is the logarithm of 10,000, −1 is the logarithm of 0.1, and so on.

Figure 3.4 shows body weights of primates plotted on a logarithmic scale. The species and raw body weights are the same as in Figure 3.3, but the species are grouped in bars based on the logarithms of their body weights rather than the raw body weights used in Figure 3.3. Logarithms are shown on the bottom axis with corresponding body weights on the top axis; note that each addition of one unit to the log of weight translates to multiplying the actual weight by 10. This way of looking at the data reveals some details that

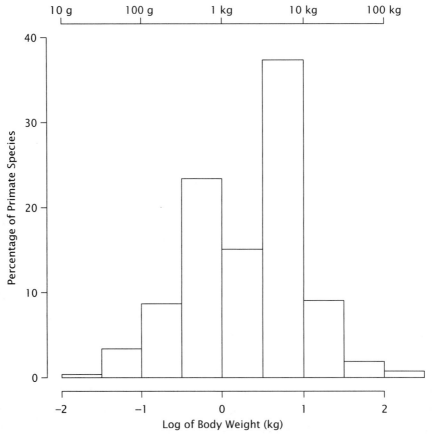

FIGURE 3.4 Mean weights of 265 species of primates from the PanTHERIA database plotted on a logarithmic scale. The scale above the figure shows actual weights corresponding to the logarithms of weight on the scale below the figure. Each interval on the upper axis multiplies weight by 10, from 10 grams to 100 grams to 1,000 grams = 1 kilogram, and so on.

aren't available in Figure 3.3. There are a few primates that weigh less than 100 grams, but not many (the left two bars in the figure), and almost 40% of primates weigh between 3 and 10 kg (the tallest bar in the figure). In addition, the distribution of weights has two peaks, from 300 grams to 1 kg and from 3 to 10 kg. These peaks are called modes, so we can see from Figure 3.4 that body weights of primate species have a *bimodal distribution*.

We began looking at data for heights and weights of individual people and then zoomed out to primates, the group of mammals that includes humans. But there are other kinds of mammals that are smaller than the smallest primates or larger than the largest primates. Smaller mammals than primates include some bats, mice, and shrews; larger ones include some carnivores, ungulates, elephants, dolphins, and whales. The range of weights for all mammals extends from 2 grams for the bumblebee bat of Southeast Asia and the pygmy shrew of Eurasia to about 160,000 kg for the blue whale (Plate 6). This compares to the range for primates from 31 grams for mouse lemurs of Madagascar to 149 kg for the eastern gorilla. Figure 3.5 shows the distribution of mean weights for all

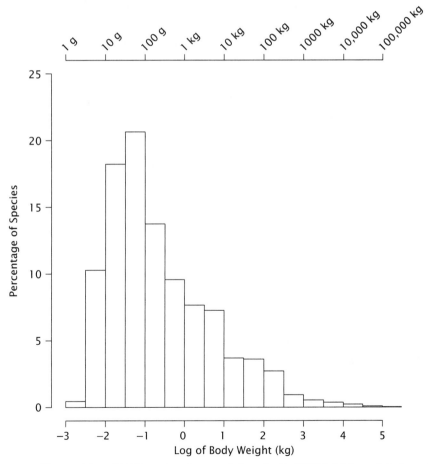

FIGURE 3.5 Mean weights of 3,542 species of mammals from the PanTHERIA database plotted on a logarithmic scale. The scale above the figure shows actual weights corresponding to the logarithms of weight on the scale below the figure.

3,542 mammal species in the PanTHERIA database in the same form as Figure 3.4 for primates. The most common weight class for mammals is from 30 to 100 grams (the tallest bar in Figure 3.5), with about 20% of species in this group. About 88% of mammals weigh less than 10 kg, although some species are much heavier than this. In fact, the mean weight of mammals is 178 kg, much greater than the median of 104 g (0.1 kg) for this highly skewed distribution of weights.

If we zoom out once more, we can compare mammals to other vertebrates like birds, lizards, turtles, and even dinosaurs, although weights of these extinct animals have to be estimated from dimensions of their fossilized bones. We began this discussion of size by considering the shortest and tallest people according to Guinness World Records, so let's compare the smallest and largest species of various types of vertebrates (Figure 3.6) and then revisit the question of why the size of an animal is one of its most important characteristics. The smallest birds are bee hummingbirds at about 2 grams, about the same size as bumblebee bats and pygmy shrews—and a US dime. Among terrestrial vertebrates, only lizards and amphibians are smaller than this. In fact, there is more than a handful of tiny animals in these groups: about 8% of lizards, 17% of frogs, and 20% of salamanders weigh less than 1 gram as adults. At the other extreme, the largest terrestrial vertebrates that ever lived were plant-eating dinosaurs called sauropods, with a maximum weight of about 80,000 kg. This is 10 times as large as living elephants and 3 times as large as the giant rhinoceros, which is now extinct. If we include marine vertebrates, however, we have blue whales that weigh about twice as much as the largest dinosaurs.

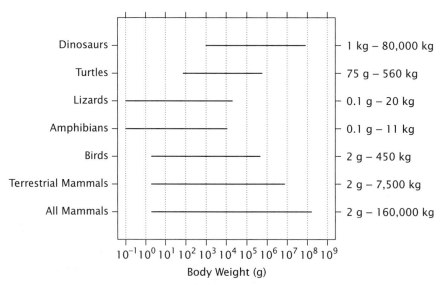

FIGURE 3.6 The smallest and largest of some types of vertebrates. Body weight is plotted on a logarithmic scale. Each horizontal line extends from the minimum to the maximum body weight for that group, and this range is shown numerically on the right side of the graph. All mammals, terrestrial mammals, and birds include species that are now extinct but represented by fossils.

It's worth taking a minute to marvel at the data summarized in Figure 3.6. Mammals have the largest range of body sizes of all animals. Body weights of mammals encompass eight orders of magnitude, from bumblebee bats and pygmy shrews to blue whales. Yet all mammals share a host of characteristics that define what it is to be a mammal—high metabolic rates, hair (even in whales, at least as fetuses though not as adults), three bones in their middle ears, and numerous other skeletal characteristics. All but three species of mammals give birth to live young; mammal mothers nurse their young with milk, even in the three species that lay eggs. How do all these common processes work in animals so different in body size?

Why Do All Birds and Mammals Weigh More Than 2 Grams?

This complex general question entails many specific questions that can't all be discussed here, but let's focus on the relationship between body size and metabolism to illustrate why size is so important for animals and why the wide range of sizes of mammals is so remarkable. Mammals and birds differ from other vertebrates in being "warm-blooded." This means that they have high metabolic rates, which enables them to have high, constant body temperatures. This metabolic pattern is called *endothermy*, meaning "heated from within." By contrast, fish, amphibians, and reptiles are *ectothermic* or "cold-blooded," meaning they gain their heat from the external environment. They have low metabolic rates and body temperatures that tend to fluctuate with temperatures of the environment.

The metabolic rate of an animal is influenced by many factors in addition to whether it's an endotherm or an ectotherm, notably its weight, temperature, level of activity, and whether it is digesting food or not. Therefore physiologists use standardized methods to make fair comparisons between different species. They measure animals at rest and long enough after they have eaten that they aren't digesting food, and they either make measurements at a common temperature or adjust measured values to a common temperature. We can quantify the differences between endothermic birds and mammals and ectothermic amphibians and reptiles based on measurements of hundreds of species summarized by Craig White and his colleagues in 2006. At body weights of 100 grams, birds and mammals have standard metabolic rates about six times greater than amphibians and reptiles. Differences between endotherms and ectotherms are somewhat greater at smaller sizes and somewhat less at larger sizes. As a frame of reference, the standard metabolic rate of a person of average size who is not exercising or digesting food is about 70 watts, similar to a typical light bulb. This can increase substantially with exercise or digestion, although not as dramatically after eating as for Burmese pythons that may increase metabolic rate by 40 times after infrequent, large meals.

The smallest mammals and birds weigh about 2 grams while many lizards and amphibians are much smaller than this as adults. Does endothermy limit how small an animal can be? To answer this question, we need to know a little more about how metabolism works, and we need to look more closely at the relationship between metabolism and body size.

Metabolism constitutes all the chemical reactions that occur in the cells of living organisms. These reactions either break down organic compounds such as carbohydrates, fats, and proteins to release energy, or use energy to make new organic compounds. Said another way, *metabolism* is processing energy to stay alive, grow, and reproduce. Metabolism happens in every cell of the body—muscle cells, brain cells, liver cells, lung cells, bone cells, not to mention leaf cells, root cells, and flower cells of plants. To stay alive, plants use sunlight as a source of energy and carbon dioxide and other raw materials to make sugars and other organic compounds, storing energy derived from sunlight in the chemical bonds that link the atoms of these compounds. Animals use oxygen together with organic compounds derived from plants or other animals that they eat in chemical reactions that release energy from the organic compounds.

Scientists can measure the metabolism of an animal in various ways, but one of the most common is to measure the rate at which it uses oxygen. A researcher places the animal in a closed chamber with air flowing through it. By comparing the amount of oxygen in the air entering the chamber with the amount in the air leaving the chamber, the researcher can calculate how much oxygen that the animal uses during a specified time period. Of course the animal must be unstressed, inactive, not digesting a recent meal, and not shivering to keep warm if the researcher wants to estimate standard metabolic rate. Also, large chambers are needed for large animals, so it's more challenging to measure metabolic rates for elephants than for mice.

Groups of researchers around the world have used methods like these to measure metabolic rates of many animals, beginning with cats, dogs, sheep, and cows in the early 1900s and gradually extending their reach to more and more groups of wild mammals, birds, reptiles, amphibians, and other animals. In 2009, Annette Sieg and five colleagues compiled data for 695 species of mammals ranging in size from the white-toothed pygmy shrew to the African elephant. We can use these data to see how metabolism relates to body size in mammals, which will help us understand how metabolism might influence minimum sizes of these endotherms compared to ectothermic amphibians and reptiles.

We used histograms to depict heights and weights of humans, weights of primates, and weights of mammals in general. In each of these cases, we were considering a single variable—height or weight—and the distributions of values of single variables can be shown clearly with histograms like Figures 3.1 through 3.5. We need different kinds of graphs to illustrate relationships between two variables, and *scatterplots* do this job effectively. Figure 3.7 is a first version of a scatterplot of metabolic rate versus body weight for mammals, using data for 695 species from Sieg and colleagues. Each point represents a single species, with the mean weight of that species plotted on the horizontal axis and its mean metabolic rate plotted on the vertical axis. Weight is shown in kilograms and metabolic rate is shown in milliliters of oxygen used by the resting animal per hour, or ml O_2/hr. These units come directly from measurements in metabolic chambers like

I described above, but metabolic rate can be translated to and from other units that may be more familiar. For example, if we wanted to compare rate of energy use by an animal and an electric appliance, we would use watts. A resting person of average size who is metabolizing at 70 watts is using about 12,500 milliliters (= 12.5 liters) of oxygen per hour. This translates to about 1,400 kilocalories per day, or a little more than two standard Whoppers from Burger King.

Figure 3.7 is not as helpful as it might be for understanding the relationship between body weight and metabolic rate because the points for most of the species are clumped near the lower left corner of the graph, with the point for African elephants far away in the opposite corner. Remember that there are data for 695 species on this graph, so points for many species overlap in the black blob near the origin of the graph in the lower left. This is similar to the problem that we had interpreting the histogram of body weights of primates in Figure 3.3, and we can solve the problem the same way—by plotting the logarithms of the data instead of the raw data. Figure 3.8 does this. We still have overlapping points because we have so much data, but we get a better sense of the

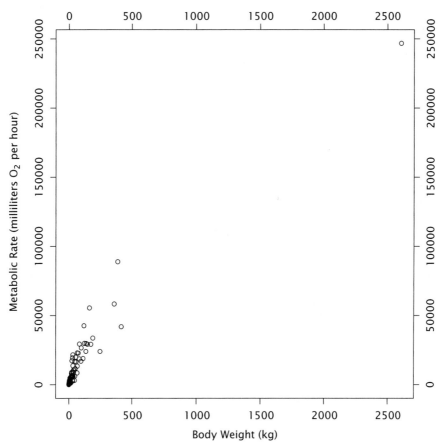

FIGURE 3.7 Total metabolic rates of 695 species of mammals in relation to their body weights. Each point represents one species.

relationship between body weight and metabolism over the whole range of weights in Figure 3.8 than in Figure 3.7.

To stay alive, not to mention grow and reproduce, animals have to supply oxygen and organic compounds to all their cells, and they have to do so at rates commensurate with their metabolic rates. This process involves digestion, respiration, and circulation. Vertebrates use gills or lungs to extract oxygen from air, digestive tracts to extract simple organic compounds from food, and blood pumped by hearts to transport oxygen and organic compounds throughout the body. Within any group of species, like birds or mammals, metabolic rate increases with body size—elephants eat more than humans, who eat more than mice (and elephants process more oxygen too, as shown in Figures 3.7 and 3.8). However, our discussion of cellular metabolism implies that we should also think about rates of metabolism at the cellular level and the fact that organisms have to supply nutrients and oxygen to their cells fast enough to keep the cells alive and functional. This is where things get interesting, because rates of metabolism *of equivalent amounts of tissue*

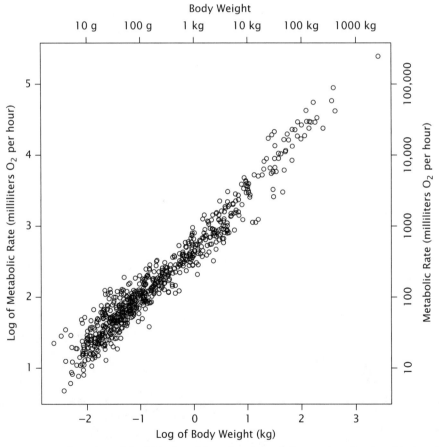

FIGURE 3.8 Data from Figure 3.7 with body weight and total metabolic rate plotted on logarithmic scales.

are greater for small animals than for large animals within a group like mammals or birds, even though *total* rates of metabolism for the whole animal increase with body size for these groups. Figure 3.9 shows this relationship between body size and metabolic rate of equal-sized chunks of tissue for mammals. I made this figure by dividing the metabolic rate for each species shown in Figure 3.8 by its weight, so the units for the vertical axis are now milliliters of oxygen *per gram* per hour.

Hummingbirds, some bats, and some shrews are very small, but no smaller than 2 grams. This means that they have very high metabolic rates at the cellular level, so they have to supply oxygen and nutrients to the cells at a rapid rate. Part of the way that they do this is to have very fast heart rates. For example, heart rates of hummingbirds and small shrews approach 1,200 beats per minute, more than 10 times typical heart rates for people at rest. In each beat, the heart fills with blood, the heart muscle contracts, blood is ejected, and the muscle relaxes. In a hummingbird or small shrew, this happens 20 times each

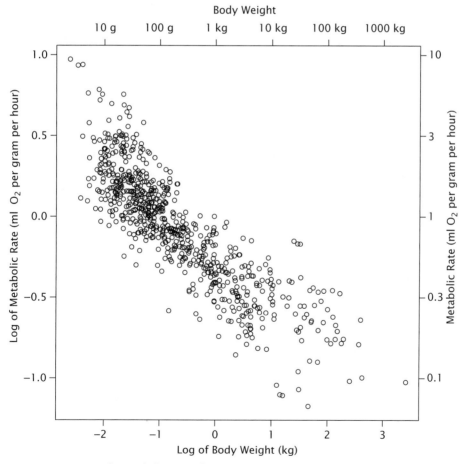

FIGURE 3.9 Mass-specific metabolic rates of mammals in relation to their body weights. Mass-specific metabolic rate is the metabolic rate of 1 gram of tissue. Each point in the graph is the amount of oxygen in milliliters used by 1 gram of tissue in one hour for one of 695 species.

second. The physiologist Knut Schmidt-Nielson suggested that it may be impossible for the complex sequence of events that occur in each heart beat to happen faster than this. Since metabolic rate at the cellular level increases as size decreases for birds and mammals, this means that a 1 gram endotherm would have an even faster heart rate than 1,200 beats per minute, which might not be feasible considering how vertebrate hearts work. But ectotherms have lower metabolic rates than endotherms, so the minimum body sizes of ectotherms aren't constrained in the same way as those of endotherms.

What about the Largest Animals?

What *about* the largest animals? Reasons for upper size limits of different kinds of vertebrates aren't as clear-cut as reasons for lower size limits, but Figure 3.6 suggests one interesting question about large animals—why are marine mammals like the great whales much larger than the largest terrestrial mammals like elephants? We can turn to the relationship of metabolism with body size to answer this question, just as we did for thinking about why there aren't any birds or mammals smaller than 2 grams. For large animals, however, the most important constraint may be whether enough food can be harvested to sustain large *total* metabolic rates.

One hint of the importance of food supply for large body size is that, on land, the largest plant eaters are larger than the largest carnivores. Elephants, giraffes, buffalo, and moose are larger than lions and tigers; herbivorous dinosaurs were larger than predatory dinosaurs. Imagine the Great Plains of North America before we eliminated native herbivores and converted the land to fields of corn and wheat. Huge herds of bison lived in the middle of vast grasslands where the next meal for each animal was a step away. The bison herds were large but moved across the landscape as they fed, so a pack of wolves had to find a herd to get its meal, select a vulnerable animal, and then attack and kill it before they could feed themselves. A similar scenario exists today for wolves and moose in the boreal forests of Canada—the moose are surrounded by their food and just have to turn around to get another meal of leaves and twigs, while the wolves have to search for a young, old, sick, or otherwise vulnerable moose to attack and kill before they can feed.

These are two examples in which herbivores have more food available than the predators that eat them, but these examples illustrate a universal pattern based on how organisms use energy to support life. The ultimate source of energy is the sun; plants convert the energy in sunlight to energy incorporated in chemical bonds of organic compounds to live, grow, and reproduce. Herbivores get energy to sustain their lives by eating plants; predators get energy by eating herbivores. But at each step of this food chain, usable energy is lost due to metabolism at the previous step. Not all of the solar energy that plants process in photosynthesis is available to herbivores because the plants use some of it in their own metabolism, and not all of the energy that herbivores get from plants is available to predators because herbivores use some in their metabolism—simply staying alive. So there is less energy available at each successive step in the food chain.

One consequence of this pattern is that maximum sizes of herbivores are greater than maximum sizes of predators—the greater energy available to herbivores can support larger body sizes.

This pattern of larger size for herbivores than carnivores breaks down for marine animals, however. The largest marine animals are baleen whales, but no whales are herbivores. Instead, baleen whales use an extremely efficient behavior called filter feeding to process enormous numbers of small, shrimplike animals called krill for food. There are 13 species of baleen whales, including bowhead whales, right whales, gray whales, fin whales, humpback whales, and blue whales, with blue whales being the largest animals of all at 160,000 kg.

Filter feeders feed by moving water over a structure that acts like a sieve to separate food particles from the water. They swallow the food, then take in more water and move this new batch of water over the filtering structure. In baleen whales, this structure consists of plates of baleen attached to the roof of the mouth (Plate 7). Baleen is made of keratin, also found in our fingernails and hair. Food particles stick to the baleen, so can be efficiently extracted from large amounts of water.

Jeremy Goldbogen and his colleagues studied filter feeding in fin whales using a combination of classical and new methods. The classical methods included examination and measurement of museum specimens; the new methods used digital devices attached to the backs of whales that recorded specific body movements during feeding bouts, called lunges. The whales feed in dense concentrations of krill. They have enormous heads that accommodate large baleen plates and they can open their mouths to almost 90 degrees to pass a large amount of water across the baleen when they make a lunge. Here is what happens when they have a mouthful of water: "During engulfment the tongue inverts into the cavum [a space between the bottom of the tongue and the walls of the mouth], retreating through the floor of the mouth and back towards the belly button, forming the large oral sac that holds the incoming seawater" (Goldbogen 2010).

Each feeding dive of the fin whales studied by Goldbogen included several lunges through a dense concentration of krill. In each lunge, which lasted only a few seconds, a whale took in more than its own body weight of water from which it filtered 10 kg of krill. It did this repeatedly, taking in enough to sustain its metabolism for a day in several hours of feeding. This amounted to about 1,000 kg of krill. Even more impressively, blue whales eat about 3,600 kg of krill per day in their summer feeding areas around Antarctica, but fast for up to eight months while they migrate to tropical waters to mate and give birth, relying on stored fat during this period.

In summary, baleen whales are very large, so they need lots of food to maintain their high metabolic rates. Unlike large terrestrial mammals, they aren't herbivores but feed on small animals a few centimeters long that occur in dense concentrations in certain parts of the ocean. Baleen whales have an integrated set of morphological, physiological, and behavioral adaptations for a highly efficient foraging method called filter feeding that makes it possible for them to gain enough energy to have very large body sizes.

Filter feeding is not uniquely related to large body size in baleen whales. Many fish are filter feeders. In fact, contrary to the general image of sharks as ferocious predators, the largest sharks are filter feeders. These filter-feeding sharks include megamouth sharks, basking sharks, and whale sharks, the largest fish of all weighing more than 20,000 kg. It may be that the efficiency of filter feeding enables animals as different as whales and sharks to attain very large body sizes.

Conclusions

My main goal in this chapter was to introduce some basic tools for dealing with sets of related observations, that is, data. Graphs are the most important of these tools, and I used two kinds of graphs to illustrate distributions of single variables and relationships between two variables. Histograms illustrated distributions of heights and weights of men and women and weights of primates and mammals in general. Scatterplots illustrated relationships between body weight and metabolic rate.

I also introduced some basic summary statistics—means, medians, ranges, variances, and standard deviations. These complement graphical representations of data. In working with graphs, I showed how transformations of data to logarithmic scales can be helpful when distributions are highly skewed.

There was less biology in this chapter than there will be in the remaining chapters, but I used examples to illustrate key points about graphing and analyzing data. These examples illustrated important biological ideas about hormones and growth in humans and about metabolism as a fundamental physiological process. They also illustrated how we can learn from comparisons in biology—between body sizes of endotherms and ectotherms and between metabolic rates of small and large mammals. We'll use the quantitative tools introduced here to interpret ideas and evidence about various biological questions in later chapters.

Questions to Ponder

1. We used deviations of individual values from the mean of a set of data to calculate the variance on page 48. These deviations are squared and then added together and divided by $n - 1$ to get the variance. I stated that without squaring the deviations, the sum would be zero. Use an example to demonstrate this. If you are ambitious, prove it!

2. Outline a study to relate diets to risk of obesity in different countries, using BMI to measure risk of obesity. You might want to return to this question after reading Chapter 5.

3. Why are the modes of the histogram of primate body weights in Figure 3.4 the bar that includes species between 300 grams and 1 kg and the bar that includes species

between 3 and 10 kg? Hint: What are the logarithms of body weight at the left ends of these bars, and what body weights do these logarithms represent?

4. Baby birds are smaller than their parents when they hatch and baby mammals are smaller than their mothers when they are born, even in bee hummingbirds, bumble-bee bats, and pygmy shrews. How do you think these babies that are smaller than their 2-gram parents survive until they reach adult size?

Experiments: The Gold Standard for Research

You've seen several brief examples of experiments in previous chapters, from tests of navigation by monarch butterflies to studies of the accuracy of humans as eyewitnesses to crimes. Experiments provide only one kind of evidence in science, but experiments are especially important because they may give clear answers to important questions. For this reason, experiments are often considered the "gold standard" for research. Therefore we need to consider experimental methods in some detail to see why experiments have such an exalted status and to learn about limitations of experimental evidence.

I'll use two extended examples to introduce you to experimental methods in biology, one from medicine and one from evolutionary biology. Medicine is a good place to start because medical researchers have a long and deep commitment to using experiments to test hypotheses. Suppose a drug company develops a new drug to treat a serious disease. This drug won't be approved for sale until it is tested in an experiment in which volunteer patients agree to be randomly assigned to one of two groups: an experimental group that receives the drug or a control group that does not. Furthermore, the drug company doesn't just have to do the experiment; it has to show significant benefits of the new drug without major adverse side effects. Would you volunteer for such an experiment knowing that you might be placed in the control group and get no relief from the debilitating symptoms of your disease? In a case like this, it seems that medical researchers are withholding a potentially valuable treatment from half of the patients who volunteer for the study. Is this ethical? Medical researchers argue that the long-term benefits of getting clear and convincing evidence that the new drug is beneficial outweigh the costs of withholding treatment from volunteers who end up in the control group. You may disagree, and the answer isn't always clear, but this is the fundamental reason why experiments like these are pervasive in tests of new drugs, new surgical procedures, and new medical treatments in general.

Two Experimental Studies of Medicinal Use of Marijuana

Laws in 23 US states and the District of Columbia allow people to use marijuana for medical purposes such as alleviating pain, yet the federal Drug Enforcement Administration lists marijuana as a Schedule I drug with no permitted medical uses. This has led to serious conflicts between federal and local law enforcement agencies, especially in California, which was the first state to legalize medical use of marijuana. Although most of the marijuana produced in California supplies a black market for recreational users throughout the country, there is a thriving business supplying medicinal marijuana within the state. Until recently, however, most of the evidence that marijuana has medical benefits came from testimonials of users. Personal stories like these can be highly persuasive but also have important limitations. Each individual is unique, so if one person reports relief from severe pain by smoking marijuana, another person with similar pain might not experience the same result. In addition, personal stories rely on an individual's subjective assessment of pain, which might not translate to another person who tried marijuana for her pain. This means that we can't generalize from a collection of personal stories, even though, as humans, we have a strong tendency to do so.

Most basic medical research in the United States is funded by the National Institutes of Health (NIH). In addition to this publically funded research, private drug companies spend large amounts of money developing and testing new drugs to treat a host of medical conditions. Neither the NIH nor drug companies have invested much money in testing the medicinal value of marijuana, although the NIH has supported research on the biochemical and physiological effects of marijuana as well as its addictive properties. In 1996, California voters passed Proposition 215 allowing use of marijuana for medical purposes, but at that time there wasn't much experimental evidence on the potential benefits or dangers of using marijuana in medicine. Therefore, the California legislature passed the Medical Marijuana Research Act in 1999. This established the Center for Medicinal Cannabis Research at the University of California, San Diego. Over the next 10 years, the legislature funded $8.7 million worth of research on medicinal use of marijuana by scientists at UC San Diego and other public institutions in California. This research program has now ended, but six studies were completed by 2012 and the results of a few others are being prepared for publication. We'll take a detailed look at two of the completed studies to learn how experiments work, why experiments are considered the gold standard of research methods, and how to critically evaluate the results of experiments.

A key feature of science is that it is a cumulative process; as Isaac Newton said, "If I have seen further it is by standing on the shoulders of giants." In the case of marijuana research, knowledge of biochemistry and physiology from previous studies provided a foundation for more rigorous tests of medicinal effects of marijuana than would otherwise have been possible. These previous studies led to identification of the main psychoactive ingredient in marijuana plants—a molecule called THC—and to an understanding of what THC does in the brain.

THC stands for delta-9-tetrahydrocannabinol, a relatively small organic compound containing atoms of carbon, oxygen, and hydrogen. THC was isolated from marijuana plants in 1964. It is an example of a plant secondary compound, that is, a chemical produced by a plant that is not involved in the basic metabolism for processing energy discussed in Chapter 3. Plants produce a huge variety of secondary compounds, many of which function as defenses against being eaten by herbivores. For example, chili peppers taste hot because of a secondary chemical called capsaicin that is most concentrated in the seeds; this deters mammals from feeding on the seeds and protects the seeds against attack by fungi. THC may have a similar function for marijuana, although this hasn't yet been established.

Between 1988 and 1995, researchers reported a series of exciting discoveries about the physiological effects of THC. They first discovered that THC binds to a molecule on membranes of cells in the brain, causing temporary changes in these cells that ultimately produce what users of marijuana experience when they get high. Researchers soon identified the gene responsible for the THC receptor molecule in the brain and then found a natural compound in brains that binds to this receptor molecule. This natural compound was named anandamide, based on the Sanskrit word for bliss, ananda. These discoveries opened up a new area of research in neurobiology to try to understand the role of anandamide and similar compounds in controlling various functions of human brains. In hundreds of studies since 1990, researchers have found that anandamide influences responses to pain and stress, memory, reproduction, and other processes. In 1996, a compound like anandamide was even found in chocolate.

Molecules of anandamide are pervasive but relatively short-lived in the brain. THC from marijuana is like a souped-up version of anandamide—it attaches more tightly and for a longer time to cell membranes. This suggests that THC might have more powerful forms of the same kinds of effects as anandamide. For example, rats treated with a compound that interferes with the anandamide system don't perform as well in standard tests of spatial memory as control rats, and in humans one common response to smoking marijuana is loss of short-term memory. Similarly, the fact that anandamide influences responses to pain provides a physiological rationale for testing marijuana for pain relief in patients.

With this biochemical and physiological foundation, the Center for Medicinal Cannabis Research was charged by the California legislature with developing a research program to test the potential medical benefits of marijuana. More specifically, their mission was to "conduct high quality scientific studies intended to ascertain the general medical safety and efficacy of cannabis products and examine alternative forms of cannabis administration." We'll consider two studies funded by the Center for Medicinal Cannabis Research that tested the efficacy and safety of smoked cannabis for pain relief.

In medicine, the phrase *high quality scientific studies* implies rigorous experiments, in particular, randomized, double-blind, controlled experiments. Both examples that follow illustrate how experiments are randomized, "blinded," and controlled and why these features of rigorous experiments are important. D. I. Abrams and several colleagues at the University of California, San Francisco and San Francisco General Hospital did the first of these studies with HIV/AIDS patients. These patients often experience a type of pain called sensory or peripheral neuropathy in their arms and legs. This can involve feelings of aching, numbness, or burning that can be exacerbated by something as simple as pulling a sheet up over bare legs.

Does Marijuana Relieve Pain for HIV/AIDS Patients?

The research group found 50 subjects for their study who were infected with HIV and reported significant pain during the week prior to the start of the experiment. This and other studies of the medicinal effects of marijuana depend on standardized but subjective assessments of pain by the subjects themselves. Abrams's group used a common method in which subjects recorded at 8:00 AM each day how much pain they felt during the previous 24 hours, with zero representing "no pain" and 100 representing "the worst pain imaginable." To be included in the experiment, a subject had to have an average pain score of at least 30 for a week. The researchers only recruited subjects who had smoked marijuana at least six times before the study began because they wanted subjects to know how to inhale and to understand the psychological effects of marijuana. About 73% of the subjects were regular users of marijuana, but these had to agree to stop before the study began.

All experiments involve comparison of two or more treatments. In the simplest case, there is an experimental treatment and a control treatment that differs from the experimental treatment in one and only one way. The goal of this study by Abrams and colleagues was to see if THC, the main psychoactive ingredient in marijuana, could reduce peripheral neuropathy of people infected with HIV. Therefore the researchers used two types of cigarettes that they got from the National Institute on Drug Abuse. One type was a regular marijuana cigarette containing on average 3.56% THC; the other was a cigarette made of material from marijuana plants from which THC had been extracted. The latter cigarettes, with 0% THC, were used for the control treatment, also called a placebo treatment in a medical study such as this. This experiment by Abrams's group fits the model of a "simple" controlled experiment with an experimental treatment and a control treatment differing in one, clearly defined way. As we will see, however, this doesn't resolve all difficulties in interpreting effectiveness of the experimental treatment compared to the control treatment.

Once subjects were recruited for the study, they were admitted to the Clinical Research Center of San Francisco General Hospital where the experiment was conducted. They spent seven days in the hospital, two days for collection of baseline data and five days for experimental treatments. During these five days, each subject smoked three cigarettes at regular times each day. Subjects recorded daily impressions of pain during their time in the hospital and for the following six days.

Half of the 50 subjects were randomly assigned to the experimental treatment group, so they smoked 15 marijuana cigarettes containing THC during the five days when treatments were given. The other 25 subjects were assigned to the control group, so they smoked 15 marijuana cigarettes with no THC. The subjects ostensibly didn't know to which group they belonged, nor did the "research staff [who] monitored patients during smoking sessions, weighed the cannabis cigarettes immediately before and after they were administered to patients, and returned all leftover material to the pharmacy."

I've just described three features of this study that are key elements of all rigorous experiments: control, randomization, and blinding (the last being especially important in medical experiments with human subjects). This was a controlled experiment because researchers compared responses of patients to two treatments that differed in one factor that was the focus of their research—the presence or absence of THC in marijuana cigarettes. It was also a double-blind experiment: neither the subjects nor the researchers who collected the data knew which treatment each subject received. Of course, some members of the research team ultimately had to know the status of each subject to analyze the data, but this information was stored in coded form and not available to the analysts until all the data were collected and recorded. The important point was that the researchers who interacted with the patients during the experiment couldn't influence the patients' records of their pain in a biased way because these researchers didn't know what the patients should expect.

In theory, both patients and researchers were blind to the treatments, but in practice, I wonder about the patients in this particular study. The researchers only recruited subjects with some prior experience smoking marijuana because they didn't want to teach them how to inhale. But this means that all subjects would know the feeling of being high. If THC is the main cause of a marijuana high, then it seems likely that subjects could have guessed whether they were in the experimental or control group based on whether they got high or not. In one respect, making cigarettes out of marijuana plant material from which THC has been extracted seems like a clever way to design a control treatment for this study; in another respect, it seems flawed because it should be easy for subjects to guess their treatment. Question 2 at the end of this chapter asks you to think about possible solutions to this problem.

Randomization is the most subtle of the key elements of effective experimental design, so let's consider the purpose of randomization using this study as an example. I'll set the stage for this by telling you a little more about the participants in the study. All were adults, with an average age of about 48. Forty-five percent of the subjects were Caucasian, 38% were African American, and the remainder were Latino or Asian. On average, the subjects had been infected with HIV for 15 years, but there was quite a bit of variation in time of infection. Fifty-six percent of the subjects were using other medications to control pain; 44% were not. Suppose the researchers had asked the subjects to pick their own treatment group. What if they found that younger subjects were more likely to choose the THC treatment while older subjects were more likely to choose the placebo treatment? Subjects exposed to the THC treatment might experience less pain than subjects exposed to the placebo, but this could be due to the younger ages of the THC subjects rather than any ability of THC to reduce pain. We would say that treatment (THC vs. placebo) is *confounded* with age in this experimental design.

It's fairly obvious that letting subjects choose their own treatment would be a poor strategy for a study like this. What if, instead, the researchers assigned subjects to treatments nonrandomly? For instance, they might assign subjects using other pain medications to the control group and subjects not using other medications to the THC group. Suppose subjects in the THC group reported similar levels of pain as subjects in the control group. This could mean that THC reduces pain to the same extent as other medications, supporting the argument that smoking marijuana has medicinal value comparable to standard medications. But it might be that subjects in the THC group weren't using other medications *because* they weren't experiencing as much pain as those using other medications. Treatment with THC or placebo is confounded with use of other pain medications, so the results of a study using this design would also be inconclusive.

The main purpose of randomization is to protect against conscious and unconscious bias in assignment of subjects to treatment groups. Without randomization, it's possible for measured or unmeasured variables to be confounded with a treatment factor and thus account for different responses of subjects in the treatment and control group. I showed how measured variables such as age or use of other pain medication could be confounded with treatment by THC or placebo in this study, leading to inconclusive results. Unmeasured variables can be even more insidious. Suppose the researchers assigned the first 25 volunteers for the study to the THC treatment and the last 25 volunteers to the placebo treatment. Late volunteers might have been late in signing up because they were more stressed than early volunteers. The researchers didn't measure stress levels of their volunteers, but this could be confounded with pain levels. By nonrandom assignment of subjects to treatments, the researchers would have created an inconclusive study. Even worse, this design would have been compromised by the unmeasured variable of stress, making it more difficult to recognize the problem.

Volunteers participating in this study kept daily records of pain that they felt in their arms and legs for seven days before entering the hospital and for two more days of acclimation to the hospital environment before they started smoking THC or placebo cigarettes. During this period before the experiment started, average pain ratings were about 60% of maximum for subjects in both groups (Figure 4.1). Pain ratings for both groups dropped during the treatment phase, but they dropped more for volunteers who smoked marijuana containing THC than for those who smoked marijuana from which THC had been extracted (Figure 4.1). After volunteers left the hospital and stopped using either active or placebo cigarettes, the average pain rating for those in the THC treatment gradually rose over the next week to the same level as the average for those in the placebo treatment (Figure 4.1).

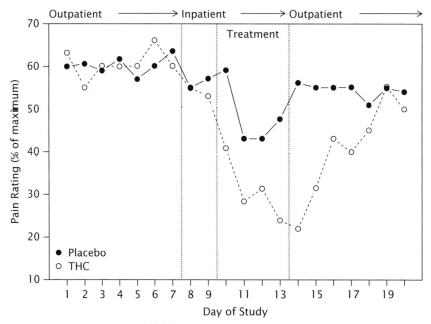

FIGURE 4.1 Effects of smoked marijuana on pain experienced by HIV patients studied by D. I. Abrams and colleagues at the University of California, San Francisco. Volunteers recorded their pain levels daily for seven days prior to hospital admission, then for two days as inpatients before treatments were started, then for five days during treatment with either marijuana cigarettes containing THC or placebo cigarettes with no THC, and then for six days as outpatients without treatment. Points are medians (see Chapter 3) for the 25 volunteers in each group. The leftmost dotted line indicates hospital admission at noon on day 7, the next dotted line shows when the first marijuana cigarette was smoked at 2 pm on day 9, and the rightmost dotted line shows when the last cigarette was smoked at 2 pm on day 13.

Figure 4.1 appears to show that smoking marijuana can alleviate a type of pain called peripheral neuropathy that is experienced by many HIV patients, although patients who smoked placebo cigarettes had some benefit as well. What about side effects? Subjects in the THC group reported higher levels of anxiety, disorientation, confusion, dizziness, and sedation (calmness) than subjects in the placebo group, while there were no differences in feelings of paranoia or nausea. As you know from reading the fine print on the information that comes with drugs prescribed by your doctor, many drugs have side effects, but these don't necessarily outweigh the benefits of taking the drugs. In the case of smoking marijuana to relieve pain, the researchers argued that side effects were relatively minor. Patients rated side effects as none (scored as zero), mild (1), moderate (2), and severe (3). The highest average rating by patients smoking THC cigarettes was 0.54 for sedation, which means that patients thought the treatment slightly increased their calmness (zero would represent no change, 1 would represent a mild increase, and 0.54 is about halfway between zero and 1). You might argue that, if anything, increased calmness would be a beneficial side effect. The next highest average rating by patients smoking THC cigarettes was 0.25 for anxiety. This should clearly be considered a negative side effect, but the average value of 0.25 implies that most patients reported no anxiety with the rest reporting only mild anxiety.

Abrams and colleagues concluded that "smoked cannabis was well tolerated and effectively relieved chronic neuropathic pain from HIV-associated sensory neuropathy." How convincing is this conclusion? Does this experiment provide definitive evidence that smoking marijuana can alleviate a common type of pain experienced by HIV patients? Think about Questions 2 to 4 at the end of this chapter as you evaluate this experimental study. Similar questions can be asked about many experiments, although some experiments are more conclusive than others. However, few experiments have such direct implications for public policy as this one. Recall that the federal government has a very different attitude about medicinal use of marijuana than several individual US states. If the results of this and similar experiments with medicinal marijuana are foolproof, should that change the terms of the debate between federal and state law enforcement agencies? Will it? What if these experiments with medicinal marijuana are flawed?

How Does Smoking Marijuana Influence Induced Pain in Healthy Volunteers?

The Center for Medicinal Cannabis Research funded another study that had some basic similarities to the study that we've just discussed but also some important differences. Like the study of HIV patients in San Francisco, this new study by Mark Wallace and seven colleagues at UC San Diego was a randomized, double-blind, controlled experiment. Wallace and colleagues measured pain in the same way, by asking subjects to rate pain on a scale from zero to 100. They recruited 15 volunteers who had smoked marijuana within six months prior to the study but agreed not to do so for 30 days before the start of the experiment. In this study, however, the volunteers did not have HIV and in fact were generally healthy.

The San Diego researchers wanted to test the effects of marijuana on induced pain rather than pain associated with HIV or some other disease. Although the motivation for the study was to learn something about use of marijuana for medical purposes, the researchers thought they could control pain more rigorously by using a standardized method to induce it in healthy people rather than relying on the variable levels of pain that HIV patients or other ill people might feel. As you know from watching movies, there are many methods of causing pain—heat, cold, sharp objects, and so on. In this study, Wallace's team injected volunteers with 100 micrograms (0.0001 grams) of capsaicin under the skin of each arm. This produced a burning sensation, similar to tasting chili peppers, which are the source of capsaicin.

Besides using healthy volunteers, the San Diego researchers tested marijuana cigarettes with four different levels of THC: 0% (the placebo), 2%, 4%, and 8%. This enabled them to test for a dose-response relationship between exposure to THC (the dose) and experience of pain (the response). The San Francisco researchers had used only two treatments in their study of HIV patients: a placebo control and a treatment with cigarettes containing 3.56% THC. One disadvantage of this approach was that they might have missed a beneficial effect, if 3.56% THC wasn't enough to alleviate peripheral neuropathy in HIV patients. Another disadvantage was that a smaller dose might have reduced pain with fewer side effects. Of course a dose-response study is more complex to conduct than comparison of a single treatment to a placebo and may not have been feasible in the research with HIV patients.

A final key feature of the study by Wallace's team in San Diego was that they used a crossover design. This means that each subject received each of the four treatments. Treatments were separated by at least one week for each subject, and the order was randomized for each subject. For example, the first subject might have smoked a cigarette with 2% THC in his first session on July 1, an 8% cigarette on July 8, a placebo cigarette on July 18, and a 4% cigarette on July 25 while the sequence for the second subject might have been 4% on July 1, 0% on July 10, 2% on July 17, and 8% on July 26, and so forth for the remaining subjects. Note that I've picked dates to illustrate separations of *at least one week* for trials of each subject because this was how the researchers designed their study.

Subjects began each trial by smoking one marijuana cigarette as guided by a nurse. Subjects lit the cigarette, inhaled for 5 seconds, waited 10 seconds before exhaling, then repeated this three times at intervals of 40 seconds. Twenty minutes later, the nurse injected capsaicin in the right arm, and then the subject recorded pain at 0, 2.5, 5, 7.5, and 10 minutes. This process was repeated in the left arm 55 minutes after the cigarette had been smoked.

As you might expect, the pain felt by subjects in this study was greatest immediately after capsaicin injection and declined gradually over the next 10 minutes. This was true regardless of whether they smoked a marijuana cigarette 20 minutes before injection in the right arm (Figure 4.2) or 55 minutes before injection in the left arm (Figure 4.3). It was also true for all doses of THC—0%, 2%, 4%, and 8% (all lines in Figures 4.2 and 4.3 slope downward to the right as times after capsaicin injection approach 10 minutes). With 20 minutes between smoking a marijuana cigarette and the pain stimulus from capsaicin, the treatments had the same effect on pain despite their differences in potency (Figure 4.2). In particular, response to the placebo was similar to responses to cigarettes containing 2%, 4%, and 8% THC. By contrast, with 55 minutes between smoking a marijuana cigarette and capsaicin injection, there were clear differences between treatments in pain responses (Figure 4.3). Subjects who smoked cigarettes with 8% THC felt the most pain, subjects who smoked cigarettes with 4% THC felt the least pain, and subjects who smoked cigarettes with 0% or 2% THC felt intermediate levels of pain. There appears to have been a dose-response relationship between exposure to THC in smoked marijuana and pain induced by injection of capsaicin under the skin about an hour later. A low dose of 2% provided no benefit compared to the placebo, a moderate dose of 4% reduced pain, but a high dose of 8% increased pain.

The researchers also measured the concentrations of THC in the blood plasma of their subjects before injecting capsaicin into the left arm. Even though the subjects were smoking marijuana cigarettes in a standardized way under supervision of a nurse, there are several reasons why they might have taken in different amounts of THC. First,

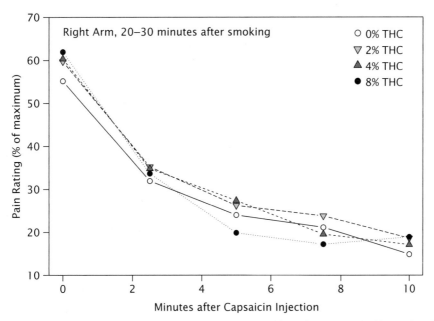

FIGURE 4.2 Effects of smoked marijuana on induced pain in healthy volunteers studied by Mark Wallace and colleagues at the University of California, San Diego. In each trial, subjects smoked one marijuana cigarette, a researcher injected capsaicin in the *right* forearm 20 minutes later, and the subjects recorded their feelings of pain on a scale of zero to 100 at 0, 2.5, 5, 7.5, and 10 minutes after capsaicin injection. Cigarettes contained 0%, 2%, 4%, or 8% THC. Points are means for the 15 volunteers for each treatment at each time after injection.

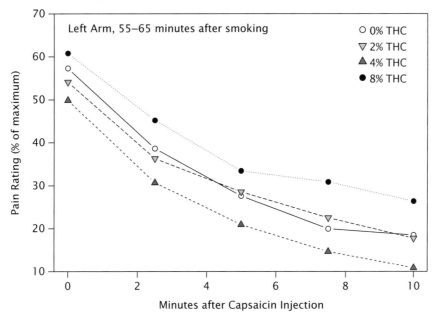

FIGURE 4.3 More effects of smoked marijuana on induced pain in healthy volunteers. In each trial, subjects smoked one marijuana cigarette, a researcher injected capsaicin in the *left* forearm 55 minutes later, and the subjects recorded their feelings of pain on a scale of zero to 100 at 0, 2.5, 5, 7.5, and 10 minutes after capsaicin injection. Cigarettes contained 0%, 2%, 4%, or 8% THC. Points are means for the 15 volunteers for each treatment at each time after injection.

the cigarettes for each treatment weren't identical. For example, those used for the low dose treatment, which ostensibly contained 2% THC, actually varied between 1.76% and 2.03% THC. Second, some subjects may not have inhaled as deeply as others, and, third, there may have been differences between subjects in how quickly or how completely the THC passed from the lungs into the blood.

There was quite a bit of variation in the relationship between THC concentration in blood plasma and pain following capsaicin injection, although higher levels of THC were generally associated with less pain (Figure 4.4). However, Figure 4.4 illustrates an important general point about interpreting scientific evidence. There is clearly a pattern in the data plotted here but also lots of variability. The pattern, or *signal*, is the negative relationship between pain and concentration of THC in blood plasma. The variability reflected in the scatter of data points is called *noise* because it interferes with the clarity of the signal showing how THC may influence pain. But this variability is also a clue that something more is going on, perhaps another compound besides THC that also influences pain.

As in the study of HIV patients, the healthy volunteers who participated in the San Diego study reported few side effects. The San Diego team also found no effects of smoking marijuana on two assessments of cognitive ability. Despite the evidence that smoking marijuana with a moderate amount of THC can reduce pain induced by a standard method of capsaicin injection in healthy people, the San Diego researchers were forthright about the limited applicability of their study to medicinal use of marijuana. They pointed out that

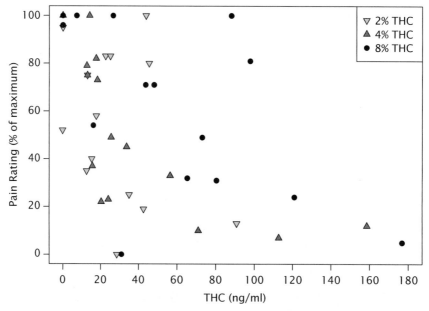

FIGURE 4.4 Relationship between concentration of THC in blood plasma and pain experienced by subjects immediately after injection of capsaicin in their left forearm. Concentration of THC was measured in nanograms per milliliter; divide by 1,000 for parts per million. Researchers drew blood to assay THC 45 minutes after subjects smoked a marijuana cigarette under controlled conditions and injected capsaicin 10 minutes later. These data extend the results shown in Figure 4.3.

sick people might respond differently to marijuana than healthy ones and that long-term use of marijuana for medical purposes might have more damaging side effects than those seen in this short-term study. Nevertheless, this study illustrates some additional features of experiments with human volunteers that we didn't see in the study of HIV patients. Notably, the San Diego team demonstrated a dose-response relationship between amount of THC exposure and experience of pain. They also collected additional data on plasma levels of THC that raised new questions for further research. Marijuana has at least 400 secondary compounds besides THC, including other cannabinoids (the group of compounds that includes THC), terpenoids (a group that includes compounds responsible for the flavors of cinnamon, cloves, and ginger), and flavonoids (a group that includes compounds found in tea, red wine, and dark chocolate). It shouldn't be too surprising that marijuana has multiple and diverse effects on human physiology. For medicinal purposes, further research will need to focus on effects of individual compounds in isolation as well as potential synergies between different compounds in this fascinating plant.

I've described two experimental tests of the ability of marijuana to alleviate pain in human subjects, but researchers also do experiments with animals that are relevant to medical uses of marijuana. Although the US Food and Drug Administration hasn't approved marijuana per se for medical use, it has approved pure THC in capsule form for treatment of nausea resulting from chemotherapy and treatment of extreme weight loss resulting from advanced AIDS. However, patients who use THC for these purposes often experience reduced attention span and loss of memory. THC also inhibits performance in a standard test of memory in mice, so Chu Chen and colleagues at Louisiana State University designed an experiment to see if they could counteract this memory loss. The researchers found that THC increases activity of an enzyme in the brain called COX-2. When they blocked COX-2 activity while giving mice THC, the mice's performance in the memory test was restored. In humans, ibuprofen is commonly used to reduce pain, and ibuprofen acts by blocking COX-2 activity. Therefore this study of mice sets the stage for new experiments with human volunteers to see if the side effects of using THC for pain control can be alleviated with ibuprofen or a similar drug. Chen and colleagues reported these results in late November 2013 as I was doing final revisions of this chapter. By the time you read this book there will undoubtedly be further advances in scientific understanding of the physiological effects and potential medicinal value of marijuana. This story about the science of marijuana illustrates the fact that most stories in science are incomplete because they describe work in progress. I hope this encourages you to continue learning about science in the years to come.

How Is Sex Determined? Observational and Experimental Evidence

If you took a biology course in high school, you probably think you know the answer to this question. We have 46 chromosomes in most of our cells. There are 22 pairs of *autosomes*, for a total of 44, plus two sex chromosomes. Members of each pair of autosomes have the same length and the same pattern of light and dark bands, representing different types of DNA (Figure 4.5). In female mammals, including humans, the two sex chromosomes symbolized by X are also paired, while males have one X chromosome and a much smaller Y chromosome. Thus females can be represented by XX and males by XY, and this chromosomal difference is ultimately responsible for all the morphological, physiological, and behavioral differences between female and male mammals (it's important to specify mammals because sex determination works differently in other animals).

One exception to the general rule that our cells have 46 chromosomes is the sex cells, or *gametes*. A special type of cell division called meiosis produces gametes with half the normal complement of chromosomes—23 in humans, 32 in horses, 20 in North American beavers. Each egg produced in an ovary of a female mammal has one X chromosome, while each sperm produced in the testes of a male has either one X or one Y chromosome.

You probably also learned in high school biology or sex education that human male ejaculate contains millions of sperm, half with an X chromosome and half with a Y chromosome. If a male and female have just had unprotected sex, the sperm will swim up the female's reproductive tract. If the female has ovulated recently or is close to ovulating, there is a reasonable chance that one of the many sperms in her reproductive tract will fertilize the egg, initiating a process that will culminate nine months later in the birth of

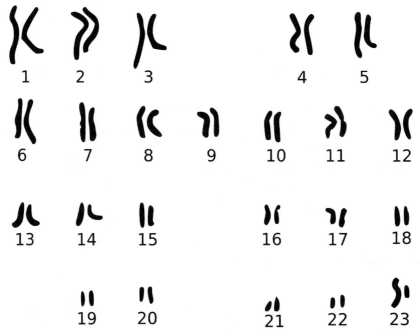

FIGURE 4.5 Karyotype of a human male. There are 22 autosomes, one X chromosome, and one Y chromosome. All cells except gametes have one pair of each autosome; the members of a pair can be recognized by their similar lengths and patterns of light and dark bands when stained (bands not shown here). The twenty-third pair includes a larger X chromosome and a smaller Y chromosome. A female's karyotype would be the same except that she would have two paired X chromosomes and no Y chromosome.

an infant. At fertilization, the 23 chromosomes in the sperm are added to the 23 in the egg, restoring the full chromosome number of 46. Since about half the sperm carry an X chromosome and half carry a Y chromosome while the egg carries an X and since fertilization is thought to be a random process, there is a 50% chance that the baby will be an XX female and a 50% chance that it will be an XY male.

Can Mothers Influence the Sex of Their Offspring? An Evolutionary Hypothesis

Despite this description of meiosis and fertilization, the sex ratio at birth is not quite 50:50 in humans, but closer to 51% males and 49% females. Many people have wondered about why slightly more males than females are born. Do Y-bearing sperm survive better or swim faster than X-bearing sperm? Are mothers pregnant with daughters more likely to have miscarriages than mothers pregnant with sons? These are examples of mechanistic explanations for sex ratios at birth in human populations, but some biologists have also thought about this question from an evolutionary perspective. Two young scientists at Harvard University, Robert Trivers and Dan Willard, asked a seemingly simple question in 1973 that inspired a huge amount of research on the evolution of sex ratios in humans and other animals. Their question was, are there circumstances in which it would be advantageous for mothers to produce more babies of one sex than the other? Trivers and Willard called their 1973 paper "Natural Selection of Parental Ability to Vary the Sex Ratio of Offspring," and they not only asked the question but proposed an explanation, now known as the Trivers-Willard hypothesis. Their paper was influential because it stimulated many different kinds of research on a great variety of animals to test their hypothesis. We'll discuss two main tests of the Trivers-Willard evolutionary hypothesis, one experimental and one observational, and a recent experimental study of a mechanism by which mothers might control sexes of the offspring that they produce. This mechanistic research grew out of some of the evolutionary studies, so these examples should reinforce the point that I made in Chapter 1 that studies of proximate (mechanistic) and ultimate (evolutionary) causation can be mutually reinforcing.

Let's begin with Trivers and Willard's question—when might it be advantageous for mothers to produce more babies of one sex than the other? What do they mean by advantageous? They were thinking in evolutionary terms, so the advantage to mothers would be measured by their genetic contributions to future generations. As discussed in Chapter 1 and Appendix 2, mutation produces genetic variation in a population. If a new form of a gene resulting from mutation contributes to increased reproductive success for carriers of that mutation, then that mutation will gradually spread through the population. In particular, one or more mutations might enable mothers to have more sons under some environmental conditions and more daughters under other conditions. If this increases the mothers' reproductive success compared to mothers without the mutations, then mutations enabling mothers to control sexes of their offspring will spread in the population.

The Trivers-Willard hypothesis builds on this basic process of natural selection. It applies specifically to species with a polygynous mating system in which dominant males mate with many females and subordinate males mate with few females or none at all. For example, Burney Le Boeuf studied a highly polygynous mammal called the northern elephant seal (Plate 8a) on Año Nuevo Island off the coast of central California. These large mammals spend two to three months each year on the island and the rest of the time at sea. Females give birth to one pup a few days after they arrive, nurse the pup for about a month, then mate, and return to the sea to feed. Le Boeuf found that females could produce at most 10 offspring between sexual maturity at age 4 and death at age 14, while the most successful male produced at least 170 offspring by being the dominant male in the colony for three successive years. In any one year, only 14% to 35% of males mated, and most males died without having any opportunity to breed.

In a polygynous mating system, physical condition is likely to affect the relative reproductive success of males more than females. A large, well-nourished, healthy male can dominate other males, thereby gaining more opportunities to mate than a small, hungry male. In species with multiple births, females in good condition may have larger litters than females in poor condition, but females generally don't get shut out of breeding altogether unless their health is very poor. By contrast, subordinate males may not have any offspring at all.

In developing their hypothesis, Trivers and Willard asked how females in good physical condition and in poor physical condition might maximize their genetic fitness if they had some control over the sex of their offspring. Based on two assumptions, they came up with different predictions for these two types of females. Suppose the physical condition of the mother influences the physical condition of her offspring at birth and the physical condition of a newborn influences its condition when it grows to adulthood. In other words, Trivers and Willard assumed that a mother in good condition would have offspring in good condition that would grow up to be adults in good condition. Conversely, if the mother was in poor condition, her offspring would be born in poor condition and would remain in poor condition as adults.

What does this mean for a female in good physical condition? By having sons instead of daughters, she will pass more genes to future generations because her sons will likely be dominant males and do more mating than sons of mothers in poor condition. This means that our mother in good physical condition will have more grandoffspring through her sons than through her daughters.

What about a female in poor physical condition? If she produces sons, they will likely also be in poor condition, so might not mate at all once they become sexually mature. Her daughters, however, will be able to mate. Therefore a mother in poor physical condition will have more grandoffspring through her daughters than through her sons.

By this reasoning, the Trivers-Willard hypothesis predicts that females in good physical condition should produce relatively more sons than females in poor physical condition. The reasoning only works for species with a polygynous mating system, so the prediction shouldn't be tested in species with other kinds of mating systems, like monogamy. The hypothesis also depends on two key assumptions—that a mother's physical condition influences the physical condition of her offspring at birth and that the condition of newborns influences their condition as adults.

An Experimental Test of the Trivers-Willard Hypothesis with Opossums

Steven Austad and Mel Sunquist described the first experimental field test of the Trivers-Willard hypothesis with a mammal in a short paper published in 1986. They worked with common opossums (Plate 8b) in Venezuela. Their experimental design was pretty simple compared to the marijuana experiments we discussed above, but the execution of the experiment in nature was challenging.

Opossums give birth to large litters of tiny and poorly developed young. The gestation period is only 13 days but the young spend 60 to 70 days nursing in a pouch on the mother's belly before independence. Austad and Sunquist attached radio transmitters to 42 virgin females before the beginning of breeding. By studying virgins, Austad and Sunquist didn't have to worry that their results were confounded by prior breeding experience of the females.

Common opossums are solitary, and, once mated, the females are responsible for all parental care. Although the variability in mating success among males is unknown, some males have large home ranges that overlap smaller home ranges of several females, consistent with a polygynous mating system. Therefore Austad and Sunquist thought it was appropriate to test the Trivers-Willard hypothesis with common opossums.

Austad and Sunquist experimentally manipulated the physical condition of some of their female subjects by giving them extra food. Specifically, they tracked them to their sleeping dens using radio telemetry and gave them sardines and cat food every other day throughout the breeding season. The extra food was placed at the openings of the dens at dusk, just before the opossums began their nightly activity period. The researchers randomly assigned 19 of the female opossums wearing radio transmitters to the experimental group getting extra food and the other 23 to a control group with no extra food. This is a simpler control treatment than in the marijuana studies, where subjects in the control group smoked placebo cigarettes. Question 9 at the end of the chapter asks you to think about some potential pitfalls of this approach.

Despite extra food, females in the experimental group had litters of the same size (7.86 on average) as females in the control group. But offspring of fed mothers grew faster in the pouch than offspring of control mothers. Offspring were marked at 10 to 15 days after birth so they could be identified after they left the pouch, and Austad and Sunquist periodically trapped the population at their study site to assess survival of these animals. Opossums whose mothers had been fed sardines and cat food were more than twice as likely to be captured as opossums whose mothers hadn't received extra food, suggesting that young in the first group survived longer. Recall that the Trivers-Willard hypothesis depends on two assumptions—that mothers in good condition produce offspring in good condition and that good condition at birth carries over into adulthood. The faster growth rate and greater survival probability of opossums whose mothers were in the experimental group are consistent with these assumptions.

Most of the opossums studied by Austad and Sunquist produced two litters, for a total of 33 litters for females in the experimental group and 36 litters for females in the control group. Females that received extra food during the breeding season had a total of 149 sons and 110 daughters, while females without extra food had 138 sons and 145 daughters. This translates to 58% male offspring for experimental mothers and 49% male offspring for control mothers, consistent with the Trivers-Willard hypothesis that females of polygynous species in good condition should produce relatively more sons than females in poor condition.

We need to think a little more deeply about this conclusion, however. Suppose female opossums can't really control the sexes of their offspring, but instead these sexes result from random fertilization of eggs by X- and Y-bearing sperm, produced in equal numbers by fathers. Could this random process yield 49% sons for a group of mothers like the control group in this study? Could the same random process yield 58% sons for another group of mothers like the experimental group? We should hesitate to accept the results of Austad and Sunquist's study as evidence supporting the Trivers-Willard hypothesis unless we can show that these results were unlikely to occur by chance alone.

Based on the logic outlined in Box 4.1, I designed a computer experiment to estimate the likelihood that the numbers of sons and daughters produced by female opossums studied by Austad and Sunquist could have been due simply to chance. Control

---

BOX 4.1 A Coin-Flipping Experiment to Illustrate How Chance Can Influence Events

Flip a coin eight times. How many heads did you get? I asked my computer to repeat this set of flips 10 times, and got 4, 6, 4, 2, 5, 3, 5, 3, 4, and 2 heads in these 10 trials. This made a total of 38 heads, which was 48% of the 80 total flips. If sex determination of fertilized eggs is completely random with a 50% chance that each resulting embryo will be male, then coin flipping should be a good model for predicting sex ratios in litters of opossums. In other words, 10 litters, each with eight babies, might have 4 + 6 + 4 + 2 + 5 + 3 + 5 + 3 + 4 + 2 = 38 total males, just as 80 flips of a coin produced 38 heads. Of course, if I repeated my computer experiment, I'd probably get a different result. Indeed I did—26 total heads on my second try, 48 on my third, 40 on my fourth, and 37 on my fifth.

Let's pursue the coin-flipping model a little further. What kind of evidence would it take to convince you that your coin was unbalanced, causing heads to come up more often than if it was a fair coin? In my first experiment I never got more than six heads or fewer than two in a set of eight flips. To get a bigger sample of possible outcomes, I repeated the experiment 10,000 times. The most common result was four

(Continued)

BOX 4.1 Continued

heads in eight flips, but this only happened 27% of the time. While three, four, or five heads occurred in 70% of trials, there was a 15% chance of getting more than five heads (Box Figure 4.1). Only 0.4% of trials, however, produced eight heads. Assuming my computer is unbiased and I didn't make a programming mistake, this implies that you should be highly skeptical about the fairness of your coin if you got all heads in eight flips. In particular, if a friend handed you a coin and offered a bet that it would come up heads eight times in a row, you might ask to do a series of test flips first.

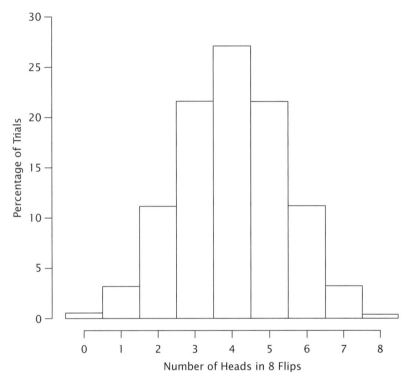

BOX FIGURE 4.1 Computer simulation of a coin-flipping experiment. In each of 10,000 trials, the computer mimicked flipping a coin eight times and recorded the number of heads in those eight flips. This histogram shows the percentages of the 10,000 trials in which zero, 1, 2, ..., 8 heads occurred.

mothers produced 36 litters ranging in size from 2 to 12. For each of these litters, I simulated how many offspring were males and how many were females just as if I were flipping a coin. Then I added up the number of males and females and compared these to the actual totals of 138 males and 145 females produced by control mothers. I repeated this process 10,000 times. I won't list all 10,000 results, but the first 10 trials produced 136, 138, 146, 138, 134, 148, 122, 137, 133, and 150 males. The numbers of females for each trial were just the total number of offspring in the 36 litters minus the number of males, for example, 283–136 = 147 females for the first trial.

Finally I asked what proportion of these trials using the coin-flipping model produced more males than the opossum mothers actually produced. For the control mothers, 6,902 computer trials produced more males than the 138 males actually born. This implies that, even if mothers had no control over the sexes of their offspring, the probability of getting 138 or more sons in the 36 litters produced by unfed mothers was 69% (Figure 4.6).

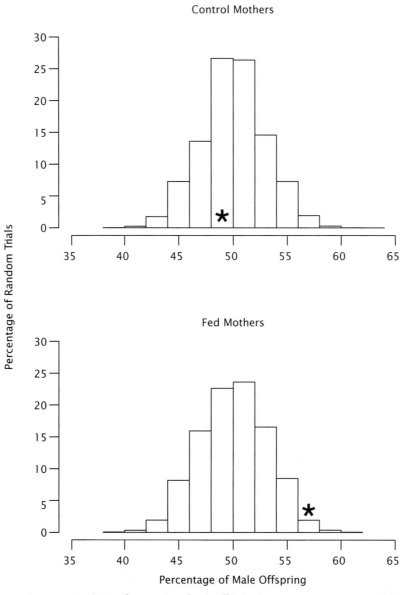

FIGURE 4.6 Computer simulation of percentage of male offspring born to common opossums in Venezuela assuming random sex determination with a 50% chance that each offspring is a male. Fed mothers were given sardines and cat food to improve their physical condition throughout the breeding season; control mothers didn't receive extra food. I did 10,000 random trials for each group. The asterisks show the percentages of male offspring observed by Austad and Sunquist in their experiment.

The story is quite different for the experimental mothers, who bore 149 males and 110 females in their 33 litters. In this case, fewer than 1% of the 10,000 computer trials produced more males than the actual proportion of 58% (Figure 4.6). The hypothesis that sexes of babies born to female opossums fed extra food are determined solely by random fertilization of eggs by equal numbers of X- and Y-bearing sperm seems quite unlikely to be true. Instead, fed females were more likely to produce males than expected by chance, a robust result consistent with the Trivers-Willard hypothesis.

We've discussed several questions to ask in evaluating experimental evidence. Do the experimental treatments make sense? Are the controls appropriate? Is the sample size sufficient? Can the results be generalized beyond the particular subjects used in the experiment? The last few paragraphs introduced another important question—can chance alone account for the results? This question is important for all experiments and for other kinds of evidence in science too. It's important for the marijuana experiments discussed in the first part of this chapter as well as for Austad and Sunquist's experiment on opossums in Venezuela, but the technical aspects of answering this question are more complex for the marijuana studies than for the opossum study. Suffice it to say that the main results of the marijuana experiments might have been due to chance but with probabilities less than 5%.

An Observational Test of the Trivers-Willard Hypothesis with Wild Horses

Elissa Cameron and three colleagues tested the Trivers-Willard hypothesis in a different way and with a very different kind of mammal. They studied a population of about 400 wild horses (Plate 8c) in the Kaimanawa region of New Zealand from 1995 through 1997. Wild horses are descendants of domesticated animals that are found in western North America and elsewhere in the world. They have been wild and free-ranging for many generations and are subject to the same ecological conditions as other large hoofed animals; for example, they don't get extra food or protection from predators, as do domesticated animals.

Cameron's group did not do an experiment, as Austad and Sunquist had done with opossums. Instead they relied solely on frequent observation of the horses over multiple years. They identified some individuals by unique natural markings and others by freeze brands placed on rumps of the horses. Because horses are large animals living in a fairly open habitat in New Zealand and because the researchers spent many hours in the field, they were able to collect extensive and detailed data on social behavior and reproduction that they used to test the Trivers-Willard hypothesis.

Wild horses have a polygynous mating system in which some adult males have harems containing several females while other adult males live in bachelor bands. Harem males don't necessarily monopolize all mating with females in their harems, but they mate with most of these females in most years. Bachelor males may have no offspring unless they are able to take over a harem after spending one or more years in a bachelor band. The gestation period of female horses is about 11 months, and mothers never have more than one offspring at a time. By making frequent observations of the population, the researchers knew within a few days when offspring were born. They then used back-dating to estimate the dates of conception.

To test the Trivers-Willard hypothesis, Cameron's group had to assess the body condition of mothers. They used binoculars or a telescope to make a visual estimate of condition on a scale from zero to 5 at intervals of 0.5. This is a standard method for horses, and earlier researchers had reported that visual estimates were highly correlated with direct measures of body fat. At the extremes, a horse with a score of zero would be emaciated and have ribs showing and a horse with a score of 5 would appear obese—which would be good for the horse because it would mean that she had eaten enough high quality food to store a substantial amount of fat.

Cameron's group assessed body condition of females frequently throughout the study. Once offspring were born, the researchers reviewed their records and retrieved condition ratings of mothers at three times: conception, halfway through gestation, and one month before birth. They found a strong relationship between condition measured at conception and offspring sex, with mothers in poor condition more likely to produce daughters and mothers in better condition more likely to produce sons (Figure 4.7). By looking at their data in more detail, they found additional evidence consistent with the

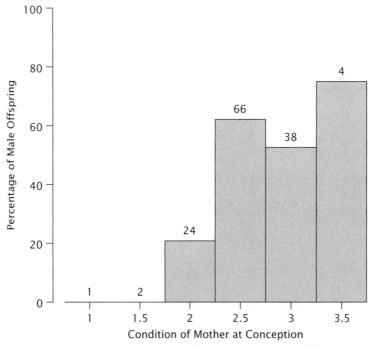

FIGURE 4.7 Relationship between physical condition of female horses in New Zealand and percentages of males born to those females. Numbers above bars are total offspring born. No adult females with condition ratings less than 1 gave birth; mothers with condition ratings of 1 and 1.5 had three total offspring, all females; mothers with condition ratings of 2 had 24 offspring, of which 5 were males (21%); and so on. No adult females had condition ratings greater than 3.5.

Trivers-Willard hypothesis. During the three years of the study, there were 25 females that had a male offspring in one year and a female offspring in another year. The researchers found that these mothers were in better condition when their sons were conceived than when their daughters were conceived.

After doing this research on wild horses, Elissa Cameron wondered about the generality of relationships between physical condition of mothers and sex ratio of offspring. Researchers have tested the Trivers-Willard hypothesis in many different species, and Cameron reviewed 87 such tests in non-human mammals. About 40% of these tests showed a positive relationship between the physical condition of mothers and the proportion of males among their offspring as predicted by the hypothesis—not a particularly impressive result. However, when Cameron subdivided the data based on when the condition of mothers was measured, she found a dramatic pattern. When condition was measured at conception, 74% of tests were consistent with the Trivers-Willard hypothesis, compared to 41% and 5% for condition measured during gestation and at birth of the young, respectively.

We've seen several lines of evidence that mothers in good physical condition produce more sons and mothers in poor physical condition produce more daughters, as predicted in 1973 by Trivers and Willard. Based on Elissa Cameron's studies of wild horses and other mammals, conception is the time at which a mother's condition is most likely to influence sex of her offspring. Cameron wondered how this might work and found research by others on cow embryos in culture dishes that provided a clue. The research with cow embryos was motivated by economic considerations; for example, a cow is more valuable to a dairy farmer than a bull. This research showed that adding glucose to culture media favored the early development of male embryos but inhibited the early development of female embryos. If this *in vitro* experiment reflects what happens in nature, the results suggest a mechanism for the Trivers-Willard evolutionary hypothesis since blood glucose levels of live cows are likely correlated with their physical condition.

How Can a Mother's Condition at Conception Influence Sex of Her Offspring? An Experimental Test with Mice

With this foundation, Cameron and three colleagues designed an experiment to test the hypothesis that manipulating blood glucose in laboratory mice at the time of conception could influence sex ratios of their subsequent litters. They used 38 female mice—18 treated to have reduced glucose levels in their blood and 20 controls. None of the females had mated or given birth prior to the experiment. The researchers used a compound called DEX to reduce glucose levels in the treated mice. DEX reduces circulating levels of glucose by partially blocking transport of glucose from the digestive tract into the blood. In keeping with principles of good experimental design, mice were assigned randomly to DEX and control treatments. Sample sizes were slightly different because two of the DEX females didn't conceive.

Before beginning DEX and control treatments, the researchers took a few drops of blood from a vein in the tails of females to measure glucose. Then they had to make a choice about how to supply DEX to treated females. One possibility was to inject the females with DEX, but they worried that stress from the injection might affect the sex ratio of offspring produced. To avoid this potentially confounding factor, the researchers added DEX to the water bottles of treated females instead of injecting them. This method, however, gave them less control over how much DEX each female received, simply because different females drank different amounts of water. This example illustrates one of many choices that must be made in designing any experiment—injecting DEX might have compromised the results by causing the mice to be stressed, but supplying DEX in water bottles caused variability in the strength of the experimental treatment received by different mice, adding noise to the results.

After taking the initial blood sample and adding DEX to water bottles of treated females, the researchers placed males in cages containing pairs of females for three days. Then they removed the males, took another small blood sample from females, and separated the females into individual cages. Females in the experimental treatment had DEX in their drinking water only while males were present, so any effect of DEX on blood glucose of mothers happened very near the time of conception. The researchers expected that DEX would decrease the condition of mothers by reducing glucose, while control mothers would not experience a change in condition at conception.

Cameron and her colleagues found that the change in blood glucose during the three days in which a mouse mated and became pregnant was a better predictor of the sex ratio of her offspring than the absolute level of glucose in her blood (Figure 4.8). In interpreting Figure 4.8, note the large variation between individuals, especially those in the control treatment. About half of these mice had higher blood glucose levels after conception than before while half had lower levels after conception. Most of the mice that drank water containing DEX lost blood glucose, as expected, but one of these actually gained glucose (Figure 4.8). Despite individual variation in responses to the treatments, however, Figure 4.8 shows clearly that mothers whose blood glucose increased at the time of conception produced litters with more sons than mothers whose blood glucose decreased at this time.

This result of a laboratory experiment with mice led Cameron to reconsider her earlier data on wild horses in New Zealand. In her original analysis, she had shown that mothers in better physical condition at conception produced relatively more sons (Figure 4.7). However, since she and her colleagues had measured condition of mothers frequently, there were many cases in which they could estimate the change of condition of mothers from approximately 20 days before conception to 20 days after conception. There were 29 mothers that lost condition and 49 mothers that gained condition during this time period. Only 3% of the mothers that lost condition gave birth to a son 11 months later, while 80% of the mothers that gained condition gave birth to a son. Cameron and Linklater (2007) argued that this pattern "shows the most extreme variation [in sex ratio at birth] in mammals ever reported."

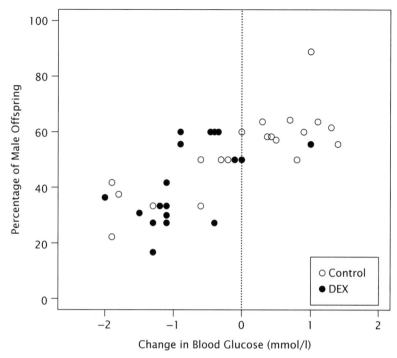

FIGURE 4.8 Percentage of male offspring in litters produced by mice in relationship to change in blood glucose around the time of conception. Researchers induced changes in blood glucose by adding DEX to the water bottles used by some females to inhibit transport of glucose from the digestive tract to the blood. Control mice did not receive DEX. Concentration of blood glucose is measured in millimoles per liter of blood, where 1 millimole = about 6×10^{20} molecules.

Conclusions

We discussed two main topics in this chapter—the potential medicinal value of marijuana and the potential ability of mothers to influence the sexes of their offspring. For the latter topic, we considered both why and how mothers might do this, that is, ultimate and proximate aspects of causation, just as we did for migration of monarch butterflies in Chapter 1. Despite fundamental differences between the stories about medical marijuana and sex determination in opossums and horses, they shared a common theme in the use of experiments to answer key questions in both cases. My main purpose in telling both stories was to illustrate some questions that you can ask about any experiments as you are thinking about whether these experiments provide convincing evidence for answering a question.

Both stories are incomplete. I described in some detail two of several experimental studies of the effects on pain of smoking marijuana, but we are a long way from a complete understanding of the short- and long-term costs and benefits of using marijuana in medicine. I also described in detail only three studies of sex determination in mammals, two tests of an evolutionary hypothesis in opossums and wild horses and one test of a physiological mechanism in mice, but researchers have done hundreds of other studies of mammals, birds, reptiles, fish, insects, and even plants. Biologists have proposed alternatives to the Trivers-Willard evolutionary hypothesis that apply to species that don't have a polygynous mating system, and mammalogists have proposed mechanisms for sex determination at conception that involve levels of stress hormones rather than blood glucose.

If you're interested in learning more about medical use of marijuana or sex determination in animals, I hope this chapter has provided some tools to help you think critically about what you hear or read. If someone makes a claim about medical marijuana or sex determination, ask yourself what evidence supports that claim. Is the evidence experimental? If so, were the methods appropriate? Did the controls make sense? Were there factors that the experimenters didn't consider that might have confounded the results? Could the results be due simply to chance? Were the assumptions justified? We'll broaden our perspective beyond experiments in Chapter 5 to include evidence from comparative and correlational research, but you should ask these kinds of questions when you evaluate all kinds of evidence about scientific hypotheses.

Questions to Ponder

1. Why do you think neither the NIH nor drug companies have invested much money to study the medicinal effects of marijuana? Do you think the reasons are the same or different for this federal agency and private drug companies?

2. The researchers who studied the effects of THC on pain experienced by people infected with HIV compared regular marijuana cigarettes to cigarettes made with marijuana plant material from which THC had been extracted. This comparison raised some questions about whether subjects were truly blind to their treatment. Is this a serious problem for the study? Why or why not? Is there a different control or a different way of doing the study that might resolve this problem?

3. Until about 2000, most of the evidence that marijuana could alleviate pain was anecdotal. For example, individuals with HIV might tell their friends (or tell the world on Facebook) that they felt much better after smoking pot. On page 65, I described the limitations of such subjective evidence to set the stage for explaining why rigorous controlled experiments are considered the gold standard in medical research. Yet the San Francisco and San Diego researchers who studied pain in volunteers who smoked THC or placebo cigarettes collected data by asking their subjects to estimate the average daily intensity of their pain on a scale from zero (for no pain) to 100 (for the worst pain imaginable). What are some key differences between this kind of subjective evidence and assorted testimonials of individuals posted on the Web? What would be the advantage of measuring pain objectively, if there were a systematic way to do so?

4. Figure 4.2 shows that different doses of THC have *similar* effects on pain induced 20 minutes after smoking one marijuana cigarette, but Figure 4.3 shows that different doses have *different* effects on pain induced 55 minutes after smoking the cigarette. However, Figures 4.2 and 4.3 also differ in where the pain was induced by capsaicin

injection—in the right forearm in Figure 4.2 but in the left forearm in Figure 4.3. In other words, site of injection is *confounded* with time between smoking a marijuana cigarette and test of the pain response. Can you imagine a reason why the different results shown in Figures 4.2 and 4.3 might be due to the location of the injection rather than the time elapsed between smoking marijuana and the injection? Even if not, should the researchers have been concerned about this possibility? How could they have controlled for this potential confounding between site of capsaicin injection and elapsed time between smoking marijuana and testing the pain response?

5. Figure 4.4 contains 44 points that relate to just three of the points in Figure 4.3. Based on the methods described in the text and the captions for these figures, which three points in Figure 4.3 relate to the points in Figure 4.4? Why do you think there are 44 points in Figure 4.4 instead of 45? Why does the vertical scale for Figure 4.4 extend from zero to 100 while the vertical scale of Figure 4.3 extends from 10 to 70?

6. Figure 4.3 suggests that an intermediate dose of THC (4%) relieves pain from capsaicin injection but a higher dose of 8% increases pain compared to a placebo injection. Figure 4.4, by contrast, suggests that higher concentrations of THC in the blood after smoking a marijuana cigarette are associated with lower levels of pain, although this is a noisy relationship. Can these results be reconciled? If so, how; if not, what might account for the inconsistency between the patterns in Figures 4.3 and 4.4? More specifically, is there any evidence in Figure 4.4 compatible with the increased sensitivity to pain associated with smoking a marijuana cigarette containing 8% THC? What if other secondary compounds like cannabinoids, terpenoids, or flavonoids occur at higher concentrations in cigarettes with 8% THC than in cigarettes with lower concentrations of THC?

7. A group of researchers met in a closed meeting in Texas in November 2012 to discuss potential new methods of male contraception, from developing a hormone-based pill like the birth control pills used widely by women to inventing special underwear that would suppress sperm production by increasing the temperature of the testes. Little research on male contraception has been done since the 1950s, when prisoners in Oregon volunteered for a study of a compound called WIN 18,446. This drug caused sperm of the volunteers to be stunted and feeble but produced few side effects in the subjects. In addition, effects of the drug on sperm were reversible, an important feature of a contraceptive. When researchers began trials in the general population, however, many subjects reported vomiting, sweating, headaches, and blurry vision. The scientists eventually discovered that WIN 18,446 produced these detrimental side effects by interacting with alcohol. The prisoners hadn't experienced side effects because they weren't allowed to drink. What are the implications of results like these for tests of medicinal effects of smoking marijuana?

8. An addiction psychiatrist named Ed Gogek wrote an OpEd column in the *New York Times* on November 7, 2012, opposing medicinal use of marijuana. He stated that most pain patients are female while about 70% of people who hold cards for use of

medical marijuana in Arizona and Colorado, the only states that report such data, are male. He further cited a National Survey on Drug Use and Health that reported that 74% of adult abusers of marijuana were male. He inferred from these data that "most medical marijuana recipients are drug abusers who are either faking or exaggerating their problems." Evaluate this argument.

9. In their experiment with opossums, Austad and Sunquist defined their experimental treatment as tracking females to their sleeping dens every other day and placing sardines and cat food near the openings of these dens at dusk and defined their control treatment as not doing these things. Females in both groups wore radio collars and were tracked and recaptured monthly to examine their litters. Besides extra food for the experimental females, how else did the experimental and control groups differ? Could these other differences account for the differences in sex ratios of offspring between the groups observed by Austad and Sunquist? If your answer is yes, give a plausible scenario for how this might happen; if no, explain why not.

10. I described two main lines of evidence supportive of the Trivers-Willard hypothesis about sex determination of offspring in polygynous animals, results of an experiment with common opossums and observational data for wild horses. What are the strengths and limitations of these two lines of evidence? Which is more convincing and why?

Resources for Further Exploration

In this chapter, we discussed how and why parents might influence the sexes of their offspring in opossums, wild horses, and mice. How about humans? The evidence is much less clear for our own species, although many researchers have asked questions about nonrandom sex determination and analyzed copious amounts of data to try to answer these questions. You won't be surprised that this research is highly controversial. Here are reports of three studies of humans that will stimulate your thinking when you read them:

1. In addition to her work on horses and mice discussed in the chapter, Elissa Cameron wrote a paper with Fredrik Dalerum entitled "A Trivers-Willard Effect in Contemporary Humans: Male-Biased Sex Ratios among Billionaires" (*PLoS One* 4:e4195 [2009]). Cameron and Dalerum concluded that "billionaires have 60% sons," substantially greater than the proportion in the general population. This excess of sons seems to be mainly determined by wealth of the father, however. Do Cameron and Dalerum make a convincing case that their results are consistent with the Trivers-Willard hypothesis? Why or why not?

2. Alan S. Miller and Satoshi Kanazawa published a book in 2008 entitled *Why Beautiful People Have More Daughters*, which summarized and extended several papers by Kanazawa—"Engineers Have More Sons, Nurses Have More Daughters," "Big and Tall Parents Have More Sons," and "Violent Men Have More Sons." Kanazawa's work

got a lot of attention, but Andrew Gelman and David Weakliem criticized it on statistical grounds in "Of Beauty, Sex and Power" (*American Scientist* 97:310–316 [2009]). Although some of the argument made by Gelman and Weakliem may be difficult to understand without statistical training, pages 80–83 of this chapter should give you a foundation for appreciating their basic points. The sections of the Gelman and Weakliem paper entitled "The 50 Most Beautiful People" and "Why Is This Important" express their argument in more concrete and less technical form than other parts of their paper. Do Gelman and Weakliem make a convincing case against the claims by Kanazawa? If you read the paper by Cameron and Dalerum on billionaires described in Question 1, how does the argument of Gelman and Weakliem influence your interpretation of the conclusions of Cameron and Dalerum? Do Gelman and Weakliem undermine all tests of the Trivers-Willard hypothesis, including the experimental and observation studies of opossums and wild horses discussed in this chapter?

3. Allen Wilcox and Donna Baird described a dramatic increase in the proportion of male births in Cuba under economically stressful conditions of the mid-1990s in "Natural versus Unnatural Sex Ratios—A Quandary of Modern Times" (*America Journal of Epidemiology* 174:1332–1334 [2011]). Wilcox and Baird propose a quite different explanation for this phenomenon than the Trivers-Willard evolutionary hypothesis or the mechanistic hypothesis about how the condition of mothers might influence sex of their offspring discussed in the text. Indeed, the Cuban results seem opposite to what Trivers and Willard would predict because economic stress would likely cause nutritional stress in mothers, which should have caused them to produce more daughters. Do Wilcox and Baird make a convincing case for their explanation of what happened to sex ratios at birth in Cuba in 1995–1996? If so, how can this be reconciled with the fact that women in the Netherlands during the severe famine of World War II also produced a disproportionate number of sons?

Correlations, Comparisons, and Causation

Although the possibility of simply discovering something new is a strong motivator for scientists, we also seek deeper understanding. In Chapter 1, I not only described how Fred Urquhart finally discovered the wintering area of migrating monarch butterflies after many years of searching but also explained that this discovery led to further questions about how and why monarchs migrate such long distances. These were questions about causation—the environmental triggers, physiological mechanisms, and adaptive value of long-distance migration. In fact, causation is the holy grail of scientific research. Think about cancer. We may know a lot about the incidence of various kinds of cancer, about what people are most vulnerable, about the typical course of the disease, even about the effectiveness of different treatments. But until we fully understand the causes of lung cancer or breast cancer or brain cancer, our understanding is only superficial and our ability to devise successful treatments is limited.

In previous chapters, we discussed several examples of research to probe the causes of various biological phenomena. This research included both experiments and purely observational studies. For example, Trivers and Willard proposed an hypothesis about the evolutionary cause of biased sex ratios at birth that Austad and Sunquist tested experimentally with opossums and Cameron tested using field observations of wild horses. These observations led Cameron to propose an hypothesis for the physiological cause of biased sex ratios at birth that she tested experimentally with lab mice.

As you saw in Chapter 4, experiments are often considered the gold standard for scientific research. This belief is especially prominent in medicine, as illustrated by the program established by the California legislature to support rigorous experimental tests of the medicinal value of marijuana. But as you also saw in previous chapters, scientists use additional kinds of evidence to answer questions and test hypotheses. These other kinds of evidence can generally be described as observational, in contrast to results of experiments, but observational evidence takes various forms. Chapter 2 included several examples of single, isolated observations or anecdotes. These kinds of observations are mostly useful for suggesting new hypotheses about causation but not for testing such hypotheses. In this chapter we'll focus on relationships between observations that lead to comparative or *correlational* evidence, which may be used—with due diligence—for testing causal hypotheses. In later chapters we'll broaden our perspective by seeing how biologists put different kinds of evidence together to study the causes of interesting phenomena in biology.

Take a look back at Figure 1.4 showing the relationship between parasite loads of monarch butterflies and distances between their summer range and overwintering location. This is an example of a correlation, as is Figure 3.9 showing the relationship between metabolic rate and body weight of mammals. Figures 4.4 and 4.8 also show correlations. As you learned in Chapter 3, these types of figures are called scatterplots because points are "scattered" within the frame of the graph. The points may represent individual organisms, like the mice in Figure 4.8; or groups of organisms in a population, like the monarch butterflies in Figure 1.4; or species, like the mammals in Figure 3.9; or something else (what do the points represent in Figure 4.4?) Whatever units of observation that the points represent, a scatterplot shows a relationship between two variables measured in those units, like change in blood glucose around the time of conception on the horizontal axis and percentage of male offspring on the vertical axis for the mice in Figure 4.8.

We'll mostly discuss relationships that can be represented on scatterplots in this chapter, but it's worth noting that there are many other ways of illustrating correlational data. For example, Figure 1.5 combines line graphs and bar graphs to show relationships between frequencies of births and food quality for wildebeest and zebras in the Serengeti. How about Figure A2.1? Do these photos illustrate correlations?

Storks and Babies

In 1988, Helmut Sies wrote a letter to the prestigious journal *Nature* reporting a strong relationship between the number of breeding pairs of storks and the birth rate in Germany between 1965 and 1980 (Figure 5.1). The complete text of Sies's letter was "SIR—There is concern in West Germany over the falling birth rate. The accompanying graph might suggest a solution that every child knows makes sense."

Figure 5.1 shows data for the stork population and birth rate in West Germany for seven years. This is a typical scatterplot with a clear overall pattern—the more storks, the more babies were born. Scientists use the *correlation coefficient*, symbolized by r, as a measure of the strength of this pattern; $r = 0.98$ for the data in Figure 5.1 (see Box 5.1).

When Sies described this pattern as one that "every child knows makes sense," he was saying that this correlation supports a causal hypothesis—that storks deliver babies. You might have believed this up until a certain age, just as you might have believed in Santa Claus, but you probably abandoned the stork hypothesis after you got some sex education.

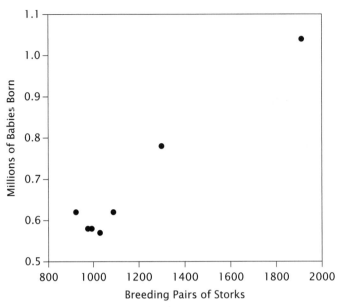

FIGURE 5.1 Birthrate in Germany was correlated with the population of storks between 1965 and 1980 ($r = 0.98$). Each point represents one of the seven years of data available.

BOX 5.1 What Do Correlated Data Look Like?

Figure 5.1 shows a correlation between numbers of storks and babies in West Germany as reported by Helmut Sies in 1988. There is a very strong relationship between these two variables ($r = 0.98$) although there were only seven years of data. Box Figure 5.1 illustrates four hypothetical correlations of various strengths for comparison. I used a computer program to generate these data. You can imagine that each panel of this figure shows hypothetical stork population sizes plotted on the horizontal axis and hypothetical birth rates plotted on the vertical axis, or you can imagine any other pair of variables of interest to you. One hundred points are shown in each panel. The correlation coefficient for the upper left panel is 0.89, pretty close to the maximum possible value of 1 but not as high as the value of 0.98 for the actual data on storks and babies. The correlation coefficient for the upper right panel is 0.49, while that for the lower left panel is –0.47. The correlation coefficient for the lower right panel is zero, indicating two variables that are completely independent of each other.

The maximum possible value of r is 1 for points lying on a single straight line with a positive slope. What is the minimum possible value? What is the fundamental difference in the pattern of points in the upper right and lower left panels? The pattern of points for the actual data in Figure 5.1 differs from the patterns for simulated data in Box Figure 5.1 because there are fewer points of actual data and these data have a higher correlation coefficient than any of the simulated data in Box Figure 5.1.

(Continued)

BOX 5.1 Continued

How else does Figure 5.1 differ from the upper left panel of Box Figure 5.1? Does this engender skepticism about interpreting Figure 5.1 as evidence that storks contribute to human births?

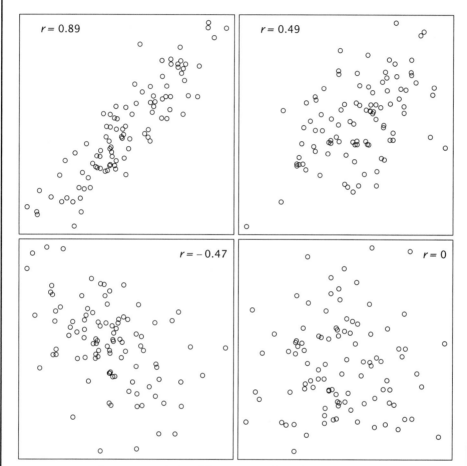

BOX FIGURE 5.1 Computer simulation of four patterns of correlated random data. Each panel displays one pattern, with correlation coefficients, r, shown near the top of each panel.

In fact, Sies wrote his letter to *Nature* to illustrate the fallacy of the common belief that correlation implies causation, as shown in a different way by Randall Munroe in Figure 5.2.

We know enough about sexual reproduction to reject the hypothesis that storks deliver human babies despite the strong pattern shown in Figure 5.1, but what accounts for this pattern? Sies actually graphed his data in a different way than Figure 5.1 that doesn't show the correlation as clearly but offers a clue about the reason for the pattern (Figure 5.3). In Figure 5.3, we see that the positive relationship between storks and

FIGURE 5.2 Relationship between correlation and causation as depicted by Randall Munroe (http://xkcd. com/552/, accessed January 17, 2014).

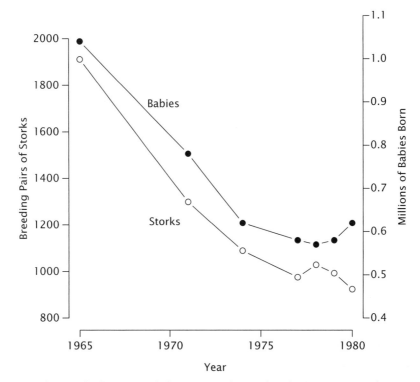

FIGURE 5.3 Changing birthrate (●) and changing population of storks (o) in Germany from 1965 until 1980 as plotted by Sies in 1988. This figure shows the same data as Figure 5.1 in addition to the years in which birthrates and stork populations were recorded.

babies is really limited to data for three years: 1965, 1971, and 1974. After 1974, this pattern breaks down. The four later years plotted on the right side of Figure 5.3 correspond to the four points in the lower left corner of Figure 5.1, which actually show a *negative* relationship between number of babies born and population of storks from 1977 to 1980.

This story about storks and babies illustrates one reason why correlations may give misleading evidence about causation: we may be ignoring other influential variables in focusing on a relationship between two variables of interest such as the numbers of storks and babies in Germany. In this case, time is tied to both of these numbers. Storks declined dramatically from 1965 to 1974, possibly as a result of habitat loss or other environmental changes. Births declined during this period as well, perhaps in response to changing economic circumstances. Changes in both storks and births were much smaller after 1974, so the overall pattern seen in Figure 5.1 was due mainly to changes before 1974. There wasn't a direct causal relationship between storks and human births, but these changed in the same direction between 1965 and 1974 due to coincidence. See Questions 1 and 2 at the end of this chapter for additional provocative data on storks and babies in Europe.

Cell Phones and Brain Cancer: A Correlational Study

We're too sophisticated to believe that storks have anything to do with babies, but other correlations in contemporary life provide more plausible suggestions of causation. One example is illustrated by a lawsuit filed in 1992 by David Reynard against NEC and GTE Mobilnet. Reynard's wife died of a rare type of brain cancer, and Reynard claimed that radiation from using a NEC cell phone on the GTE network contributed to growth of his wife's cancer. The court ruled against Reynard but acknowledged that its conclusion might have been different if more evidence had been available.

In 1990, one year after Susan Reynard was diagnosed with astrocytoma, there were about 12 million cell phone subscriptions worldwide. This number increased to more than 6 billion in 2011, and during the intervening two decades many researchers looked for evidence that use of cell phones contributes to brain cancer. It's difficult to imagine an experiment with human volunteers to study this relationship, so most of the evidence is correlational. I'll describe some of this evidence, but let me begin with one warning—as of January 2014, we don't have a conclusive answer to the question of whether use of cell phones contributes to brain cancer. Like the study of medicinal use of marijuana described in Chapter 4, this story of cell phones and brain cancer is a work in progress; as you learn more about future research, you may discover a different ending than presented here.

Reynard's lawsuit was based on the fact that use of cell phones produces radiation and some kinds of radiation damage living tissue, so we need to understand some basic things about radiation to see if there is a plausible connection between cell phones and cancer. The most familiar type of radiation in our environment is visible light. Starting in about 1800, scientists discovered other forms of radiation, ranging from radio waves through microwave and infrared radiation that carry less energy than visible light to ultraviolet radiation, X-rays, and gamma radiation that carry more energy than visible light. These forms of radiation constitute the electromagnetic spectrum (see Appendix 3 for an application to visual perception by animals).

There are two general categories of radiant energy, ionizing radiation and non-ionizing radiation. These categories are defined based on the effects of radiation on atoms and molecules, including the atoms and molecules in living tissue. Ionizing radiation can remove electrons from atoms and break bonds between atoms. This means that ionizing radiation can directly damage biological molecules such as DNA, causing mutations and cancer.

The potential biological effects of non-ionizing radiation like the microwave radiation emitted by cell phones are less clear than the effects of ionizing radiation. Although non-ionizing radiation doesn't directly damage DNA, it heats tissue by causing atoms and molecules to move faster or vibrate. By increasing the motion of molecules in cells, this heating may indirectly damage these molecules. The question is, can this ultimately cause or contribute to cancer? After 20 years of explosive growth in use of cell phones, we have lots of data, but it still isn't clear whether this risk exceeds the known risk of talking on a cell phone while driving.

In 2011, Steven Lehrer and two colleagues at the Mt. Sinai School of Medicine in New York described a relationship between use of cell phones and incidence of brain tumors in the United States. They used data on brain tumors diagnosed in 2000–2004 for 19 states as reported to the Central Brain Tumor Registry of the United States. These 19 states were the only ones to submit data to the central registry. The data included both malignant and non-malignant tumors affecting the brain and central nervous system. Lehrer's team reported a strong positive relationship between number of brain tumors and number of cell phone contracts in these states (Figure 5.4). The correlation coefficient for these data was a very impressive 0.98.

Does the striking correlation between cell phone contracts and brain tumors in 19 US states imply that use of cell phones causes brain cancer? There are several reasons to be skeptical of this conclusion. First, we don't actually have data on *use* of cell phones but rather on the numbers of cell phone *contracts* in the various states. Some people with contracts may use their phones for frequent, lengthy conversations every day, and others may use their phones only occasionally. To further complicate things, users may differ in how much talking and texting that they do. These differences could be important because holding a phone against an ear exposes the brain to microwave radiation while holding it for texting does not. Yet another complication is that some users may use a headset with Bluetooth technology that minimizes radiation exposure.

Using number of contracts as an index of cell phone use in each state has additional problems. Some users may have multiple contracts for different phones, yet only use one phone at a time. Other contracts may cover multiple users, such as parents and their children. A California resident may go off to college in Nevada but keep her phone number and contract in California.

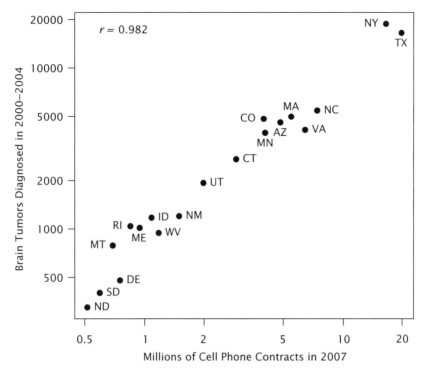

FIGURE 5.4 Relationship between number of cell phone contracts and number of brain tumors in 19 US states. Cell phone contracts and brain tumors are plotted using logarithmic scales because distributions of these variables are skewed; review discussion of Figures 3.7 and 3.8 for explanation.

Considering these problems, why did Lehrer's team—and other researchers—use cell phone contracts or subscriptions rather than data on actual use of cell phones in their analyses? They could have done a more rigorous test of the hypothesis that cell phone use contributes to brain cancer if they had looked at use directly, rather than relying on contracts as an indirect index of use. One reason for this roundabout approach is that usage data are proprietary to the companies that supply cell phone service and aren't generally made available for research purposes. If these companies released usage data for their customers that could be linked to medical records of these individuals, it would be a serious violation of the customers' privacy rights.

Despite these caveats about whether number of cell phone contracts in a state accurately represents number of individuals exposed to significant microwave radiation from talking on cell phones, there must be an approximate relationship between these variables. For example, Figure 5.4 shows about 16 million contracts for New York and 1.5 million for New Mexico, and it seems likely that cell phone usage in New York greatly exceeded usage in New Mexico as well. So the striking relationship between contracts and brain tumors in 19 states probably isn't completely undermined by using contracts as an indirect measure of cell phone usage.

A concise way of summarizing the ideas in the last two paragraphs is to say that Lehrer's team *assumed* that number of contracts for cell phone service in a state was approximately related to usage of cell phones in that state. Recognizing and identifying assumptions is one of the most important things for you to do in evaluating research. Sometimes researchers state their assumptions clearly and justify them fully. Sometimes they don't, in which case it's especially important for you to ferret out the hidden assumptions and decide whether they are acceptable.

Lehrer's team made another assumption that may be more problematic for inferring a causal relationship between cell phone use and brain cancer. Figure 5.4 shows the relationship between brain tumors diagnosed between 2000 and 2004 in 19 states and cell phone contracts in these states in 2007. Is there any way using a cell phone in 2007 could cause someone to have a brain tumor three to seven years earlier? We can be pretty certain this is impossible. In fact, despite all the complexities of analyzing causation, one thing we know for sure is that causes must precede their effects.

We can go even further with this example because cancer generally takes a long time to develop. If we were trying to explain cancer incidence in 2004, we would probably want to use cell phone data from at least 10 years earlier. Contracts by state in 1994 might be a good measure, because this is a few years after use of cell phones began to accelerate but 10 years before 2004. Apparently Lehrer's team didn't have access to these earlier data, so they implicitly assumed that numbers of contracts by state in 2007 were directly proportional to numbers in 1994.

The story about storks and babies at the beginning of this chapter foreshadows yet another problem with the analysis by Lehrer and his colleagues—a problem that might be considered a fatal flaw in their attribution of some brain cancers to cell phone use. Take another look at Figure 5.4. The two states with the most cell phone contracts and cases of brain cancer are New York and Texas, shown in the upper right corner of the figure. North Dakota, South Dakota, and Delaware have the fewest cell phone contracts and cases of brain cancer and appear in the lower left corner. New York and Texas have much larger populations than North Dakota, South Dakota, and Delaware, so naturally the former two states have more cell phone contracts and brain cancer than the latter three. What about the other states? What are the relationships between population size, cell phone contracts, and incidence of brain cancer?

Figure 5.5 shows the relationship between population size and number of cell phone contracts in the 19 states. The correlation coefficient is 0.998, about as close to a perfect linear relationship as possible with data from the real world. The relationship between population size and incidence of brain cancer (Figure 5.6) is almost as tight, with a correlation coefficient of 0.983.

Just as we did for storks and babies, we need to consider whether the correlation between cell phone contracts and cases of brain cancer demonstrated by Lehrer's team was a byproduct of another variable. For storks and babies, the extra variable was year—parallel declines from 1965 to 1974 in population size of storks and number of human births (Figure 5.3) produced an apparent association between storks and babies (Figure 5.1) without a real causal connection between these variables. Similarly, a

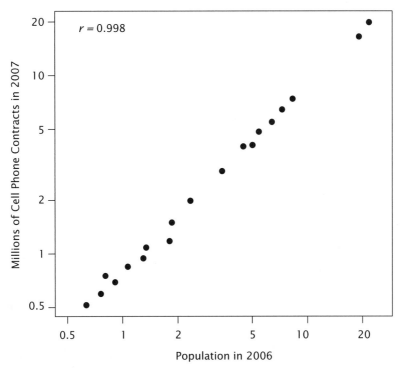

FIGURE 5.5 Relationship between population size and number of cell phone contracts in 19 US states, with both variables plotted using logarithmic scales.

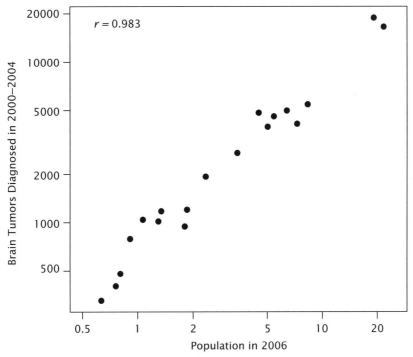

FIGURE 5.6 Relationship between population size and number of brain tumors in 19 US states, with both variables plotted using logarithmic scales.

correlation between number of cell phone contracts and cases of brain cancer in 19 US states does not necessarily imply that cell phones cause brain cancer. Instead, the correlation may have arisen from a third variable, population size, that was closely related to both cell phone contracts and incidence of brain cancer in these states.

We can test this idea by dividing the number of cell phone contracts and cases of brain cancer in each state by the population sizes of the states and making a scatterplot of these *per capita* data (*per capita* means "per head" in Latin, i.e., "per individual"). This is one way of factoring out the effect of population size on both cell phone contracts and incidence of brain cancer and getting a clearer picture of the direct relationship between cell phones and brain cancer. Figure 5.7 shows this relationship. Cell phone contracts per person varied between 0.65 and 0.93 in the 19 states, while cases of brain cancer varied between 0.0005 and 0.0011 per person, or 5 and 11 per 10,000 people as shown in the figure. More important, the correlation between cell phones per person and brain tumors per 10,000 people was only 0.18. Contrast Figure 5.7 with Figure 5.4, where the correlation between *total* number of cell phone contracts and *total* number of brain tumors in the states was 0.98.

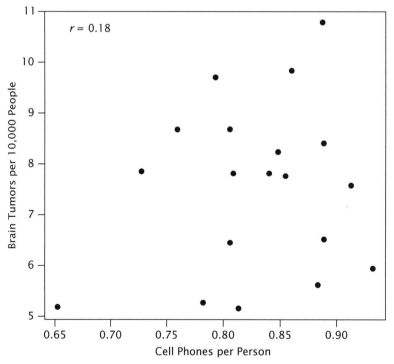

FIGURE 5.7 Relationship between number of cell phone contracts per person and number of brain tumors per 10,000 people in 19 US states. Since numbers of cell phone contracts and brain tumors are highly correlated with population sizes of the 19 states (Figures 5.5 and 5.6), this figure shows the relationship between an index of cell phone use and incidence of brain cancer more directly by factoring out the effect of population size on each variable.

We're fundamentally interested in whether an *individual's* risk of brain cancer is related to her use of cell phones (although we don't have data on use of cell phones, so we use contracts as an index of use). Figure 5.4 obscures the potential effect of cell phone use on an individual's risk of brain cancer by showing total contracts and cases of brain cancer in 19 states that differ greatly in population size. Figure 5.7 shows individual risk directly on the vertical axis, but there's no clear pattern in arrangement of points for the 19 states. The correlation coefficient for these data is much smaller than for Figure 5.4, which used total numbers of cell phone contracts and cases of brain cancer rather than *per capita* values. But the correlation coefficient for Figure 5.7 *is* positive, at 0.18. Does this suggest the possibility of a significant positive relationship between the number of cell phone contracts per person and the cases of brain cancer per 10,000 people that might be more apparent if we had data for all 50 states?

We can answer this question using a method similar to the method that we used in Chapter 4 to examine the sex ratios of young produced by opossum mothers given extra food while pregnant. For the data on *per capita* cell phones and brain cancer in Figure 5.7, I wrote a computer program to randomly rearrange the number of cell phone contracts and cases of brain cancer in the 19 states and then divided these numbers by the actual populations of the states to get hypothetical *per capita* values. This created a random association between cell phones and brain cancer with a correlation coefficient of 0.60. I repeated this 10,000 times and found that the median correlation coefficient for these random data was 0.18, exactly the same as for the real data in Figure 5.7. This means that the correlation between *per capita* cell phone contracts and *per capita* risk of brain cancer is no different than what would be expected purely by chance.

Cell Phones and Brain Cancer: A Comparative Study

Steven Lehrer and two colleagues described a relationship between use of cell phones and risk of brain cancer in the United States and suggested that people limit use of cell phones pending further research. I dissected this study in detail to illustrate the pitfalls of inferring causation from correlation, just as we saw in the story about storks and babies. There have been many other studies of a possible relationship between cell phone use and brain tumors, but virtually all of these studies were like Lehrer's in relying on comparative observations or correlations rather than experiments. This shouldn't be too surprising, even though experiments are the gold standard for research in medicine (Chapter 4). You might be able to imagine a short-term experiment with human volunteers to test the effects of radiation from cell phones on brain physiology, but this would provide only indirect evidence because brain cancer develops over a long period of time. Suppose it was feasible to design a long-term experiment to study development of brain cancer in humans using cell phones. Would it be ethical to do this experiment? A few researchers have done long-term experiments with animals, although their relevance may be questioned. Nonetheless, there are various ways of doing purely observational research in addition to simply examining correlations between variables as done by Lehrer's group, and some of these alternatives may provide more insight.

One comparative method used often in medicine is called a *case-control study*. In this method, researchers identify two sets of people: those who have a particular disease (cases) and those who don't (controls). The researchers then gather information on past exposure to potential causes of the disease for both cases and controls. If some past exposure differs for cases and controls, this provides evidence that the exposure contributed to the disease. For example, one kind of evidence that smoking causes lung cancer came from case-control studies—people with lung cancer (cases) were more likely to have a long history of heavy smoking than controls without lung cancer.

In 1998, the International Agency for Research on Cancer initiated a large-scale case-control study of the effects of cell phone use on brain tumors. This study was called the INTERPHONE study, and the INTERPHONE Study Group published their results in the *International Journal of Epidemiology* in 2010. The INTERPHONE study actually included 14 separate case-control studies in 13 countries, 8 in Europe plus Australia, Canada, Israel, Japan, and New Zealand. The 14 studies were designed similarly, though not identically. In particular, all used matched controls, meaning that each person with a brain tumor who agreed to participate in the study was matched for statistical analysis with a control of the same sex, age (±5 years), and general place of residence within the country where the study was conducted.

As of December 2013, the INTERPHONE study was the largest published case-control study of a possible link between cell phone use and brain cancer. The research team focused on the two most common types of brain tumors, gliomas and meningiomas (Senator Ted Kennedy died of a malignant glioma in 2009). The researchers compared cell phone use for 2,708 cases of glioma and 2,409 cases of meningioma to cell phone use for the matched controls for these two groups. They interviewed all individuals about their past use of cell phones, asking whether they used cell phones at all, when they started, how many calls they made, and how much time they spent using phones. Of course people could only provide rough estimates of these variables, especially their cumulative number of calls and call times, since some individuals had been using cell phones for more than 10 years.

Medical researchers use *odds ratios* to summarize results of case-control studies. Odds ratios express risks of disease or death in relation to factors that might influence the chance of getting a disease or dying (see Box 5.2). For example, the INTERPHONE researchers computed the odds of having meningioma or glioma for regular users of cell phones (those who averaged at least one call per week for six months or more) and for nonusers (those who averaged fewer calls than this or didn't use a cell phone at all) and then calculated the ratio of these odds. A ratio greater than 1 meant that regular users had greater risk than nonusers; a ratio less than 1 meant that regular users had less risk

---

BOX 5.2 What Are the Chances of Dying in a Maritime Disaster like the Sinking of the Titanic?

When the *Titanic* sank in 1912, 752 of the 1,184 adult passengers died. How did the probability of death differ between males and females? Among males, 650 died and 132 survived, so the odds of dying were 650/132 = 4.92. The picture was quite different for females: 102 died and 300 survived, so the odds of dying were 102/300 = 0.34. Therefore the odds ratio was 4.92/0.34 = 14.5; that is, adult male passengers were almost 15 times as likely to die when the *Titanic* sank as adult female passengers.

This dramatic difference in odds of dying between males and females on the *Titanic* occurred because the crew directed women and children to board lifeboats first, and male passengers generally stood back to facilitate this (the odds of dying were similar for children as for adult females). This, in turn, has been attributed to a long-standing principle of "women and children first" in rescue efforts following disasters. However, in 2012 Mikael Elinder and Oscar Erixson analyzed 17 other maritime disasters between 1852 and 2011 and found that men survived better than women and children in these cases. They concluded that "human behavior in life-and-death situations is best captured by the expression 'every man for himself.'"

---

than nonusers. They adjusted odds ratios based on the country in which a study was conducted, sex, age, and educational level. These adjustments were done to control for differences between samples in these variables, for example, if there were relatively more females in the Australian sample than in the Norwegian one.

The overall results of the INTERPHONE study suggested that, if anything, cell phone use decreased the risk of the two main types of brain tumors. For meningioma, the odds ratio for regular users compared to nonusers was 0.79; for glioma the odds ratio for regular users compared to nonusers was 0.81. Stated differently, regular users of cell phones were 79% as likely to have meningiomas as nonusers and 81% as likely to have gliomas as nonusers. But these odds ratios raise another question. If users and nonusers of cell phones had exactly the same risks of contracting these two types of brain tumors, odds ratios would equal 1. How likely is it that the observed odds ratios of 0.79 and 0.81 were due simply to chance, and there was no real difference in risk of meningioma or glioma for users and nonusers?

This question can be answered by considering *confidence intervals* for the odds ratios. The authors of the INTERPHONE study reported a 95% confidence interval of 0.68 to 0.91 for the meningioma odds ratio. This means that we can be fairly confident that the odds ratio was between 0.68 and 0.91 and not outside this interval; specifically, there's only a 5% chance that the odds ratio was less than 0.68 or greater than 0.91. More important, it's unlikely that the odds ratio was as large as 1, which would indicate equal vulnerability to meningioma for users and nonusers of cell phones. Instead, regular users were less likely to have meningioma than nonusers.

For glioma, the 95% confidence interval for the odds ratio was 0.70 to 0.94, suggesting that cell phone users were less likely than nonusers to have this type of brain tumor as well. These results were surprising to the INTERPHONE researchers and to scientists who read the report of their research. The INTERPHONE study was based on the premise of a plausible mechanism by which use of cell phones might increase risk of brain tumors. Although cell phones only release non-ionizing radiation, this radiation might have indirect biological effects because it slightly increases the temperature of nearby tissue. With extensive exposure, these indirect effects might include cancer. But how could use of cell phones *decrease* cancer risk, as the INTERPHONE study implies? No one, before or since, has proposed a plausible mechanism for this result.

The INTERPHONE team took a much more detailed look at their data than described in this general summary. For example, they asked participants to estimate the total amount of time that they had talked on cell phones since beginning to use them. The researchers then divided participants into 10 groups based on their usage estimates—less than 5 total hours, 5 to 13 total hours, and so on, up to more than 1,640 total hours. The researchers computed odds ratios with 95% confidence intervals for each of these groups, as shown in Figure 5.8 for glioma (the pattern for meningioma was similar). For 8 of the 10 groups, the odds ratio was less than 1, implying that risk of glioma was less for individuals in these groups than for nonusers of cell phones. For people who claimed to have used cell phones for a total of 13 to 31 hours, the odds ratio was slightly greater than 1, but the confidence interval overlapped 1, so this slightly greater risk of glioma for users at this level may have been due to chance. However, for people in the highest category of use—more than 1,640 total hours—the odds ratio for glioma was 1.40, far greater than for any other category of use.

This is a peculiar result. If use of cell phones causes an increased risk of glioma, then we might expect gradually increasing odds of glioma for individuals spending more and more time talking on cell phones. Instead, odds ratios were less than or equal to 1 for up to 1,640 total hours of use, suggesting a protective effect of cell phones, and only greater than 1 for more than 1,640 hours of use. Ignoring the possible protective effect for a moment, we might attribute elevated risk of glioma for the highest category of use to a threshold response. Maybe using cell phones is safe up to a point, at least as far as brain cancer is concerned, but becomes risky beyond a certain level. Other known carcinogens lack such threshold responses but show steady increases in risk as exposure increases, but perhaps non-ionizing microwave radiation differs from this general pattern.

There were other strange aspects of these results. The researchers subdivided their data for high-intensity users (more than 1,640 total hours) into short-term users (1–4 years of use), medium-term users (5–9 years), and long-term users (10 or more years). Long-term and medium-term high-intensity users were somewhat more likely to develop glioma than nonusers (34% and 28%, respectively), but short-term high-intensity users were almost four times as likely to develop glioma as nonusers. Since brain cancer tends to develop slowly over a long period of time, this is opposite to the expected pattern of higher odds for long-term users.

The INTERPHONE researchers considered several possible sources of bias that could explain both the apparent protective effect of cell phone use at low to moderate levels and the detrimental effect at the highest level. The most important bias is a general problem with many case-control studies—subjects may not have accurate memories of their exposure to a risk factor. For example, subjects with a disease (cases) may overestimate their exposure to a potential risk factor in trying to account for their misfortune, while controls may have more accurate memories of exposure or even underestimate their exposure. The INTERPHONE team had two ways of testing this possible recall bias: (i) some volunteers at some locations used modified phones that recorded actual calling data, and (ii) some service providers gave records of cell phone use to the researchers. These data confirmed a tendency for cases to overestimate their usage. As a dramatic example, 10 subjects claimed an average of more than 12 hours of cell phone conversation per day. The researchers thought this much use daily was highly unlikely. If so, the fact that all of these subjects had a brain tumor would artificially inflate the odds ratio for the highest usage category of more than 1,640 total hours (Figure 5.8).

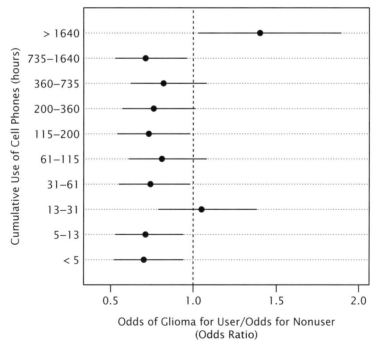

FIGURE 5.8 Odds ratios for glioma comparing users of cell phones at various levels and nonusers as reported by the INTERPHONE Study Group. Filled circles show odds ratios and solid horizontal lines show 95% confidence intervals for these ratios. The vertical dashed line at an odds ratio of 1 represents equal odds for users and nonusers.

The INTERPHONE team concluded that both of their main results—decreased incidence of two common brain tumors in low-intensity cell phone users compared to nonusers and increased risk of these tumors in high-intensity users—could be attributed to various sampling problems and biases. Despite the large scale of this study, it did not give a definitive answer to the question of whether microwave radiation from cell phones contributes to brain cancer.

The INTERPHONE study was the largest case-control study of the effect of using cell phones on brain cancer but not the only observational study of this relationship in humans. In 2012, a large group of European researchers reviewed all of these studies, concluding that their analysis "did not support a causal relationship between wireless phone use and the incidence of adult cancers in the areas of the head that most absorb RF energy from the use of wireless phones. There are insufficient data to make any determinations about longer-term use (≥ 10 years)." Another way to approach this problem is to ask whether the incidence of brain tumors has been greater in the 25 years since cell phones were introduced than it was before this time, or whether tumors have increased as use of cell phones has skyrocketed. Researchers have looked at such data for several countries and found no evidence of either pattern. Nevertheless, in May 2011 the International Agency for Research on Cancer classified the non-ionizing radiation released by cell phones as "possibly carcinogenic," although the US National Cancer Institute disagreed with this classification.

Finally, it's worth mentioning that a few researchers have done experiments in which laboratory animals have been exposed to the type of radiation emitted by cell phones. So far, none of these experiments has produced clear evidence of increased frequency of tumors in experimental groups compared to control groups. However, the National Toxicology Program has begun a long-term experiment with mice in which cell phone radiation will be activated and inactivated for 10 minute periods for 20 hours a day (no one knows whether mice would make calls typically lasting 10 minutes if they could do so voluntarily). This experiment should be completed by 2015. Combined with further comparative and correlational studies of people, we may have a fairly definitive answer about the risk of brain cancer from using cell phones before too long.

Lead and Crime Rates

Most of us are concerned about crime in various ways. We don't want to be victimized, nor do we want our families, friends, or neighbors to be victimized. Beyond these personal concerns, we probably would also like to see less crime in our cities, our country, or even globally. Even if we don't personally know victims of a mass shooting in another part of our own country or of a terrorist attack by a suicide bomber in another part of the world, we have an innate emotional response to reports of these heinous crimes.

What actions can individuals and societies take to reduce crime rates? Answering this question requires thinking about causes of crime. These causes are no doubt numerous and intertwined in complex ways. They involve sociology, economics, psychology—and biology, which is one reason why I'm writing about causes of crime in this book. The more important reason is that this story illustrates how comparative and correlational data can provide strong evidence of causation. You may have gotten the impression from learning about the tenuous relationship between cell phones and cancer that correlations (Figure 5.4) and comparisons (Figure 5.8) aren't worth much for understanding causation. Here's an example of a relationship between crime rates and exposure of children to lead in the environment that suggests the opposite. By understanding the differences between these examples, you'll be able to evaluate future research on these and other questions more effectively.

Humans have used lead for millennia. Today, we use lead in making many products, from building materials to batteries to ceramic glazes for pottery. We've also known since at least the time of the Roman Empire 2,000 years ago that lead is poisonous. A wide range of case reports, epidemiological studies, and animal experiments have demonstrated the biochemical and physiological effects of lead and the consequences of these effects for disease and death. Nervous tissue is particularly sensitive to lead, so children with their developing brains are especially vulnerable to lead poisoning. For this reason, two important uses of lead have been banned in developed countries. In the United States, for example, lead-based paint was used widely for painting houses from 1875 through the 1920s before its use began to decline in the 1930s. Leaded gasoline was introduced in the 1920s and its use increased dramatically as use of lead paints declined, until the Environmental Protection Agency began a program to phase out the use of leaded gas in the 1970s.

Crime rates differ in different regions and fluctuate over time. There are many thoughts about causes of this variation—is it related to differences in economic circumstances, drug use, law enforcement, the proportion of young males in a population, availability of guns, or severity of punishment? In the United States as a whole, the rate of violent crime increased from about 200 per 100,000 people in 1964 to 760 per 100,000 in 1992 and then began to decline, reaching a low of 386 per 100,000 in 2011. Why did violent crime increase by almost 400% in 30 years and then decrease by 50% in the next 20 years?

Rick Nevin is an economist affiliated with the US National Center for Healthy Housing. He knows quite a bit about exposure of children living in older houses to lead paint and about evidence that this exposure can lead to reductions in IQ. After World War II, however, the main source of lead in the environment was leaded gasoline. When cars burn leaded gas, some lead enters the atmosphere and then is taken up by people when they breathe. Exposure to tetraethyl lead from gasoline increased from 1941 to 1971 and then decreased as the mandate to eliminate leaded gasoline took effect. As a consequence, the average concentration of lead in the blood of people in the United States decreased by 78% between 1976 and 1991.

Nevin wondered if there might be a relationship between the increasing and then decreasing exposure to lead and the increasing and then decreasing rate of violent crime in the United States. Of course you could probably think of other variables that show similar patterns of change after World War II. As noted by Kevin Drum, for example, "Sales of vinyl LPs rose in the postwar period too, and then declined in the '80s and '90s" (*Mother Jones*, Jan/Feb 2013). For lead and crime, however, there was a plausible connecting mechanism because of the well-established relationship between exposure of children to lead and their IQ. Nevin hypothesized that reduced IQ due to lead exposure in early childhood might cause elevated crime rates when these children, especially males, become young adults. When he looked at the data, he found that the pattern for lead exposure matched the pattern for violent crime in the United States quite closely (Figure 5.9) as long as he used a 23-year lag time between lead exposure and crime rate. This makes sense when you figure that kids who had been exposed to lead as infants would be in their prime years for committing crimes 20 to 25 years later.

You know enough by now to be skeptical of reaching a conclusion about causation from a single correlation like that shown in Figure 5.9. This didn't work for storks and babies in Figure 5.1, nor for cell phones and brain tumors in US states in Figure 5.4, so why should it work here? One way to test the credibility of a pattern like this is to look for other examples. If the pattern is consistent across different scales, then it begins to look more like a real causal relationship.

Nevin first looked back in time. He was able to obtain estimates of the use of lead in paint and gasoline from 1914 to 1980 from records of the US Geological Survey and to

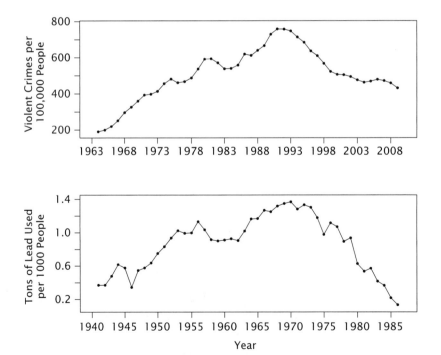

FIGURE 5.9 Relationship between use of tetraethyl lead in gasoline and rate of violent crime in the United States 23 years later.

use the early data for paint to extrapolate back to 1880. The National Center for Health Statistics compiled data on murder rates from 1900 to 1988, so Nevin examined the relationship between lead used in paint and gasoline, the two main sources of environmental contamination affecting children, and rates of murder through most of the twentieth century. With a lag time of 21 years, there is an impressive relationship between lead exposure and murder rate (Figure 5.10). Both variables have two peaks, and the peaks and valleys of changes in lead exposure and murder are closely matched. The first peak of lead exposure was due to increased use of lead-based paint before World War I; the second peak was due to increased use of leaded gasoline until the early 1970s when leaded gas was gradually phased out.

Nevin then extended his study to eight other developed countries. For this analysis, he used estimates of average lead levels in blood of preschoolers. In all nine countries (including the United States), there was a close correlation between rise and fall of estimated lead levels in preschoolers and crime rate 19 years later.

Jessica Reyes was another researcher who became interested in this problem while a graduate student at Harvard University in the late 1990s. She learned that US states differed in rates at which use of leaded gasoline decreased in the 1970s and consequently in rates of decrease in exposure to lead particles suspended in the air. She then found corresponding differences between states in rates at which crime rates declined in the 1990s. In 2012, Howard Mielke and Sammy Zahran extended the analysis that Nevin had done

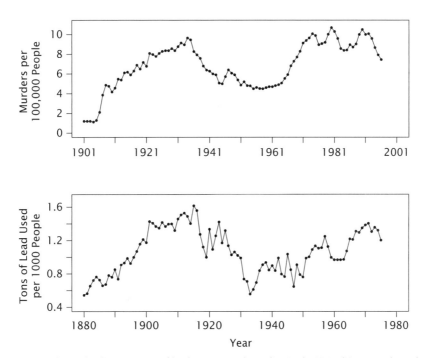

FIGURE 5.10 Relationship between use of lead in paint and gasoline in the United States and murder rate 21 years later.

at the national level and Reyes had done at the state level to six individual cities in the United States, from New Orleans in the south to Minneapolis in the north to San Diego in the west. Mielke and Zahran found as close a relationship between amount of lead in the atmosphere and rate of aggravated assault in these six cities as Nevin had found for the United States as a whole. In all six cities, the peak rate of assault occurred about 22 years after the peak level of atmospheric lead. Mielke has also examined concentrations of lead in the soil of different neighborhoods in New Orleans, where he lives, and found that this correlated with crime rates in these neighborhoods. It's difficult to imagine that increases and decreases in another variable, like sales of vinyl long-playing records, followed the same pattern as changes in crime rates in nine different countries, all 50 US states, six individual cities in the United States, and even different neighborhoods in New Orleans.

Which is a more dangerous place in which to live, a large or small city? Most of you would probably say that a large city, despite all its advantages, is more dangerous. Indeed, murder rates in US cities with more than a million people have generally been greater than murder rates in smaller cities. From 1991 to 2005, however, the average murder rate in cities of more than a million declined sharply, but the murder rate in smaller cities declined modestly, if at all, so that now there is no difference between cities with 100,000 residents and cities with more than 1 million residents. This pattern might be explained by greater changes in socioeconomic factors, law enforcement, or drug use in large cities than in small ones, although there's no evidence of such differences. Lead is another matter. Large cities have more vehicles, more traffic, more congestion, and so higher atmospheric levels of lead. Therefore eliminating the use of leaded gasoline had a bigger effect on air quality in large cities than in small ones. As for the examples discussed in the previous paragraphs, we should expect a time lag of about 20 years between reduction in exposure of young children to lead and reduction in crime rates, which is what happened in cities with more than a million residents.

The original correlation that Nevin found between lead exposure and violent crime in the United States is far from an isolated pattern; instead, it seems to be everywhere that we look. Although none of these data are experimental, their consistency across different temporal and spatial scales makes it difficult to dismiss them as artifacts. The various analyses discussed in the last several paragraphs give increasing credibility to the causal hypothesis that exposure of infants to lead influences crime rate 20 years later. This is much like the evidence that smoking causes lung cancer—there is not only the general fact that smokers are more likely to get lung cancer than nonsmokers but also the facts that the risk is greater for people who have smoked for a long time and for those who smoke more often and lower for those who smoke filter-tipped cigarettes and for those who quit. There is another kind of evidence, however, providing even stronger support for a causal relationship between lead exposure of infants and later crime rates (there's a parallel kind of evidence for smoking and lung cancer that I won't discuss here). This evidence is still correlational, because we're dealing with a problem that can't be studied experimentally. It involves a detailed look at the mechanisms that connect lead exposure and crime.

I mentioned that Nevin's initial work was motivated by a relationship between early exposure to lead and IQ of children. This relationship has been well established in multiple studies, and the US Centers for Disease Control and Prevention has regularly reduced the recommended tolerable concentration of lead in children's blood. In 2012, the US Centers for Disease Control and Prevention announced that there is no concentration known to be safe for children, and it recommended that children younger than 6 with lead levels greater than 5 micrograms per deciliter (µg/dl) be monitored for adverse effects of this exposure to lead (2.5% of children in the United States have levels greater than 5 µg/dl).

Exposure to lead in childhood not only adversely affects IQ but may also increase the risk of attention deficit/hyperactivity disorder and perhaps aggressiveness and impulsivity of children—a combination of traits that could contribute to later criminal activity. In fact, long-term studies of individuals have shown that high concentrations of lead in blood during childhood are correlated with greater likelihood of being arrested for violent crimes later in life. To this point, I've mentioned this relationship as a plausible but hypothetical reason why childhood exposure to lead could be correlated with higher crime rates in countries, states, cities, and neighborhoods, but now we see that data for individuals tracked over long periods of time document the mechanism for this correlation.

How does lead affect brains? A group of researchers at the University of Cincinnati started a long-term study of 300 children born in low-income neighborhoods of the city in 1979. The researchers measured concentrations of lead in these children from birth to 6½ years of age. When the children had grown to young adulthood (age 19 to 24), the researchers asked if they would be willing to undergo MRI scans to determine effects of early exposure to lead on brain development. There were two main effects. Higher lead exposure was associated with reduced gray matter in the prefrontal cortex of the brain and with reduced production of myelin. The prefrontal cortex is involved in aggression control and other so-called "executive functions" like attention and verbal reasoning. Myelin, which forms the white matter of the brain, is involved in connections between brain cells, that is, communication between different areas of the brain. The researchers also found that lead had greater detrimental effects on the brains of males than of females.

In summary of this example, multiple correlations suggest a causal relationship between exposure to lead early in childhood and crime rates 20 to 25 years later. Other correlations document the mechanisms by which this might occur, from effects of lead on developing brains to ultimate consequences for behavior. Of course this doesn't mean that exposure to lead is the only cause of crime, which remains a social problem with a complex web of causation. Nor does it mean that every child who grows up with levels of lead in his blood of 5 or 10 or even 20 micrograms/deciliter is destined to be a criminal. Quite the contrary, the vast majority will not follow lives of crime. It's also not the case that we've completely solved the problem by banning lead paint and leaded automobile fuel, because lead remains in the soil and gets resuspended in the atmosphere each summer when the soil warms up and dries out. Neighborhoods within cities differ in levels of lead in the soil (Plate 9), with poorer neighborhoods often having higher levels, so children are exposed to different amounts of lead depending on where they live, with some of the consequences discussed in previous paragraphs.

Conclusions

Observational data in the form of comparisons and correlations are important kinds of evidence in science. Many interesting and important questions can't be answered by doing experiments, like the question of what causes variation in crime rates. In other cases, we may be able to do experiments with animals to complement observational studies of humans, as in the study of whether exposure to radiation from cell phones increases risk of brain tumors. As you'll see in later chapters, some kinds of questions about ecology and evolution can be studied experimentally but others cannot. Methods available depend on scale—things that happen in days, weeks, or a few years or in areas up to the size of small lakes or patches of forest are amenable to experimentation; things that happen over centuries or millennia or at larger spatial scales often are not.

This chapter illustrated some pitfalls in interpreting comparative and correlational data as well as some features of studies using these kinds of data that can facilitate persuasive inferences about causation. Criticism of observational studies may be trumped by a consistent pattern of correlations at multiple scales and especially by evidence of a plausible mechanism for how one factor influences another. Experimental studies also have pitfalls, as we'll see in future chapters. It's rare that a single piece of evidence—one key experiment or one dramatic correlation—answers a question with finality. Instead, we usually have multiple pieces of evidence—some individual observations, some comparisons and correlations, and maybe some experiments. Each piece has both limitations and strengths; we must reach a conclusion based on our assessment of the overall weight of evidence. I'll develop this theme with several examples in later chapters.

Questions to Ponder

1. Figure 5.3 shows changes in the stork population and human birthrate in Germany from 1965 to 1980. We might also test the hypothesis that storks deliver babies by seeing whether there is a correlation between numbers of storks and birthrate in different countries of Europe. Table 5.1 shows data for 17 countries in 1990. Draw a scatterplot of the relationship between stork populations and birth rates in these 17 countries. Use pencil and paper, Excel, a graphing calculator, or a web applet. Is the human birth rate correlated with the population of storks in these countries? Could the children's tale about storks delivering babies be true after all? If not, how can you explain these results? (Hint: Look at the other differences between the countries shown in the table.)

TABLE 5.1 Areas, numbers of storks and people, and numbers of babies born in 17 European countries in 1990.

Country	Area (km²)	Pairs of Storks	Millions of People	Thousands of Babies Born
Albania	28,750	100	3.2	83
Austria	83,860	300	7.6	87
Belgium	30,520	1	9.9	118
Bulgaria	111,000	5,000	9.0	117
Denmark	43,100	9	5.1	59
France	544,000	140	56	774
Germany	357,000	3,300	78	901
Greece	132,000	2,500	10	106
Holland	41,900	4	15	188
Hungary	93,000	5,000	11	124
Italy	301,280	5	57	551
Poland	312,680	30,000	38	610
Portugal	92,390	1,500	10	120
Romania	237,500	5,000	23	367
Spain	504,750	8,000	39	439
Switzerland	41,290	150	6.7	82
Turkey	779,450	25,000	56	1,576

2. The possibility that storks might actually assist in delivering human babies has remained fascinating to researchers in the years since Helmut Sies sent his short letter to *Nature* in 1988. In 2004, a group of researchers published additional data on babies born in Berlin and storks nesting in the rural province of Brandenburg that surrounds Berlin (no storks nest in Berlin itself). Figure 5.11 shows the results reported by these researchers. There is no clear relationship between total births in Berlin and the stork population in Brandenburg, but there is a correlation between home deliveries and the stork population. What might account for this correlation?

3. Steven Lehrer, Sheryl Green, and Richard Stock reported a strong correlation between numbers of cell phone contracts and brain tumors in 19 US states (Figure 5.4). In a comment on the paper by Lehrer's team, Mathieu Boniol and two French colleagues showed that number of brain tumors in these states was also strongly correlated with number of hospitals and with number of beauty salons. Lehrer replied, "Prolonged use of dark-colored permanent hair dyes [in beauty salons] may have a relationship to the genesis of brain tumors . . . and many forms of cancer, including brain tumors, have been attributed to medical radiation [in hospitals]" (Lehrer 2011:435). Does this rebuttal by Lehrer support the original claim of Lehrer, Green, and Stock that use of cell phones may contribute to brain cancer? Why or why not?

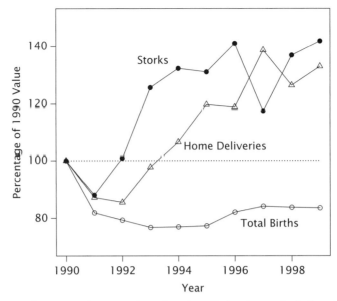

FIGURE 5.11 Pairs of nesting storks near Berlin and number of babies born in Berlin from 1990 to 2009. In 1990, there were 959 pairs of storks, 3,604 births, and 897 home deliveries. Data plotted for subsequent years are percentages of these 1990 values.

4. A case-control study such as the INTERPHONE study is a retrospective study because researchers identify cases of people with a disease and matched controls and then look backward to study possible differences in exposure to risk factors, often by asking the subjects to recall their use of cell phones, history of smoking, dietary habits, and so on. By contrast, a prospective study enrolls volunteers before any of them get a disease and then collects data on their exposure to various risk factors until some get the disease of interest and some do not. What are the advantages and disadvantages of prospective and retrospective studies? Could researchers use a prospective study to test the hypothesis that use of cell phones increases the risk of brain tumors? Why or why not?

5. Siddhartha Mukherjee (*New York Times*, April 13, 2011) described three general types of relationships between risk factors and disease—both the risk factor and the disease are rare, both are common, or the risk factor is common and the disease is rare. In the first case, a correlation between a rare risk factor and a rare disease is fairly convincing evidence that the risk factor causes the disease. For example, as early as 1775 Sir Percivall Pott found that the incidence of scrotal cancer was much greater in chimney sweeps than in the general population. Few men are chimney sweeps and few men get scrotal cancer, so Pott's discovery suggested that exposure to some toxin that accumulated in chimneys was a likely cause of scrotal cancer.

It's more difficult to determine causation in the other two cases. Smoking and lung cancer were both common in the 1930s, but it wasn't until careful case-control studies were done in the 1950s and 1960s that medical researchers established smoking as a cause of lung cancer. Cell phones and brain cancer illustrate the third type of relationship—an extremely common environmental factor and a relatively rare disease. Why does the relationship between a very common environmental factor and a rare disease produce the greatest challenges in inferring that the environmental factor contributes to increased risk of the disease? Can you think of any other examples that illustrate this dilemma?

6. As of 2012, the US Centers for Disease Control and Prevention uses a reference level for the concentration of lead in children's blood of 5 µg/dl; 2.5% of children younger than 6 in the United States have greater concentrations than this. One microgram is one-millionth of a gram, that is, 1 µg = 0.000001 g = 10^{-6}g; 1 deciliter (dl) is one-tenth of a liter. Blood volume of infants in milliliters is related to their body weight in kilograms as follows: Blood volume (ml) = 85.3 × Body Weight (kg). A typical 2-year-old girl weighs 14 kilograms. How much lead would her blood contain if the concentration was 5 µg/dl?

7. In this chapter we discussed evidence that exposure of children to lead influences rates of violent crime 20 to 25 years later. Does this evidence undermine the argument that more stringent gun controls may reduce crime? Why or why not?

Resources for Further Exploration

1. Many examples illustrate the pitfalls surrounding the relationship between correlation and causation. CrossValidated is a website for questions and answers about statistics. Many of the discussions are quite technical, but here is one called "Examples for Teaching: Correlation Does Not Mean Causation" (http://stats.stackexchange.com/questions/36/examples-for-teaching-correlation-does-not-mean-causation). What's your favorite example among those discussed at this site?

2. Correlation doesn't imply causation—or does it? Even though we can find many examples of taking a foolish leap from correlation to causation, there *is* a relationship between these two concepts. Greg Laden is a biological anthropologist who discusses a wide range of scientific topics in his blog, including what he calls "Today's Falsehood: Correlation Implies/Does Not Imply Causality" (http://scienceblogs.com/gregladen/2011/06/20/todays-falsehood-correlation-i/). Laden's conclusion to this posting is "'Correlation does not mean causation' or some variant thereof is, sometimes, a dog whistle." What does he mean? Do his arguments alter your understanding of the examples in Chapter 5? If so, how; if not, why not?

The Diverse Uses of Models in Biology

You've been introduced to the term *model* in previous chapters. In Chapter 1 I described monarch butterflies as a model system for studying biological clocks and bats as a potential new model for studying the physiology of senescence; in Chapter 4 I used coin flipping as a model for predicting sex ratios of opossums at birth. These were brief examples of a few uses of models in biology, but in fact scientists use various kinds of models in many different ways, and modeling is a key part of doing science. You've already seen examples of some of the problem-solving tools of science, from making systematic observations of nature to analyzing correlations to conducting experiments. Modeling is another of these tools, so learning more about modeling will be important in learning how science works. Some models are abstract and some involve heavy duty math, but my four examples will be concrete and light on math. My goals are to show how modeling contributes to asking and answering important questions in biology and to illustrate through these examples why you should care about modeling.

To begin with an illustration from everyday life, consider maps. You probably use your smartphone to find your way around a new place, but until recently most people used printed maps like the map of Treasure Island in Robert Louis Stevenson's book by the same name (Figure 6.1). A map is a simplified representation of part of the real world designed to be used for a particular purpose, like finding treasure on Treasure Island. This definition works for models too—they are simplified representations of parts of the real world made for a particular purpose. Models simplify reality by highlighting certain features of nature and suppressing others, helping users focus on their key questions. Maps are one kind of model, but there are many other kinds—from scale models like the plastic helicopters that my son used to build to mathematical models in the form of equations or graphs or computer programs to run your favorite apps. Models have various purposes as well, although all involve "finding treasure" in some

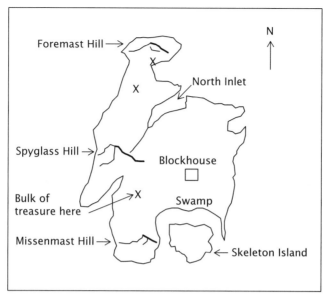

FIGURE 6.1 Map of Treasure Island, which "was about nine miles long and five across, shaped, you might say, like a fat dragon standing up, and had two fine land-locked harbours, and a hill in the centre part marked 'The Spy-glass.' There were several additions of a later date; but, above all, three crosses...two on the north part of the island, one in the south-west, and, beside this last, ... and in a small, neat hand, very different from [Captain Flint's] tottery characters, these words: 'Bulk of treasure here.'" (Stevenson 1930). Stevenson's description of this map in Chapter 6 of *Treasure Island* continued with directions to the three sites where the pirate Captain Flint had buried his treasure.

sense, either literally for an engineering model that leads to a commercially successful invention or metaphorically if we interpret treasure to mean better understanding of the world in which we live.

Two Concrete Scale Models

You know from frequent news reports that DNA carries the genetic information of living organisms. New applications of DNA technology are announced almost every month; in a few years, determining a person's complete DNA sequence will be a regular and inexpensive part of medical diagnosis. Friedrich Miescher isolated DNA from salmon sperm in about 1870 and Phoebus Levine described the chemical components of DNA in 1919, but as late as 1950 scientists were uncertain about whether the genetic material found in the chromosomes of all cells was made of DNA or protein. Finally in 1953, James Watson and Francis Crick made a key discovery about the three-dimensional structure of DNA that set the stage for tremendous progress in understanding how DNA works, both in inheritance and as a recipe for the development of organisms from embryos to adults. Watson and Crick used scale models to figure out the structure of DNA, and their discovery has become one of the great stories of modern biology.

When Watson and Crick began making models, they knew that DNA was a very large molecule with three components—phosphate groups, sugars called deoxyribose, and structures called nucleotides (the last two components give DNA its name—deoxyribonucleic acid). The nucleotides come in four varieties—adenine, thymine, guanine, and cytosine (Figure 6.2). Nucleotides are also called bases, and they are abbreviated A, T, G, and C when writing out the genetic codes represented in DNA.

Watson and Crick reported their discovery of the structure of DNA on March 25, 1953, in a very short paper in *Nature* with the famous penultimate sentence: "It has not escaped our notice that the specific pairing we have postulated immediately suggests a possible copying mechanism for the genetic material." What Watson and Crick meant was that their model showed how the DNA in any cell could produce identical copies that would be passed on through cell division to subsequent generations of cells including sperm and eggs, thus accounting for similarities of parents and offspring. How did they use scale models to reach this dramatic conclusion?

Francis Crick and James Watson began the collaboration that led to one of the most important scientific discoveries of the twentieth century in fall 1951 when Crick was 35 years old and Watson was 23. Crick was trained as a physicist but switched to biology in the late 1940s and was working on his PhD in biology at Cambridge University in 1951. Watson got his PhD in molecular biology at the University of Indiana in 1950 and then went to Europe for postdoctoral research. After spending time at the University of Copenhagen, Watson moved to Cambridge and met Crick.

The structure of DNA was a longstanding mystery that was getting a lot of attention in the late 1940s, when scientists were developing new techniques that could be applied to solving the mystery. The biochemist Linus Pauling at the California Institute of Technology in Pasadena was one prominent scientist who had turned his attention to DNA structure after making key contributions to understanding the structure of proteins. Watson and Crick were motivated in part by competition with Pauling.

Before arriving at Cambridge to work with Crick, Watson attended a meeting in

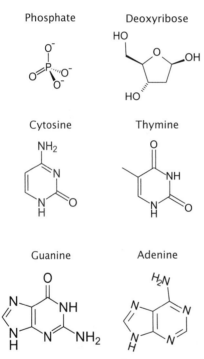

FIGURE 6.2 Components of DNA. Watson and Crick knew that DNA was a very large molecule made of phosphate, deoxyribose, cytosine, thymine, guanine, and adenine. These drawings show skeletal structures of the components. Except for phosphate, the skeletons contain five- or six-membered rings, or both in the case of guanine and adenine. A carbon atom (C) is located at each vertex of each ring; some of these have attached hydrogen atoms (H) that aren't explicitly shown in the skeletons. The other atoms that are shown are oxygen (O), phosphate (P), and nitrogen (N). How do cytosine and thymine differ? How do guanine and adenine differ?

Italy where he heard a talk by Maurice Wilkins, head of a research group at the University of London that was using X-ray photography to study the structure of large biological molecules such as DNA and proteins. In his talk, Wilkins showed an X-ray photo of DNA suggesting that DNA had a simple, regular three-dimensional structure in its crystalline form. Watson and Crick began their work by assuming that DNA has a helical shape, like a spiral staircase or coiled spring—and like some proteins, as had recently been demonstrated by X-ray studies. The regular structure of DNA seen in Wilkins's photo and the assumption that DNA was shaped like a helix led Watson and Crick to think that making models of alternative possible forms of this structure might be productive.

In the meantime, Crick, with his strong background in math and physics, had developed a mathematical model of how X-rays should interact with molecules shaped like helices. This model implied that DNA would contain two, three, or four chains of nucleotides. This was important because it limited the number of potential structures for DNA that Watson and Crick had to investigate. They built their first model using metal plates for the nucleotides and wires for the phosphate and deoxyribose groups. This model was a triple-helix comprised of a central backbone of alternating phosphate and deoxyribose groups with nucleotides attached to the outside of this backbone.

It's time to introduce one more key character in this story—Rosalind Franklin, an expert in using X-ray photography to deduce the structure of crystals who worked with Wilkins at the University of London. The three-chain model built by Watson and Crick was based in part on a talk that Franklin gave in London, so when Watson and Crick completed their model, they invited Franklin to visit them in Cambridge to examine it. Franklin quickly saw that having the nucleotides on the outside didn't allow for nearly as many water molecules to bind to the DNA as she actually found in her X-ray work. She had described this in her talk, but Watson remembered it incorrectly.

The three-chain model was disproved by Franklin's X-ray photography showing the likely water content of DNA, so Watson and Crick began experimenting with two-chain models. Their challenge was to figure out how the nucleotides could fit inside the helical molecule, with an external backbone of phosphate and deoxyribose that could accommodate the amount of water associated with DNA. Erwin Chargaff at Columbia University provided a key hint with his report on the nucleotide content of DNA from various species. All the species differed in their relative amounts of adenine, thymine, guanine, and cytosine, but in each species the amount of adenine equaled the amount of thymine and the amount of guanine equaled the amount of cytosine. After some false starts, Watson and Crick built a two-chain model with adenine paired with thymine and guanine paired with cytosine. Since adenine and guanine each have two rings and thymine and cytosine each have one ring (Figure 6.2), an adenine-thymine pair has three rings as does a guanine-cytosine pair. Therefore Watson and Crick's model was a double helix of a fixed width (Figure 6.3). The two chains in the helix are complementary; a G in one chain is always paired with a C in the other and an A is always paired with a T. This accounted for Chargaff's discovery that A = T and G = C

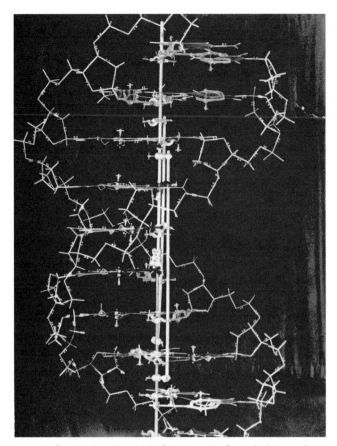

FIGURE 6.3 Photograph of original scale model of DNA made of metal plates and wires by Watson and Crick in 1953. The model was 6 feet tall.

within species. It also explained Watson and Crick's conclusion in the 1953 paper that "the specific pairing we have postulated immediately suggests a possible copying mechanism for the genetic material." When a cell divides, the DNA in each chromosome in the nucleus is unzipped, making two, complementary single-stranded molecules, each of which produces a complementary copy, so the double-stranded DNA in each of the two daughter cells is identical to that in the original cell—except for occasional copying errors, that is, mutations.

About the same time as Watson and Crick were building their scale model of DNA in England, a group of engineers was building another scale model in Sausalito, California, just north of the Golden Gate Bridge. This was a one-acre model of San Francisco Bay built in a warehouse by the US Army Corps of Engineers in response to a proposal by a developer named John Reber to build two dams in the Bay to conserve fresh water, increase the size of the deep-water harbor, and create filled land for construction. Many were skeptical of the proposal and its alleged benefits, including the Corps of Engineers. Their model included hydraulic pumps to mimic the action of tides and river flows and barriers like those described by Reber. Although "San Francisco Bay in a Warehouse" (Plate 10) is no longer used for research, it has been expanded to three acres encompassing the Sacramento-San Joaquin Delta as well as San Francisco Bay and is open to the public at the Bay Model Visitor Center (http://www.spn.usace.army.mil/Missions/Recreation/BayModelVisitorCenter.aspx).

When the Corps of Engineers tested tidal and river flows in this model of San Francisco Bay, they found that the dams resulted in pools of stagnant water rather than fresh water lakes as promoted by Reber. Adding openings in the dams to enable water flow and perhaps mitigate the creation of huge stagnant pools only made things worse by producing intense flows that would have impeded boat traffic in the Bay. The Corps of Engineers concluded that Reber's project would not produce the benefits that he touted but would severely damage the aquatic environment. Needless to say the project was never built.

San Francisco Bay in a warehouse and Watson and Crick's model of the double helix couldn't have been more different in their size, scope, and objectives. One was a model of an aquatic and terrestrial landscape of many square miles; the other was a model of a microscopic molecule inside the nuclei of cells of living organisms. One model was so big that the builders needed a warehouse in which to build it; the other fit on a tabletop. One was designed to address a very pragmatic policy question—should a development proposed for San Francisco Bay be allowed to proceed? The other model was designed to advance fundamental understanding of biology, although it certainly led to practical applications in subsequent years. One was built to test a specific hypothesis—that Reber's proposal for dams in San Francisco Bay would produce the benefits that he claimed without damaging the environment. The other was built in an attempt to discover the three-dimensional structure of DNA, not to test a preconceived hypothesis about what that structure might be. The key feature that these models shared is that both were concrete, physical, scale models, not unlike the model helicopters that my son made as a child. I used these examples as a gentle introduction to modeling, but most models that scientists develop are more abstract than this; we'll now consider two biological examples.

A Mathematical Model of the Spread of Disease

All human societies live with disease, as do all animal and plant populations. Disease has had an important role in human history, from the plague that may have killed half of Europeans in the Middle Ages, to diseases like smallpox and measles brought by Europeans to the Americas that killed millions and probably destroyed the Aztec civilization of Mexico, to HIV today. In 2004, the World Health Organization estimated that infectious diseases are responsible for about 25% of deaths worldwide each year. In recent years we've seen several examples of spillover—the transfer of pathogens from another animal species to humans. Besides HIV, these examples include Lyme disease, SARS, and Ebola (see Chapter 9 for more on the history of HIV).

Infectious diseases are caused by many types of organisms and are spread in many different ways. Viruses, bacteria, protozoa, and fungi are especially important causes of infectious disease in humans. Some diseases are spread by direct contact between people (the common cold, flu), others by contamination of drinking water or food (dysentery, cholera), and others with the aid of another organism as a vector (malaria, for example, is caused by a protozoan called *Plasmodium* that has part of its life cycle in humans and part in mosquitoes). Epidemiologists who study infectious disease use many different types of models to understand and predict the course of particular diseases. For instance, measles is caused by a virus that is easily transmitted from person to person. Cases of measles in cities typically follow a cyclic pattern, with outbreaks every one to three years when few children are vaccinated and five to seven years when vaccination is more complete. Epidemiologists have built mathematical models that reproduce these cycles.

Some models of disease are quite complex, but we will explore a very simple model that is at the core of epidemiological studies of the spread of disease. The ability of a disease to spread in a population depends on the average number of new individuals infected by carriers who introduce the disease to a population lacking it. This number is called the *basic reproduction rate* and is symbolized by R_0. If the basic reproduction rate is greater than 1, the disease will continue to spread; if the basic reproduction rate is less than 1, the disease will eventually die out in the host population. For example, suppose four unrelated people pick up a new strain of flu while visiting Hong Kong and then fly home to Minneapolis. After they pick up their baggage, they spread out to different parts of the city. If the four travelers transmit the new strain of flu to a total of eight victims, then $R_0 = 8/4 = 2$. The eight newly infected people will infect 16 more, and this strain of flu will spread rapidly in Minneapolis. By contrast, suppose two of the original carriers each transmit the disease to one new person, but the other two don't give the disease to anyone. Then $R_0 = 2/4 = 0.5$. The four original carriers transmit the flu to two new victims; these two will infect one new person, and the disease will die out.

The basic reproduction rates of human diseases depend on many factors such as rate of production of infective particles by a pathogen in a human host, duration of infectiousness, ease of transmission, density of people in an area, and density of a vector such as mosquitoes if the disease depends on the vector. Ease of transmission, in turn, depends on mode of transmission as well as environmental factors including efforts to control transmission. Pathogens spread by coughing or sneezing such as cold or flu viruses spread more easily than pathogens spread by sexual contact such as human papillomavirus (HPV) or HIV, partly because sexual contact is less frequent than encountering people with a cold or the flu (other factors such as characteristics of these viruses themselves also contribute to differences in how easily they are spread from one person to another). Condoms can reduce the ease of transmission of viruses such as HIV and, to a lesser extent, HPV. Pathogens that cause vomiting or diarrhea can spread more easily where water sources are contaminated but can be controlled by cleaning water supplies; these types of pathogens also spread more easily among preschoolers than among college students.

Many epidemiologists devote careers to elucidating how all of these factors influence the basic reproduction rates of their favorite diseases. They work with complex models consisting of multiple equations that represent these factors. We can get a sense of this process by asking a simple question that leads to a simple model represented by a single equation. Here is the question: How does vaccination influence the spread of a disease? More specifically, how many people must be vaccinated to halt the spread of a disease? Even more specifically, how many people must be vaccinated to reduce the reproduction rate of the disease below 1, since this should cause the disease to eventually die out in our population as illustrated above?

The basic reproduction rate, R_0, is the average number of new individuals infected by a carrier of a disease. If some proportion of a population is vaccinated, then the carrier will come in contact with some people who are vaccinated and some who aren't vaccinated but won't spread the disease to the vaccinated people. So this reduces transmission potential from the basic reproduction rate to a lower number, which I'll call the effective reproduction rate and which I'll symbolize by R_e. If the proportion of the population that is vaccinated is represented by p, then we can write an equation to show how the reproduction rate of the disease is reduced with vaccination:

$$R_e = R_0(1-p)$$

In this model, $1-p$ is the proportion of the population that is *unvaccinated*. Imagine our infected carrier of the flu moving about Minneapolis in his daily activities. Of the people he encounters in the coffee shop, in class, or at the gym, only the unvaccinated ones are susceptible to this strain of flu. If he encounters 100 people before returning home and crawling in bed to nurse his illness and if 80% of these people are vaccinated ($p = 0.8$), then $1-p = 0.2 = 20\%$ is the proportion who are unvaccinated. Whatever the basic reproduction rate of this strain of flu in the absence of vaccination, the effective reproduction rate will be 20% of this value because the other 80% of people our carrier encounters are protected from getting the disease. Of course this argument depends on some assumptions (see Question 2 at the end of the chapter).

We can take one more step in developing this model and ask what proportion of the population must be vaccinated to reduce the reproduction rate to 1, since this is the threshold below which the disease won't persist. This condition can be expressed symbolically as follows:

$$R_e = 1$$

Since $R_e = R_0 (1 - p)$, the effective reproduction rate equals 1 if $R_0 (1 - p) = 1$. Rearranging in two steps to isolate p gives:

$$1 - p = \frac{1}{R_0}$$

$$p = 1 - \frac{1}{R_0}$$

This last form of the equation gives us a target for our vaccination program—if the proportion of the population that we vaccinate (p) equals 1 minus the inverse of the basic reproduction rate (R_0), then the effective reproduction rate will equal 1. If we can vaccinate a higher proportion of the population than this, the disease shouldn't be able to persist. Of course we'd probably want to incorporate some margin of error and vaccinate enough extra people to allow for the possibility that our estimate of the basic reproduction rate is inaccurate.

Let's apply this model to some actual human diseases. Researchers have estimated the basic reproduction rate to be about 4 for smallpox and about 17 for whooping cough. Figure 6.4 shows these values of R_0 as well as those for polio, chicken pox, and measles. Figure 6.4 also shows the proportion of individuals that must be vaccinated to reduce the effective reproduction rate to the threshold for disease control. As described above, this value is determined by setting $p = 1 - 1/R_0$. A line connects the points representing $R_e = 1$ for the five diseases; above this line, R_e is less than 1, implying a successful vaccination program. The main pattern in Figure 6.4 is that diseases with higher basic reproduction rates require greater vaccination rates for successful control. Smallpox can be controlled if 75% of a population is vaccinated, and smallpox has been eliminated worldwide as a human disease. Similarly polio, requiring 83% vaccination for control, has been largely eliminated except for a few cases in Nigeria, Afghanistan, and Pakistan (216 cases in these three countries in 2012, 6 cases elsewhere). It's much harder to eliminate diseases like chicken pox, measles, and whooping cough, for which 90% or greater rates of vaccination must be achieved.

Are you surprised that a population as a whole can be protected from disease without everyone being vaccinated? This protection, called *herd immunity*, is a consequence of the model that we developed in the last few paragraphs. Herd immunity has important implications for public health policy and even ethical decisions of individuals. Before summarizing these issues, we should ask whether there is any evidence that herd immunity works in the real world. This is an important question because our model is a very simplified representation of reality, ignoring all of the specific factors that influence basic reproduction rate and assuming that all carriers of a disease have the same likelihood of transmitting it to others.

There is quite a bit of evidence consistent with the relationship between basic reproduction rate of disease organisms and herd immunity illustrated in Figure 6.4. A recent

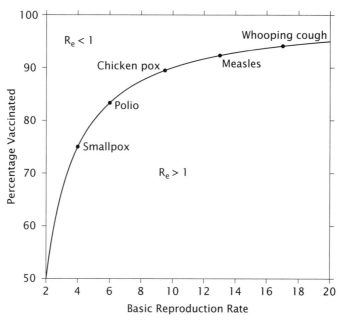

FIGURE 6.4 Potential for control of some common human diseases. The horizontal axis shows basic reproduction rate, the vertical axis shows the percentage of a population that is vaccinated, and the line shows the percentage of a population that must be vaccinated to reduce the effective reproduction rate to 1. Below the line, the effective reproduction rate is greater than 1 ($R_e > 1$) and the disease can spread if introduced to a human population. Above the line, the effective reproduction rate is less than 1 ($R_e < 1$) and the disease shouldn't be able to persist in the population. You can estimate the threshold for control of each of the five diseases shown by estimating the vertical position of the point representing that disease.

story about measles vaccination provides a good example of this evidence and, more generally, an example of a type of evidence that we haven't considered so far called a *natural experiment*. MMR vaccine is used worldwide to protect against measles as well as mumps and rubella (hence the name of the vaccine). Children are typically vaccinated before age 2, and these three viral diseases have become much less common since MMR vaccine began to be widely used. In the United States, for example, there were more than 400,000 cases of measles in most years from 1944 to 1964, when vaccination was licensed, and fewer than 10,000 cases in most years since 1964.

Based on our model, about 92% of a population should be vaccinated with MMR to control measles (Figure 6.4; critical percentages would differ for mumps and rubella because they have different basic reproduction rates than measles). In the United Kingdom in the 1990s, 90% to 92% of children were vaccinated with MMR by age 2. In early 1998, however, a physician named Andrew Wakefield reported an association between early childhood vaccination and the development of autism. Many parents were alarmed by this news, and vaccination rates dropped to 80% by 2003. Shortly thereafter, the number of measles cases in the United Kingdom increased to levels that hadn't been seen for many years (Figure 6.5). Wakefield's claim that early childhood vaccination caused autism has since been discredited, and vaccination rates in the United Kingdom gradually increased to 92% by 2012. Cases of measles remain at high levels because vaccination rates for children born between 1997 and 2003 were low, and these children are now entering high school where the disease spreads easily. It remains to be seen whether vaccination rates will stay high and measles will decline in the next few years.

This recent history of changes in MMR vaccination and corresponding changes in incidence of measles in the United Kingdom is a *natural experiment* because it shares one key feature of the designed experiments that we discussed in Chapter 4. All experiments involve comparison of different treatments. In this case, the "treatments" were rates of childhood vaccination of about 92% during the 1990s that researchers expected would control measles in the United Kingdom compared to rates below 88% from 1999 to 2009 that researchers expected would be too low to prevent the spread of measles in the United Kingdom. Unlike a designed experiment, however, these different rates of vaccination above and below the threshold for herd immunity were not planned by researchers but resulted from fear in parents of adverse consequences of vaccination for their children.

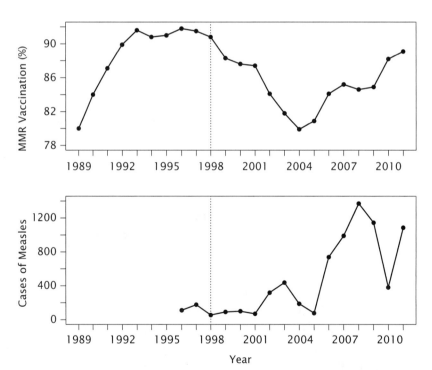

FIGURE 6.5 Percentage of children in the United Kingdom receiving measles-mumps-rubella vaccination by age 2, 1989–2011 (upper panel). Confirmed cases of measles in the United Kingdom, 1996–2011 (lower panel; data unavailable for previous years because screening methods differed). Vertical dashed line shows when Andrew Wakefield and colleagues published a paper in *The Lancet* claiming that early childhood vaccination causes autism.

Although natural experiments seem attractive because they lack the artificiality of some designed experiments, drawing credible conclusions about causation from natural experiments is tricky. Maybe the increased incidence of measles in the United Kingdom associated with decreased vaccination after 2000 was coincidental, and the real cause was a change in weather, increased immigration of vulnerable children from other countries, or any factor that differed between the last decade of the twentieth century and the first decade of the twenty-first. Although evidence like that in Figure 6.5 can be described as a natural experiment, this kind of evidence is more similar to correlational and comparative evidence than to evidence from designed experiments. Nonetheless, natural experiments can provide some evidence for evaluating hypotheses. Combined with various kinds of evidence from other diseases, this case study of measles in the United Kingdom supports our simple model of herd immunity. Ultimately, as you'll see in Chapter 8, we need to consider the weight of evidence to reach conclusions in science, where the weight of evidence includes everything from designed experiments through natural experiments to simple observations.

The concept of herd immunity is based on mathematical models of the spread of diseases in populations, but it raises difficult questions for personal decision-making and public policy. We make medical decisions for ourselves and our children on a regular basis. Some of these are heart wrenching but affect only our own families. Other decisions may have broader impacts. Consider whooping cough, or pertussis. This is a highly contagious bacterial infection that causes intense coughing that can last for weeks. Most adults can tolerate whooping cough, but infants may suffer pneumonia, seizures, or other brain disease as a complication of whooping cough. About 300,000 people worldwide, mostly children, die of whooping cough each year. Public health agencies in various countries introduced vaccines for whooping cough in the 1940s, and the effectiveness and safety of these vaccines were improved in subsequent decades. Imagine yourself as a parent of a young child. Would you get her vaccinated for whooping cough, considering the potentially dire consequences of the disease but also the fact that all vaccines have a small risk of serious side effects?

DTaP is a combination of vaccines that is highly effective against diphtheria, tetanus, and whooping cough (pertussis, hence the P). Medical personnel recommend five doses for children beginning at 5 months of age. You have a new baby and need to decide about getting her vaccinated against these and other diseases. On one hand, your doctor explains that not being vaccinated may substantially increase your daughter's risk of getting whooping cough. You do some research on the Web and verify this risk, but you also discover several anti-vaccination websites that raise concerns about childhood vaccination. These concerns range from legitimate—any vaccine can have adverse side effects in a small percentage of those vaccinated—to bogus—an earlier form of vaccine that used whole pertussis cells was thought to cause permanent brain damage in a very small percentage of cases, but this was eventually disproved. In making your decision about having your daughter vaccinated, you need to weigh the benefits of protecting her from trauma and possible death if she gets whooping cough against the potential risks of being vaccinated. So far, this seems like the same kind of decision that you might make about any other medical procedure. For example, suppose your daughter has strabismus, or lazy eye, which can result in a lifetime with poor depth perception if uncorrected. Should you opt for the noninvasive treatment of covering the good eye with a patch to force your daughter to exercise the lazy eye or a more aggressive treatment of using surgery as well as the patch, which has a better chance of success but carries more risks?

There's a key difference, however, between your decision about vaccinating your daughter and treating her for lazy eye. Your daughter can't make her own decision about her treatment for lazy eye, so you have to do it, but that decision affects only her. But your decision about vaccination may affect not only your daughter but also her friends, neighbors, classmates, and even the broader community. Here's how. Some kids in your community won't be able to be vaccinated because they're allergic to the vaccine or have weak immune systems. Others won't get vaccinated because their families don't have access to high-quality healthcare. Others won't because their parents believe vaccination is a government plot to control society. The net result may be a vaccination rate that drops below a level that can provide herd immunity and control whooping cough in a community, especially because DTaP doesn't provide protection against whooping cough for as long a time as the older form of the vaccine, so people need to get revaccinated periodically. Vaccinated children are still protected, but those who couldn't be vaccinated because of allergies or immune problems are susceptible, as are poor children who missed out on the recommended vaccination program, as are adults who were vaccinated years ago but have lost their immunity. In short, your decision about vaccination in combination with the decisions of others may contribute to the spread of a disease like whooping cough among vulnerable members of your community. Unlike lazy eye, vaccination isn't just about you and your family.

This isn't just a hypothetical example. The incidence of whooping cough has increased in many countries in recent years. In the United States, there were about 200,000 cases annually before the advent of vaccination, but only about 1,000 in 1976 when vaccination was widespread. In 2012, there were 41,000 cases and at least 18 deaths, mostly newborns in their first three months of life who weren't old enough to be vaccinated. Minnesota, Wisconsin, and Vermont had the highest incidence of whooping cough. As shown in Figure 6.4, whooping cough has a very high basic reproduction rate, so more than 90% of a population has to be vaccinated to prevent the disease from spreading. In some communities in Vermont, however, only 60% of children are vaccinated. Vermont is also 1 of 20 states that allow parents to use a philosophical or religious justification for not having their children vaccinated (all states permit medical exemptions).

George Till is a physician and a member of the Vermont state legislature. He learned from a pediatrician practicing near Burlington that only 25% of kids in a local kindergarten class were completely vaccinated with DTaP. Therefore Till introduced a bill in the legislature in 2012 to eliminate the opportunity for Vermont parents to refuse vaccination on philosophical grounds. This bill passed the state senate but its consideration was delayed in the House of Representatives. By the time it came up for consideration, an anti-vaccination movement had mobilized, lobbying against the bill on grounds that vaccination causes autism—a claim that has been thoroughly discredited. The bill failed, although Till reintroduced it in January 2013.

We began this section with an abstract mathematical model of the spread of disease. This model took the form of an equation for the basic reproduction rate of a disease organism. The model led us to the concept of herd immunity that we applied to some human diseases. In the last few paragraphs, I developed the implications of the model for pragmatic personal decisions with an ethical dimension—our decisions about vaccination may have negative impacts for other members of society. Finally, I described a connection between herd immunity, personal decisions about vaccination, and public policy. Working with models isn't just an intellectual exercise!

Predators and Prey: A Different Kind of Mathematical Model

Plants, animals, and microbes interact in a multitude of ways. In Appendix 3, you learned about mutually beneficial relationships between bees and plants, in which plants provide nectar as food for bees and bees provide pollination services to plants. In the last section of this chapter, you learned about measles and other diseases as examples of parasitic relationships between microbes and their human hosts, where the microbes benefit at the expense of the hosts. Predation is another interesting and important interaction between species. Nature shows on television often feature predation by large, charismatic predators like lions or wolves on their large, charismatic prey like wildebeest or bison. Ecologists who study predation may focus on anatomical or behavioral adaptations of predators and prey, influences of predation on the population dynamics of the prey and predators, the evolution of predator–prey interactions, or the consequences of predation for the biodiversity of natural environments. Predation has also been a theme of ecological modeling; we'll discuss a contemporary version of the first and simplest mathematical model of predation and find that it has a surprising practical consequence.

Vito Volterra was an Italian mathematician whose daughter was engaged to a fisheries biologist named Umberto D'Ancona. D'Ancona analyzed fish harvests from the Adriatic Sea during and after World War I and found that harvests of sharks and other predatory fish increased relative to harvests of cod and other prey fish during the war but declined after the war ended. Because the Adriatic was a war zone, anglers caught fewer fish in this area during World War I than before or after the war. Assuming that harvests of different species were proportional to population sizes of those species, D'Ancona and other biologists wondered why decreased human fishing should shift the abundances of different species in favor of predators. At the fish market in Fiume, for example, sharks were about 12% of fish sold in 1914, as the war was beginning. This relatively low percentage of predators in the Adriatic Sea should make sense in relation to what you learned in Chapter 3 about energy flow through food chains. However, the percentage of sharks sold in the market at Fiume increased during World War I to 36% in 1918, the last year of the war, and then decreased to 11% by 1923. Why should reduced human fishing during the war be associated with increased proportions of predatory fish in the Adriatic Sea? D'Ancona asked Volterra if he had any ideas that could account for this surprising pattern.

Volterra was not a biologist but a mathematician, so he approached the problem by building an abstract mathematical model. His model consisted of two equations, one for the growth rate of a prey population and one for the growth rate of a predator population. This model was more complicated than the one that we discussed above about vaccination and disease, partly because the predator–prey model had two equations instead of one but mostly because it involved not just algebra but also calculus. Despite this complexity, Volterra's model had some unrealistic consequences. However, predator–prey models have evolved substantially since the time of Volterra and D'Ancona, although most of these models provide the same explanation for increased abundance of sharks in the Adriatic in World War I as Volterra's original model.

Do you have a little brother or sister who plays video games, or do you enjoy video games yourself? A video game is essentially a computer program, and I'll describe a mathematical model of predation in the form of a computer program. My model is much simpler than a video game but no different in its basic structure. If you like video games, learning about this model of predation may help you appreciate your favorite games even more. More important, biologists are making more and more use of computer models to ask questions and test hypotheses, so learning how to interpret these models will help you better understand how the science of biology works.

Imagine a large meadow containing a population of rabbits and a population of coyotes (we'll return to sharks and cod in the Adriatic later). The rabbits and coyotes move around the meadow, the rabbits eating grass and the coyotes catching and eating rabbits. To keep track of individual rabbits and coyotes, I'll divide the meadow into a grid of 2,500 square cells, all the same size. I'll populate the meadow with 400 rabbits and 250 coyotes to start and keep track of the two populations for 1,300 time intervals as individuals of each species reproduce and die. The time intervals might be days, weeks, months, or years, but I'm purposely not specifying what they are because I intend this to be a general model of an interaction between predators and prey, not a model specific to coyotes and rabbits. For instance, I might apply the model to trout feeding on small, aquatic invertebrates called zooplankton in a lake. In this case I would divide the lake into a grid just like I did the meadow for coyotes and rabbits, but the cells of the grid would be smaller for the lake than for the meadow, and the time interval would be shorter as well because of the shorter lifespans of trout and zooplankton.

At the beginning of a run of the model, each coyote and rabbit has a certain amount of stored energy from previous feeding. The coyotes and rabbits are randomly distributed across the meadow, so many of the 2,500 cells don't have either predators or prey; some have only predators, some have only prey, and a few may have both predators and prey. We also start by imagining that about half of the cells have grass that rabbits can feed on. The other cells may have had grass in the past, but this grass has been depleted by previous feeding by rabbits. As you'll see shortly, the grass will eventually grow back. The upshot is a meadow that is a changing mosaic over time of grassy green cells and bare brown cells, with rabbits and coyotes moving about, giving birth, dying, and occasionally interacting.

This simulation model can be described by a set of rules for the predators, the prey, and the prey's food. Of course these rules have to be written as computer instructions, which are like equations, to run the model, but I'll spare you these details and describe the rules in words instead.

There are three rules for predators. These rules are applied to each individual predator at each time interval of the model. First, if you (as a predator) share a cell with a prey individual, catch it and eat it. This increases your energy level by 20 units. Second, you may reproduce, but only with a probability of 5%. Reproduction is simulated by picking a random number between 1 and 100; if this number is less than 5, a new predator is "born" on the same cell as its parent. This 5% chance of reproduction means that a population of 100 predators would have, *on average*, five offspring, since 5% of 100 = 5. The actual number of new predators born to 100 adults might be more or less than 5 because the model treats birth as a random process. The third rule simulates movement to an adjacent cell selected at random, with loss of 1 unit of energy and death of the predator if its energy level drops to zero. This means that a predator that has just eaten can live for 20 time steps before eating again—it gains 20 units of energy when it eats a prey individual and loses 1 unit in each subsequent time step. If the prey population is sparse enough that a satiated predator doesn't find another meal as it moves around the meadow for the next 20 time periods, it dies. Of course, these numerical specifications of the model can be adjusted; shortly I'll show you how to run the model yourself and do this.

The rules for prey are similar to those for predators. If you (as a prey individual) occupy a cell with grass, eat the grass and gain 4 units of energy and then reproduce, with a probability of 5%. Finally, move to an adjacent cell at the cost of 1 unit of energy. You might wonder why rabbits gain so much less energy from eating grass than coyotes do from eating rabbits. While coyotes can live for 20 time steps on one meal, rabbits live for only 4. This difference (loosely) reflects biology in two ways: (i) a diet of grass, with its high fiber content, is less nutritious than a diet of rabbits, but (ii) grass is more abundant, so, as rabbits move around the grid, they are reasonably likely to land on another cell with grass within four time steps.

Finally, there is a rule for regrowth of grass after grazing by rabbits. After a rabbit has eaten the grass in a cell, the cell will remain barren for 30 time steps and then return to grass. Just as with the other parameters of the model like birth rates of predators and prey, the user can adjust regrowth time to see how results change when a simulation is run.

As with any model, this model of predation is a simplification of reality. For example, I mentioned that *individual* predators and prey reproduce. Doesn't it take two coyotes, a male and a female, to make baby coyotes, and two rabbits to make baby rabbits? Of course it does! The question is whether the added complexity of incorporating sexual reproduction in the model is necessary for our uses of the model. If we were interested in genetic changes in the populations of rabbits and coyotes, we would probably want to include both sexes and simulate sexual reproduction between males and females that meet on the grid representing our habitat. But I want to use the model to learn something about population dynamics, not genetics, so ignoring sex may be a reasonable simplification.

This computer simulation model has three advantages over Volterra's original 1926 model using two equations to link populations of predators and prey. First, the computer simulation is less abstract. I hope you can visualize rabbits and coyotes (or cod and sharks) moving around their environment, eating food if they encounter it, giving birth, and dying more easily than you can visualize P for number of predators, V for number of prey, r for growth rate of the prey population, a for capture rate of prey by predators, b for birth rate of predators, and d for death rate of predators. Second, the computer simulation is more realistic than Volterra's model, despite the many simplifications of the simulation (see Question 5 at the end of the chapter). For example, Volterra's model considered only predators and prey as if they existed in a vacuum, while the computer simulation also includes grass as food for the prey. To be fair, modelers have developed analytical extensions (systems of equations) of Volterra's original model that are as realistic as our computer simulation, although still very abstract. Third, and perhaps most important, you can play with the computer model yourself, at a website called NetLogo (http://ccl.northwestern.edu/netlogo/).

Plate 11 shows the basic setup from NetLogo, with a grid of 50 × 50 = 2,500 cells, each colored green if it has grass or brown if not. The plate also shows 50 coyotes and 100 rabbits scattered across the habitat. As you run the model in NetLogo, you will see the locations of rabbits and coyotes change as they move around the meadow, you will see their numbers change as they reproduce and die, and you will see the mosaic of green and brown cells shift as rabbits eat grass, turning cells from green to brown, and as grass grows back, turning cells from brown back to green.

Although I've described this model in terms of coyotes and rabbits, I've mentioned that it could apply to any predator–prey interaction, such as trout and zooplankton in a pond or sharks and cod in the Adriatic Sea. Indeed, Volterra's original version didn't mention any particular species, just a generic predator and a generic prey species. As we discussed, Volterra was stimulated to make his model by conversation with his future son-in-law, Umberto D'Ancona, who had been surprised by an increase in relative abundance of sharks in fish harvests from the Adriatic Sea during World War I. How does Volterra's model relate to D'Ancona's unexpected observation?

Figure 6.6 is a simplified diagram of the food web involving fish in the Adriatic Sea. It's simplified because predatory fish include other species besides sharks, prey fish include other species besides cod, seabirds and other predators also eat cod and other

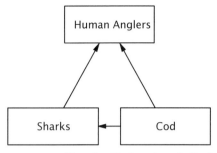

FIGURE 6.6 Very simplified version of a food web in the Adriatic Sea. Sharks catch and eat cod; human anglers catch and eat both sharks and cod.

prey fish, and prey fish eat zooplankton and other foods. In fact, Figure 6.6 is a kind of model of the situation in the Adriatic, somewhat like the maps that we discussed at the beginning of this chapter. This diagrammatic food web model doesn't include any quantitative information with which we could predict population dynamics of sharks and cod, but it does include a third species—human anglers, that function quite differently than predatory fish like sharks. The sharks are relatively specialized predators, feeding mostly on smaller prey fish, but the humans are generalized predators, feeding on both sharks and cod as well as other prey fish. As Volterra discovered from his model, predation by a generalized predator on both a prey and its specialized predator can alter the dynamics of both species. Humans harvest both cod and sharks. In doing so, we have a *direct negative* effect on the population size of cod but an *indirect positive* effect through suppressing the number of sharks that also eat cod. Volterra showed that the *net* effect of human fishing on the system was to benefit prey species at the expense of predator species. Conversely, his model showed that reduction of human fishing as occurred in World War I would benefit the predators at the expense of the prey, consistent with D'Ancona's report that the relative abundance of sharks in the Adriatic tripled during World War I.

This result of Volterra's model has become known as the *Volterra Principle* and has been applied much more broadly than just to predatory interactions between fish and humans. The Volterra Principle states that anything that harms both predators and prey indiscriminately will increase the number of prey relative to the number of predators in a habitat. Figure 6.7 illustrates the Volterra Principle using my Predator, Prey, Poison model in NetLogo. In this example, I ran the model for 500 time steps with just grass, rabbits eating grass, and coyotes eating rabbits. Then, to model a harmful effect on both predators and prey, I introduced a poison that killed both coyotes and rabbits in a small percentage of the meadow for the next 300 time steps. Finally, I removed the poison for the last 500 time steps. On average, there were 338 coyotes before poisoning started, 204 coyotes while poisoning was simulated, and 314 coyotes after poisoning ended. For rabbits, however, average numbers were 639 before poisoning, 702 during poisoning, and 677 after poisoning. In this experiment with the model, a poison that killed both coyotes and rabbits caused the predators to decrease by 40% but caused prey to increase by 9%. Notably, in this experiment I used poison in only 1.5% of the cells in the habitat. Slightly higher levels of poison caused the predator population to disappear altogether so the prey population got even bigger.

So far I've couched this discussion in terms of coyotes and rabbits. This isn't the best example to illustrate the Volterra Principle, however, because human interaction with coyotes and rabbits is quite different from human interaction with sharks and cod or

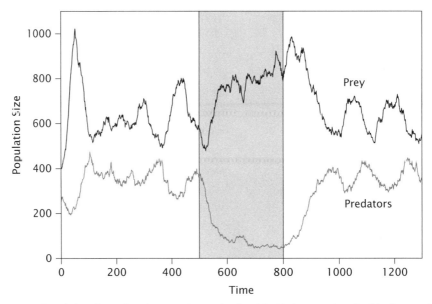

FIGURE 6.7 Population sizes of predators and prey simulated by NetLogo as described in the text. Light gray shaded area indicates the time during which a poison that killed both predators and prey was applied to 1.5% of the cells in the habitat.

other pairs of predator and prey species. The story of biological and chemical control of insect pests is a more realistic—and dramatic—illustration of the Volterra Principle.

Groves of citrus trees in California are the setting for this story. California is one of the most important agricultural states in the United States today, but the history of agriculture in California is relatively brief compared to other states. The citrus industry began in the 1880s, but by 1887 a pest insect called the cottony cushion scale was doing tremendous damage to lemon, orange, and other citrus trees. Evidence suggested that this pest had been accidentally introduced to California in 1868 on acacia trees brought from Australia. Therefore Charles Riley, head of the Entomology Division of the US Department of Agriculture, sent his assistant Albert Koebele to Australia to search for a natural parasite or predator of cottony cushion scale. On October 15, 1888, Koebele found a beetle feeding on cottony cushion scale in a garden in North Adelaide. He shipped samples of these vedalia beetles back to California in the fall and winter of 1888–1889, and these beetles were released on citrus trees. By the summer of 1889, damage by cottony cushion scale was much reduced, and the citrus industry became well established and remained successful for the next 50 years.

This was the first successful demonstration of biological control—use of a natural biological process to control a pest population. In this case, the biological process was predation, and the predator was a beetle that fed voraciously and specifically on cottony cushion scale. In the 1940s, however, a new era of chemical control of pests began with the introduction of DDT. During World War II, DDT was successfully deployed in various parts of the world to control insect vectors of diseases such as typhus, malaria, and dengue fever. After the war, farmers began using DDT to control insect pests of crops. This turned out to be disastrous for citrus, however (Figure 6.8). DDT and other chemical pesticides are indiscriminate poisons that kill not only pest species like cottony cushion scale but also predators of those pests like vedalia beetles—the Volterra Principle in action.

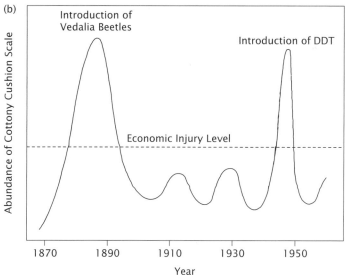

FIGURE 6.8 Cottony cushion scale with vedalia beetles on a tree branch (a); history of cottony cushion scale in California showing rapid increase and subsequent control by vedalia beetles in the late 1800s and an irruption of the pest population in 1947 associated with the introduction of DDT (b; the original source did not provide numerical values for the vertical axis).

DDT was finally banned for most uses in the United States in 1972, partly in response to Rachel Carson's documentation of its widespread detrimental effects on insects, birds, fish, and humans in *Silent Spring*, published in 1972. During the same period of time, there was a resurgence of interest in biological control, especially for pests introduced accidentally from one part of the world to another. In these cases, researchers often found that pests showed explosive population growth in the new environment because they left their natural predators and parasites behind in their native environment. This suggested that it might be possible to control pests in the new environment by introducing a predator or parasite from the native environment, especially if that predator or parasite fed exclusively on the pest species and had a high potential rate of population growth itself (see Question 6). Many scientists like to travel, and searching for potential agents of biological control provides a perfect justification for learning the natural history of exotic places. These travels and ensuing experiments resulted in many successful examples of biological control of insect pests.

Conclusions

We've explored four relatively simple models in this chapter—concrete scale models of DNA and San Francisco Bay and mathematical models of herd immunity and predator–prey interactions. The model of herd immunity took the form of an algebraic equation while the model of predator–prey interactions took the form of a computer simulation, illustrating different ways that researchers express mathematical models. The topics of these models were very different—from a fundamental aspect of the biochemistry of life, to a development proposal to alter a California environment beloved by many, to the control of human disease by vaccination, to the interactions of chemical and biological control of pests. Like all models, these four examples were simplifications of nature. All four also had important practical implications. I described these implications for the models of San Francisco Bay, herd immunity, and predator–prey interactions; Watson and Crick's DNA model was a key step in understanding biochemistry, physiology, and genetics that eventually led to the new field of molecular medicine.

How do models relate to the other tools of science that we've discussed so far—observations, comparisons, correlations, and experiments? All science is motivated by curiosity, a search for treasure as I described in the beginning of this chapter (Figure 6.1). This search for treasure may lead to wonderful new discoveries, like the winter home of monarch butterflies described in Chapter 1, to better understanding of causation, or both. So there is a unity of purpose in science, but a diversity of methods to satisfy our curiosity about the natural world and how it works. There is not *a* scientific method; instead, different questions are answered with different combinations of observations, comparisons, correlations, natural experiments, designed experiments, and models. So modeling complements the other tools that scientists use. The best way to explain this may be to revisit two of the examples in this chapter, stepping back from the details of the models to the questions that motivated the models and the other methods that were used to answer these questions.

The fundamental purpose of vaccination is to protect individuals who are vaccinated from contracting a disease, but even without a model of herd immunity, it should be fairly obvious that vaccination may control the spread of disease even if some people aren't vaccinated. What does our model expressed in the equation $p = 1 - 1/R_0$ add to this basic understanding of herd immunity? To review, p in this equation is the proportion of a population that is vaccinated and R_0 is the basic reproduction rate of the disease organism; if p is greater than $1 - 1/R_0$, the model predicts that the disease will be controlled. Without the model, we have only a qualitative understanding of what influences the spread of a disease; with the model, we can make quantitative predictions about the level of vaccination required for control. In addition, the model focuses the attention of epidemiologists on estimating the basic reproduction rates of disease organisms, since estimating R_0 is the key to controlling diseases. In other words, the model encourages observational, comparative, correlational, and other research to get accurate estimates of R_0, not that this research is necessarily easy or straightforward.

Our model also leads to specific predictions about herd immunity (Figure 6.4) that can be tested by comparing the responses of diseases like smallpox and whooping cough to vaccination campaigns and by natural experiments such as the shift in acceptance of MMR vaccine in the United Kingdom following the claim that this vaccine led to autism (Figure 6.5). Finally, our quantitative model of herd immunity crystallizes some practical and ethical issues that parents should consider when deciding about vaccinating their children.

I must emphasize that this is the simplest possible model of disease spread, depending as it does on many assumptions, especially that the basic reproduction rate of the disease organism is constant. Epidemiologists use more complex and more realistic models to make more specific predictions about how rapidly diseases will spread and how vaccination programs will work. For example, early in the history of the HIV epidemic Roy Anderson and Robert May used models to predict how rapidly HIV would spread in various countries. In a 1992 review in *Scientific American*, they wrote, "Mathematical models . . . sometimes produce results that upset simple intuition." Their modeling suggested that educational programs that caused women to change sexual partners *less* often might cause *faster* initial spread of HIV in the general population, although fewer infected individuals overall in the long term. This prediction followed from the assumption that such educational programs for women in the general population might cause sexually active men to visit prostitutes more frequently. In their model for HIV, unlike ours for measles, R_0 was different for different subpopulations depending on extent of sexual activity with different partners.

Our NetLogo model has similar relationships to empirical research on predation as our model of herd immunity has to empirical research on the spread of disease. Figure 6.6 is a conceptual, "boxes-and-arrows" model of interactions between cod as prey, sharks as specialized predators, and human anglers as generalized predators that harvest both cod and sharks. This model is like the qualitative understanding of herd immunity described three paragraphs above. In the case of Figure 6.6, a net beneficial effect of human anglers on cod is certainly possible, although not obvious because the net effect depends on the relative strength of predation by the anglers on the sharks and the cod. Formulating this conceptual model as a quantitative model using NetLogo and generalizing from human anglers and fish to predators, prey, and a poison that kills both predators and prey leads to a demonstration that an indiscriminate poison can indeed favor the prey population (Figure 6.7), although this result may depend on specific values of parameters of the model such as the birth rates of predators and prey.

Models of predation are connected to the real world in two other important ways that illustrate how modeling complements other scientific tools. First, Volterra didn't just decide out of the blue to make a mathematical model of predation. Instead, he was inspired to make his model (which we mimicked in NetLogo) by a surprising observation by D'Ancona about fish harvests in the Adriatic Sea during and after World War I. Second, although agricultural researchers in California did the first experiments with biological control long before Volterra built his model, Volterra's model helped researchers understand why the use of DDT on citrus crops in the 1940s had effects opposite to those that were intended (Figure 6.8). This in turn contributed to expansion of biological control efforts and eventually to integration of biological with chemical control in more rational programs of integrated pest management.

Predator–prey models illustrate one final point about effective use of scientific tools. Volterra's original model was the simplest possible model of population dynamics of interacting predators and prey. Volterra didn't pay any attention to individual predators and prey in his model, and mathematical biologists who made this model more realistic by including such things as food for the prey and satiation of foraging predators also ignored individuals and just focused on total population sizes of the two species. Our NetLogo model, by contrast, followed individual rabbits and coyotes as they moved around the environment, fed, gave birth, and died. All of these models, however, despite their differences in complexity and basic structure, predicted the Volterra Principle that anything that harms both predators and prey indiscriminately will increase the relative number of prey. This should enhance our confidence that the Volterra Principle is a robust concept with broad applicability. More generally, we can increase our confidence in models by getting similar results from different types of models just as we can increase our confidence in empirical research by getting similar results from different observational and experimental studies. You saw an example of this in the story of lead and crime in Chapter 5.

Modeling is the most difficult tool in the scientific toolbox to understand, so I used fairly simple examples to introduce you to this process. Yet some complex models have very important implications for decisions that you will have to make as a citizen about policies that affect all people. How can you deal with models of global climate change, for example, that involve hundreds of equations and months of time to analyze using supercomputers? I offer some advice in Appendix 4, and discuss the empirical evidence relating to climate change in this appendix and in Chapter 10.

Questions to Ponder

1. In introducing the idea of models in this chapter, I began with two explicit biological examples from earlier chapters—monarch butterflies as a potential model system for studying biological clocks and bats as a potential model system for studying the physiological mechanisms of senescence. In fact, biologists have a long history of using model organisms to study various topics. In genetics, Gregor Mendel used pea plants to discover the principles of inheritance, later researchers used fruit flies to study how inheritance was influenced by the dance of the chromosomes in cell division, and still later workers used bacteria and viruses to home in on the structure and function of DNA. Cell biologists use a roundworm called *Caenorhabditis* to study how brains work; medical researchers test new surgeries on dogs. Why do you think scientists use organisms such as these as model systems to ask certain kinds of questions? What are the benefits and limitations of working on model organisms?

2. In describing the effect of vaccination on the reproduction rate of a disease organism, I imagined a carrier of the flu contacting 100 people in his daily activities. With no vaccination and a basic reproduction rate of 10, he will pass the disease to about 10 of these people (not exactly 10, because R_0 is the average number of new infections that result from one infected person and transmission involves an element of chance). If 80% of the population is vaccinated, however, we expect our carrier to pass the disease to only two of the people whom he encounters. This analysis depends on two key assumptions: that infectiousness is the same for all carriers and that susceptibility is the same throughout the population. These assumptions imply that the basic reproduction rate of the disease is a fixed value for all carriers and that the effective reproduction rate is the same throughout the area occupied by a population, like the city of Minneapolis in this example. Describe a scenario in which these assumptions don't hold true. What might be the consequences for spread of the disease?

3. If you were a member of the Vermont legislature, would you vote for or against George Till's bill to require parents to follow the prescribed vaccination program for their children unless they qualify for a medical exemption (i.e., a philosophical exemption would no longer be available)? Why or why not?

4. D'Ancona's data for sales of fish at the Fiume market on the Adriatic Sea showed that sharks increased from 12% at the beginning of World War I to 36% at the end of

World War I and then decreased to 11% in 1923. As described in the text and illustrated in Figures 6.6 and 6.7, these results may be attributed to the Volterra Principle. What else might account for these changes in relative abundance of sharks harvested during World War I, considering the increased risk of fishing in the Adriatic while fighting was underway?

5. One simplification of the Predator, Prey, Poison model in NetLogo is that I ignored sexual reproduction by coyotes and rabbits. I justified the assumption that all individuals can reproduce on page 134. Describe some other simplifying assumptions of the model. Which of these can you justify, as I did for the absence of sexual reproduction, and how? Which assumptions do you think are most critical for the results of the model and so should be modified to make the model more realistic?

6. An insect species from Asia is accidentally introduced to rangeland habitat in North America and becomes a damaging pest on native vegetation. Researchers travel to the Asian home of the insect pest to search for a predator that can be brought back to North America and used for biological control in areas where the pest has become established. Successful biological control agents have high rates of population growth and feed exclusively on the pest species that they are intended to control. What might happen if the prospective biological control agent in this example has a high rate of population growth but doesn't feed exclusively on the intended pest species?

Resources for Further Exploration

1. We used software called NetLogo to model predation and develop the Volterra Principle about the effects of a generalized predator or indiscriminate poison on the interaction between a specialized predator and its prey. You can run sample NetLogo models in your Web browser or download NetLogo from the Center for Connected Learning and Computer-Based Modeling at Northwestern University and install it on your Windows, Mac, or Linux personal computer (http://ccl.northwestern.edu/netlogo/). Our model is called "Predator, Prey, Poison" and is available at http://ccl.northwestern.edu/netlogo/models/community/; you can use it to test how variations in predator and prey birth rates and energy requirements, rates of grass regrowth, and intensity of poisoning influence results of simulations. For example, what if there is no poisoning that kills both predators and prey? Do populations of the two species cycle indefinitely, or do they reach relatively stable levels that are maintained for a long time? You can also experiment with many other interesting sample models in NetLogo in fields ranging from math to biology to psychology to art.

2. Three of the sample models available with NetLogo simulate the spread of diseases in human populations: AIDS, Virus, and EpiDEM Basic. By studying these models (load a model and read the text at the Info tab) and experimenting with them, you can learn about more realistic models of disease spread than the basic model we discussed in this chapter.

3. I introduced our NetLogo model of predation and the Volterra Principle by asking if you like video games. Mary Flanagan and her students and colleagues at Dartmouth College have developed a game called POX: Save the People that involves stopping the spread of a disease by vaccinating people, thereby illustrating the principle of herd immunity. Alternative versions of POX are POX: Save the Puppies and ZOMBIEPOX. These games are available for various platforms at http://www.tiltfactor.org/store/.

PLATE 1 Great Pyramid of Giza, by Nina Aldin Thune (a); arctic tern protecting her nest, by Oddur Benediktsson (b); Harris's sparrows, by Richard Crossley (c); migratory ungulates in the Serengeti of East Africa, with zebras and wildebeest in the foreground (d).

PLATE 2 The life cycle of monarch butterflies. Adult, by Armon (a); late-stage caterpillar after shedding its skin, by Ryan E. Poplin (b); chrysalis (pupa) in which metamorphosis to an adult butterfly occurs, by Armon (c); identification tag on the wing of a monarch, by Jim Gagnon (d).

PLATE 3 *Anchiornis huxleyi*, a 150-million-year-old dinosaur about the size of a chicken, painted by Michael Digiorgio. Colors of this specimen were determined by Li et al. as described in the text. This specimen had no tail, so the tail shown here is based on another specimen of the same species found earlier, although the colors of the tail aren't known.

PLATE 4 Chromolithograph of an ivory-billed woodpecker, by Theodore Jasper (a); photo of a male pileated woodpecker, by Dick Daniels (b).

PLATE 5 Dr. Jaime Guevara-Aguirre with some Ecuadorean villagers with Laron Syndrome in 2009.

PLATE 6 One of the smallest mammals, the bumblebee bat, by Medhi Yokubol (a); skeleton of the largest mammal, the blue whale, at the Canadian Museum of Nature, by D. G. E. Robertson (b); one of the smallest primates, the pygmy mouse lemur of Madagascar, by Bikeadventure (c); the largest primate, the eastern gorilla, by F. Löcker (d).

PLATE 7 Baleen plate on one side of a partially open mouth of a right whale.

PLATE 8 Three northern elephant seals—a large male mating a much smaller female with her pup from the previous year beside her—by Mike Baird (a); a common opossum in Caracas, Venezuela, by Juan Tello (b); a band of wild horses in Theodore Roosevelt National Park, North Dakota, by the National Park Service (c).

PLATE 9 Amounts of lead in the soil in different neighborhoods of New Orleans mapped by Howard Mielke and Christopher Gonzales. Lead is measured in milligrams per kilogram of soil, which is equivalent to parts per million.

PLATE 10 Scale model of San Francisco Bay in a warehouse in Sausalito, California (partial views). The top photo shows the Golden Gate Bridge at the entrance to San Francisco Bay with some of the infrastructure of the facility in the background. The bottom photo shows the Bay Bridge across San Francisco Bay from Oakland at the lower left across Yerba Buena Island to the city of San Francisco at the top. Both photographs taken by Michael Weisberg, University of Pennsylvania.

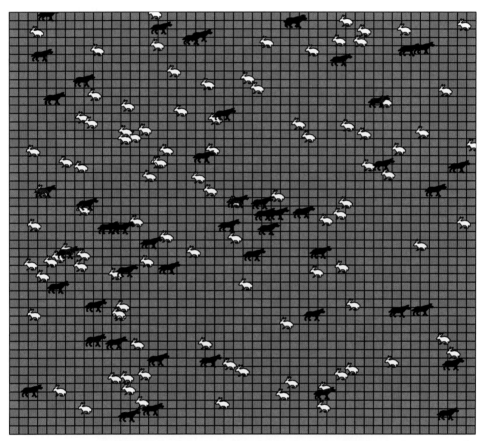

PLATE 11 Coyotes and rabbits distributed on a meadow at the beginning of a simulation using the Predator, Prey, Poison model in NetLogo. The meadow consists of a grid of 2,500 patches—some green because they have grass that can be eaten by rabbits and others brown because they have been fed on recently.

PLATE 12 Photograph of two domesticated zebra finches, by Luis Miguel Bugallo Sánchez (a); photograph of yarrow in Scott's Valley, California, by Dvortygirl (b).

PLATE 13 Starfish at the edge of a bed of mussels with seaweeds, by Dave Cowles (a); sea otter grooming itself, by Mike Baird (b); killer whales, by the National Oceanic and Atmospheric Administration, (c); Steller sea lions, by G. Frank Peterson (d).

PLATE 14 Sea urchins carpeting the ocean floor at Santa Cruz Island, Channel Islands National Park, by the National Park Service (a); kelp forest at Santa Barbara Island, also in the Channel Islands off the coast of southern California, by Chuck Kopczak (b).

PLATE 15 Photographs of the same site in the Lamar Valley of Yellowstone National Park taken in 1997, two years after reintroduction of wolves (a) and 2010, 15 years after reintroduction (b). The new plant growth in the foreground of the 2010 picture is willow shrubs.

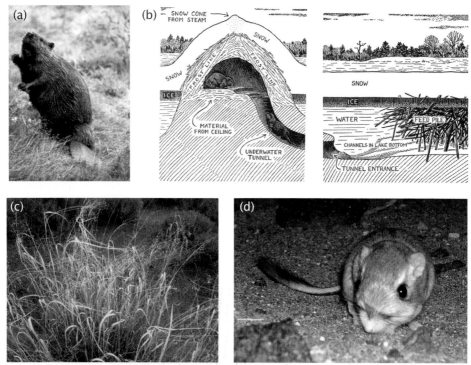

PLATE 16 North American beaver, by Ross Haley (a); diagram of a beaver lodge with a food cache of a large pile of sticks anchored in mud at the bottom of the pond nearby (b); Indian ricegrass, by Matt Lavin (c); Merriam's kangaroo rat, the primary rodent that harvested and stored seeds of Indian ricegrass at our study site, by Ned Dochtermann (d).

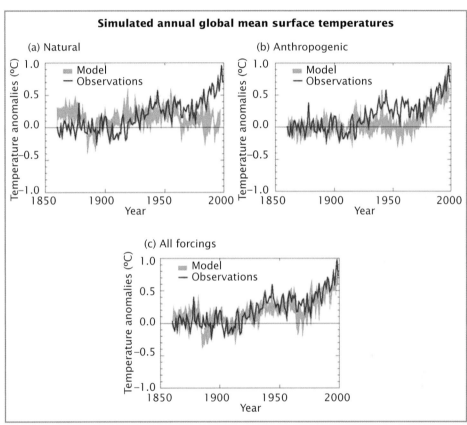

PLATE 17 Average annual temperature of Earth's surface from 1860 to 2000 (red lines) compared to projections of three sets of global climate models: models with only natural forcings such as changes in solar radiation and volcanic activity (a), models with only anthropogenic forcings such as release of greenhouse gases into the atmosphere (b), and models with both natural and anthropogenic forcings (c). The gray bands show the range of projections for four models in each case. The temperature anomaly shown on the y-axes of these graphs is the deviation from the average temperature between 1880 and 1920 in degrees Celsius (multiply these values by 1.8 for degrees Fahrenheit).

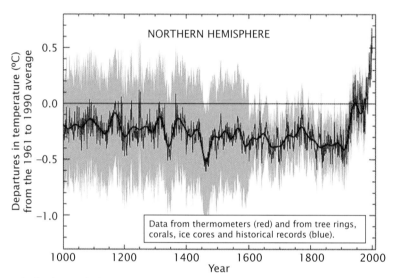

PLATE 18 The hockey stick showing estimated temperatures for the Northern Hemisphere for the last 1,000 years. The blue line shows proxy data from tree rings, corals, ice cores, and historical reports, the black line is a smoothed average of these proxy data, the gray region represents the 95% confidence interval on these proxy data, and the red line is the instrumental record. Note the overlap in the proxy and instrumental data from the mid-1800s through the mid-1900s, which was used to validate the proxy data.

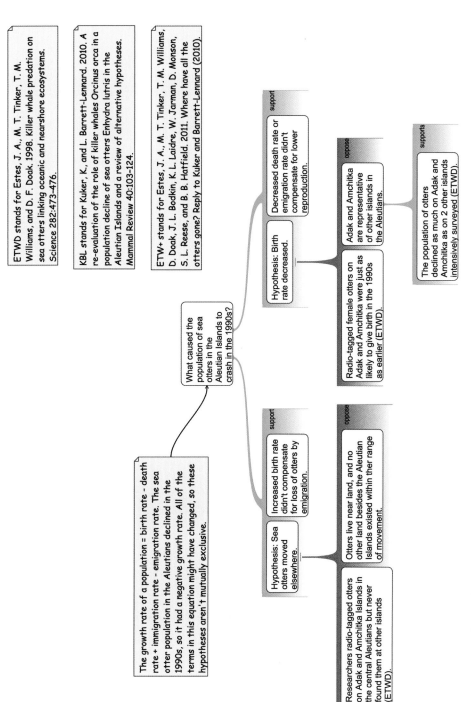

PLATE 19 Partial map of arguments supporting and opposing alternative explanations for the population crash of sea otters in the Aleutian Islands in the 1990s. This part of the map considers the hypotheses that sea otters declined because emigration increased or birth rate decreased. I made this map and those in Plates 20 to 23 with Rationale™ software.

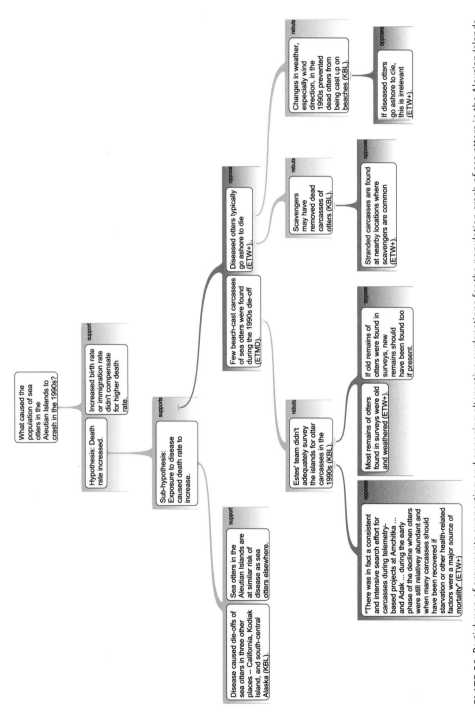

PLATE 20 Partial map of arguments supporting and opposing alternative explanations for the population crash of sea otters in the Aleutian Islands in the 1990s. This part of the map considers the hypothesis that sea otters declined because death rate increased due to disease.

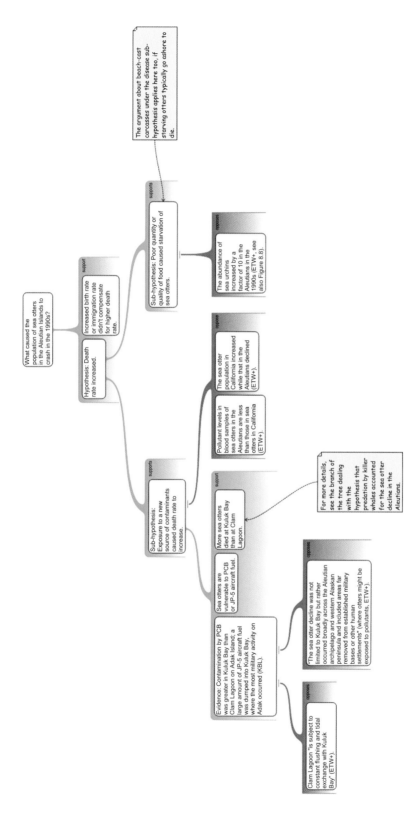

PLATE 21 Partial map of arguments supporting and opposing alternative explanations for the population crash of sea otters in the Aleutian Islands in the 1990s. This part of the map considers the hypotheses that sea otters declined because death rate increased due to exposure to contaminants or starvation due to lack of food.

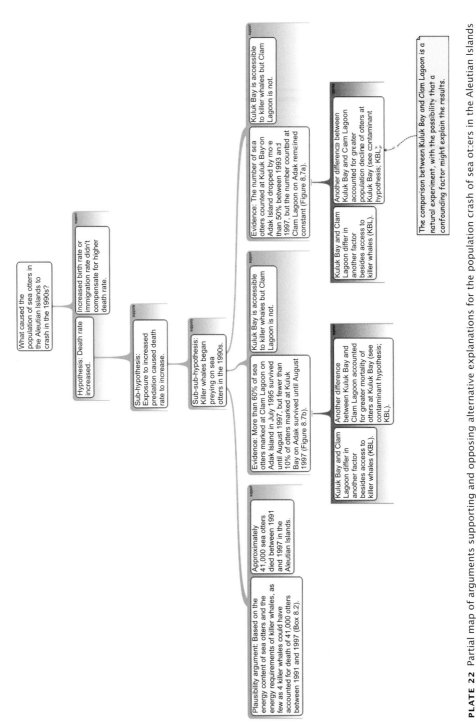

PLATE 22 Partial map of arguments supporting and opposing alternative explanations for the population crash of sea otters in the Aleutian Islands in the 1990s. This part of the map considers the hypothesis that sea otters declined because death rate increased due to predation by killer whales.

The following text appears within the argument map boxes:

What caused the population of sea otters in the Aleutian Islands to crash in the 1990s?

Increased birth rate or immigration rate didn't compensate for higher death rate.

Hypothesis: Death rate increased.

Sub-hypothesis: Exposure to increased predation caused death rate to increase.

Sub-sub-hypothesis: Killer whales began preying on sea otters in the 1990s.

Plausibility argument: Based on the energy content of sea otters and the energy requirements of killer whales, as few as 4 killer whales could have accounted for death of 41,000 otters between 1991 and 1997 (Box 8.2).

Approximately 41,000 sea otters died between 1991 and 1997 in the Aleutian Islands.

Evidence: More than 60% of sea otters marked at Clam Island in July 1995 survived until August 1997, but fewer than 10% of otters marked at Kuluk Bay on Adak survived until August 1997 (Figure 8.7b).

Kuluk Bay is accessible to killer whales but Clam Lagoon is not.

Another difference between Kuluk Bay and Clam Lagoon accounted for greater mortality of otters at Kuluk Bay (see contaminant hypothesis; KBL).

Kuluk Bay and Clam Lagoon differ in another factor besides access to killer whales (KBL).

Evidence: The number of sea otters counted at Kuluk Bay on Adak Island dropped by more than 50% between 1993 and 1997, but the number counted at Clam Lagoon on Adak remained constant (Figure 8.7a).

Kuluk Bay is accessible to killer whales but Clam Lagoon is not.

Another difference between Kuluk Bay and Clam Lagoon accounted for greater population decline of otters at Kuluk Bay (see contaminant hypothesis; KBL).

Kuluk Bay and Clam Lagoon differ in another factor besides access to killer whales (KBL).

The comparison between Kuluk Bay and Clam Lagoon is a natural experiment, with the possibility that a confounding factor might explain the results.

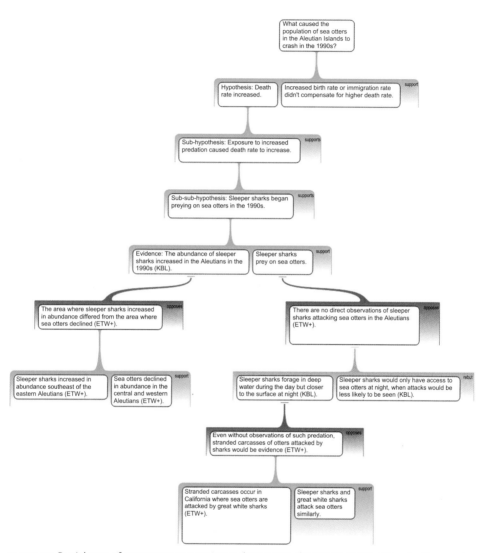

PLATE 23 Partial map of arguments supporting and opposing alternative explanations for the population crash of sea otters in the Aleutian Islands in the 1990s. This part of the map considers the hypothesis that sea otters declined because death rate increased due to predation by sharks.

Genes, Environments, and the Complexity of Causation

Scientists are motivated by curiosity and rewarded by discovering new things about the natural world. As seen in the story of monarch butterflies in Chapter 1, there are several different dimensions of discovery, from learning something new about where an animal or plant lives or what it does to learning how it works or its evolutionary history. When scientists ask the latter two questions, they move from descriptive research to an analysis of causation. In the case of monarch butterflies, for example, Fred Urquhart had to describe migration routes and find the winter home in central Mexico before he and others could study the environmental, physiological, and evolutionary causes of the remarkable long-distance migration of this species.

There are two reasons why scientists want to understand causation and you should too. The first reason is simply the intellectual satisfaction of gaining a deeper appreciation of how the world works. Have you ever explored the website HowStuffWorks? You might do so simply to satisfy your curiosity. For example, three of the most popular postings on April 13, 2013 were "How PlayStation 4 Works," "How Foot Binding Worked," and "What Causes Mental Illness?" These and other postings discuss causation to a greater or lesser extent. Sony is promoting PlayStation 4 as more powerful than its predecessor—what makes this possible? According to HowStuffWorks, PlayStation 4 has "the equivalent of a pretty powerful desktop computer—but one that's geared towards optimizing graphics-intensive gameplay." The posting on the old Chinese practice of foot binding explains how binding the feet of young girls caused changes in bone structure resulting in strongly arched feet only 3 inches long when the girls matured. The posting on mental illness begins honestly with "we don't know what causes mental illness," and then briefly discusses the contributions of genes and environment to mental illness.

The second reason for wanting to understand causation is more pragmatic. Without a good understanding of the causes of mental illness, developing treatments is at best a trial and error process; a better understanding of causation may lead to targeted treatments that are more effective. Thinking about how to predict the future provides another example. Climatologists keep trying to improve their models to make more reliable projections of future climate (Appendix 4); investors would be ecstatic if they could more accurately project future earnings of various kinds of investments. Causal models have more promise for accurate forecasts of future conditions than simple extrapolations of past trends (Box 7.1).

BOX 7.1 Understanding Causation and Predicting the Future

The simplest way to project a trend into the future is to extrapolate past data. This is tricky with something like stock prices, which fluctuate over time (Box Figure 7.1). Extrapolation may be more tempting where there is a clear directional trend in past data, as there is for estimates of the total number of people on Earth (Box Figure 7.2a). The human population increased slowly but relatively steadily from about 170 million 2,000 years ago to 720 million in 1800 but much more rapidly after 1800, to about 7 billion in 2010. If we zoom in on the period from 1800 to 2010, we see that the rate of increase accelerated even more after 1930 (Box Figure 7.2b). Extrapolating this trend produces estimates of 14 billion people in 2050 and 32 billion in 2100 (see Question 1 at the end of this chapter). Yet the United Nations Population Division

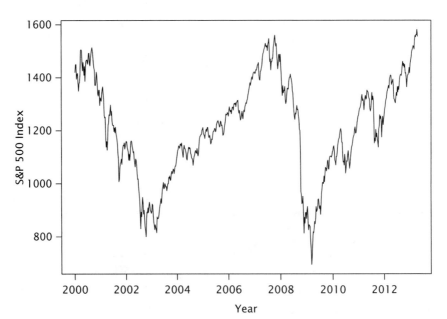

BOX FIGURE 7.1 Value of the Standard & Poor's Index of 500 U.S. stocks from January 2000 to April 2013.

(Continued)

BOX 7.1 Continued

estimates a likely population of only 10 billion in 2100. Why is this much less than our extrapolation of 32 billion? The reason is because the United Nations uses a model in which changes in population size are driven by changes in fertility rates and death rates, which in turn are influenced by socioeconomic factors in different regions of the world. This model is based on the causes of population growth at two levels—the immediate causes (birth and death rates) and underlying causes such as health, education, and prosperity that influence birth and death rates. As discussed in Appendix 4, this doesn't guarantee the United Nations projection to be true, but it focuses our attention on changes in socioeconomic conditions that influence fertility and mortality factors and might cause the population in 2100 to be substantially higher or lower than 10 billion.

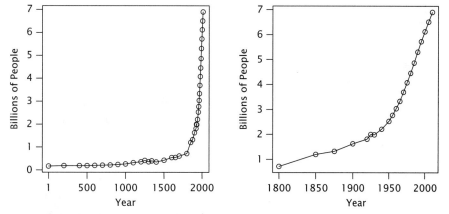

BOX FIGURE 7.2 Estimated human population of Earth from 1 to 2010 (left panel) and from 1800 to 2010 (right panel, which is a blow-up of the right-hand portion of the left panel).

The Problem of Nature and Nurture and the Complexity of Causation

Long before people learned about genes, chromosomes, and DNA, we speculated about influences of nature and nurture on human traits. In *The Tempest*, Shakespeare has Prospero describe Caliban as "a devil, a born devil, on whose nature Nurture can never stick" (Act IV, Scene 1). According to Miranda, on the other hand, "good wombs have borne bad sons" (Act I, Scene 2). Three seventeenth- and eighteenth-century philosophers offered different views of human nature. Thomas Hobbes argued that "the life of man [in a state of nature is] solitary, poor, nasty, brutish and short," while Jean-Jacques Rousseau believed that "nothing can be more gentle than him [man] in his primitive state." Both Hobbes and Rousseau emphasized the role of innate nature in human personality, although from different perspectives. By contrast, John Locke proposed that our nature is a *tabula rasa*, or blank slate: "Let's suppose the mind to be, as we say, white paper void of all characters, without any ideas. How comes it to be furnished? . . . To this I answer in one word, from experience."

These and other Renaissance and Enlightenment writers, however, didn't view nature and nurture as opposing forces influencing human traits. According to Evelyn Fox Keller, "the aim of education [for Shakespeare] is to 'nurture nature'" (Keller 2010:18). Keller makes a persuasive case that Francis Galton was the first writer to identify nature and nurture as separate factors, in his 1874 book *English Men of Science: Their Nature and Nurture*. Galton had diverse interests, ranging from anthropology to statistics and beyond. He invented the correlation coefficient for describing the relationship between two variables (see Chapter 5). Galton also was the first person to do systematic quantitative research on twins, and studies of twins remain one of the most important methods for learning about genetic and environmental contributions to human traits. Galton's separation of the effects of nature and nurture on human traits also remains influential today, sometimes to the detriment of true understanding of the complex interplay of genes and environments in the development of these traits.

How can we learn about influences of genes and environments on characteristics of humans, other animals, and organisms in general? This is a question about genetic and environmental factors as causes of our morphology, physiology, behavior, susceptibility to disease—indeed, all aspects of our lives. It is one of the most important questions in biology. It is also one of the most challenging questions. It would be simpler if we could imagine putting all of our traits into two piles, one of genetically determined traits and another of environmentally determined traits—height and susceptibility to breast cancer in the genetic pile and weight and susceptibility to lung cancer in the environmental pile, for example. In fact, however, all traits of all organisms are influenced by both genes and environments, so our challenge is to understand how these two types of causes interact as a tomato seedling or human baby develop to maturity. This is the complexity of causation in understanding the problem of nature and nurture.

Traits with a "Simple" Genetic Basis

Some traits of organisms appear to have a simple genetic basis, but even these traits can be influenced by environmental conditions. Traits that have a simple genetic basis are called Mendelian traits after Gregor Mendel, the Austrian monk who discovered the principles of inheritance by experiments with pea plants in the 1850s and 1860s. Mendel studied seven traits involving the color, size, and shape of peas, pods, and the plants themselves. These aspects of physical appearance are part of the *phenotype* of pea plants (phenotype comes from the Greek root *phanein*, meaning to show, as do such interesting and diverse words as phenomenon, diaphanous, and epiphany). For example, some peas were yellow and others were green. Mendel crossbred plants that produced yellow peas with plants that produced green peas and then counted the number of offspring with yellow and green peas. He repeated this process for additional generations and concluded that traits like these were determined by discrete factors that were still present in plants even when not expressed. We now know that these factors are different forms of a gene, called *alleles*, that occur at a particular spot, called a *locus*, on a particular chromosome in the cell nuclei of pea plants. The chromosomes are paired, so each individual has two alleles for each gene, one on each of the paired chromosomes where that gene is located. The alleles may be the same—both Y for yellow seeds or both G for green seeds—or they may be different—one Y and one G. Therefore a pea plant may have one of the following *genotypes* at the locus for pea color—YY, YG, or GG. Mendel discovered that both YY and YG plants produced seeds with a yellow phenotype, while GG was the only genotype that produced green seeds. In other words, the yellow allele is *dominant* to the *recessive* green allele, just as in humans the allele for brown eyes is dominant to the allele for blue eyes.

A few aspects of human appearance may have a simple genetic basis, although eye color no longer belongs in this category. Most human traits with a simple genetic basis are diseases including some forms of blindness and deafness, Huntington's disease, cystic fibrosis, albinism, and Tay-Sachs disease. Phenylketonuria (PKU) is an interesting example because it illustrates a role for environmental conditions even when genetic causation seems straightforward. PKU is caused by a recessive mutation in a gene that codes for an enzyme involved in the metabolism of phenylalanine, a constituent of proteins that is present in many foods. Since PKU is caused by a recessive mutation, a person must have two copies of the defective gene, one from each of his parents, to be afflicted with PKU. If untreated, PKU causes severe mental retardation because of accumulation of toxic amounts of phenylalanine. However, PKU can be treated with a diet lacking phenylalanine, resulting in normal brain development. Therefore newborns are routinely diagnosed for PKU so treatment can be started before irreversible brain damage has occurred. About 1 in 15,000 babies in the United States is born with PKU, although almost all are diagnosed early and treated successfully. Now that the genetics, biochemistry, and physiology of PKU are understood, it is no longer strictly a genetic disease but a disease that only occurs under particular genetic *and* environmental circumstances—an infant would have to have two copies of a defective gene at the PKU locus *and* not be given the appropriate controlled diet to suffer the consequences of PKU. In short, the mental retardation that in the past was associated with PKU is now determined by an interaction between a person's genes and environment.

Learning from Twins

Most traits of organisms are influenced by many genes, so can't be analyzed like the colors of peas or phenylketonuria in people. Francis Galton pioneered the study of twins to learn about the genetics of complex human traits. How does this work?

Humans produce twins in two different ways. Occasionally when a woman ovulates, she releases two eggs instead of one. If sperm fertilize both of these eggs, the mother becomes pregnant with fraternal twins who share the same proportion of genes on average (50%) as two siblings with the same parents but born at different times. Fraternal twins are also called *dizygotic* twins because they result from two separate fertilized eggs, or zygotes.

Alternatively, a woman may release one egg at ovulation that is fertilized by one sperm, but this zygote divides in half after the first few cell divisions of development. If each of the resulting embryos completes development, the mother will give birth to *monozygotic* twins who share 100% of their genes. Monozygotic twins are often called identical twins, but this terminology is misleading—they become unique individuals by interacting with their environments in different ways. The environments of monozygotic twins even differ during gestation—most such twins share one placenta but have different amniotic sacs, 18% to 30% have different placentas as well as amniotic sacs, and only 1% to 2% have the same placenta and amniotic sac.

Twinning in humans is relatively rare; in the United States, for example, twins represent about 3% of all births. About two-thirds of twins are dizygotic, and one-third are monozygotic. While the frequency of monozygotic twins is the same in all human populations that have been studied, the frequency of dizygotic twins varies between populations, highest in central Africa and lowest in Latin America and Asia.

In many other mammals, multiple births are the typical pattern rather than the exception. When female mammals such as cats or dogs or rats or mice have litters of two, five, eight, or more, it's usually because several eggs were released when the mothers ovulated and many or all of these eggs were fertilized. Armadillos are interesting exceptions because they regularly produce four or more genetically identical offspring in each litter. Few people study armadillos, and they aren't as easy to keep in captivity as other model organisms used for biological research like fruit flies, but armadillos would be an intriguing source of information about the interactions of nature and nurture in the development of individuals.

Monozygotic Twins Reared Apart

What can we learn from twins about how genes and environments influence human traits? One approach to answering this question is to study monozygotic twins adopted into different families and raised separately from an early age. In 1979, Thomas Bouchard started the Minnesota Study of Twins Reared Apart to collect data on these types of twins. Bouchard was motivated by learning about a pair of monozygotic twins who were reunited at age 39 after being adopted into separate families when they were just a month old. No doubt coincidentally, both twins were named Jim by their adoptive parents. Both had been married twice, first to Lindas, then to Bettys. They smoked the same kinds of cigarettes and drank the same kinds of beer. According to Bouchard, when he first met the two Jims, "Both had fingernails that were nibbled down to the end. And I thought, no psychologist asks about that, but here it is, staring you in the face" (Miller 2012:54). The similarities between these monozygotic twins raised apart went on and on, encompassing appearance, health, and personality. So Bouchard sought out other pairs of twins who were raised separately and brought them to Minnesota for extensive interviews, measurements, and tests to search for consistent similarities and differences between genetically identical people exposed to different environments.

The basic premise of studies of twins reared apart is that similarities between members of a monozygotic pair must reflect genetic causation because genes are the same while environments are different. Yet despite the popularity of stories like that of the Jim twins, there are serious limitations of this approach to teasing out effects of nature and nurture on human traits. First, some similarities between twins reared apart may be coincidental. For example, the Jim twins were born in 1939 and both married women named Linda and then women named Betty. We don't know how old their wives were, but let's assume they were comparable in age to the two Jims, so born in the late 1930s or early 1940s. According to the Social Security Administration, Betty was the second most popular name given to girls in the 1930s and Linda was the second most popular name for girls in the 1940s. Even if Jim and Jim shared some unknown genes that led them to favor women named Betty and Linda, the popularity of these two names for girls born at a time when they could be future wives for the two Jims makes it possible that their wives shared the same names purely by chance.

This discussion of Jim, Linda, and Betty times two illustrates a deeper problem with using stories about identical twins reared apart to demonstrate the importance of genetics for human characteristics. When we hear such stories, we are captivated by the remarkable similarities of such twins. I didn't even mention, for example, that both Jims had dogs named Toy. But the human interest of these stories isn't commensurate with their biological importance. Both Jims smoked the same brands of cigarettes and beer—amazing! But these were very popular brands, so this too may have been coincidental. We ought to be more interested in whether genes contribute to becoming addicted to nicotine or alcohol (I'm setting the stage for a general point here, not inferring from their preference for Miller Lite that the Jims were alcoholic).

This isn't the only story of remarkable similarities between monozygotic twins reared apart, but all of these stories are anecdotal evidence, like some of the observations discussed in Chapter 2. Anecdotes are fascinating but their value in answering questions in science is limited. To get closer to answering our questions about nature and nurture, we need to have a context in which to interpret anecdotal observations. This context is provided in part by gathering multiple observations using a systematic protocol, as Bouchard tried to do by recruiting additional pairs of twins reared apart for his study. As we discussed in Chapter 3, rigorous science depends on moving from individual observations, anecdotes, or case studies to sets of multiple observations collected systematically, that is, data. But another challenge of studying twins reared apart is the limited number of such twins that exist and can be identified and recruited for research. Bouchard's team in Minnesota continued their study for 20 years, during which time they found 137 pairs of twins who were adopted and raised separately. Only 81 of these pairs were monozygotic. By contrast, large numbers of twins are raised together, and many parents are willing to enroll these twins in twin registries that exist in various countries. Although not as straightforward as studying monozygotic twins adopted by different families, comparing the extent of similarity of monozygotic and dizygotic twins raised together is another way that researchers can disentangle contributions of nature and nurture to human traits. This involves correlation analysis, which you first saw in Chapter 5.

Correlations of Height between Twins

What are the relative contributions of nature and nurture—genes and environments—to variation in human traits? This is a pretty vague, general question. One tactic that scientists use to tackle this type of question is to convert it to a more specific question. If we have a specific question, we can figure out what kind of data to collect and how to analyze the data to answer the question. If the specific question accurately represents the general question and if the assumptions needed to answer the specific question can be justified, then we gain some insight about an answer to the general question. Working through one example of estimating genetic and environmental contributions to a specific human trait won't resolve the nature–nurture question for all traits, but it may give you a better appreciation of the challenges of thinking about nature and nurture.

I'll use human height to illustrate this process. Height is an easily recognizable trait, unlike IQ or some aspect of personality. Height can be measured precisely and without bias. Finally and most important, height is obviously a variable trait in every human population. If we gather together a group of people, we can scan the group and see the variation in height. With a tape measure and notebook, we can tabulate everyone's height. Individuals in our group will differ in many other ways too, but some of these will be difficult to measure. For instance, we would need to collect blood samples to record everyone's blood type.

We can now start to refine our general question about nature and nurture as follows. We have a group of people who differ in height. These people differ genetically and in the environments in which they grew up. How do these genetic and environmental differences contribute to the differences in height?

Your height is part of your phenotype, together with all other aspects of your physical appearance—weight, eye color, hair color, and so on. Your phenotype also includes your biochemistry, physiology, and behavior. So your blood type (A, B, AB, or O) is part of your phenotype, as is the time that it would take you to run 100 meters, as is your vocabulary. The phenotype includes all observable and measurable characteristics of an organism *except* its unique genetic code, which is its genotype.

We are homing in on a specific version of the nature–nurture question. We can measure variability in a phenotypic trait such as height among individuals in a population, and we want to see how that variability relates to genotypic and environmental variability in the population. We discussed the measurement of variability in Chapter 3, where I described the variance as a statistic used for this purpose. I developed a formula for the variance on page 48 and used histograms to illustrate variation in heights of men and women in the US Army in Figure 3.1. The variances in height measured in centimeters (cm) were 40.4 for women and 44.6 for men.

Unfortunately, we can't get much closer to answering our question with these data. We don't know what specific genes influence heights of people or even how many genes influence height, so we have no way to estimate the genetic variability in this population, much less the variability in environmental factors that these individuals experienced during growth to their adult heights. We need another approach to make further progress, like a running back in football switching direction to bypass a tackler and continue down the field. This approach involves using data for twins. The Mid-Atlantic Twin Registry (MATR) is a research program based at Virginia Commonwealth University that seeks twins for studies of the role of genetic factors in health and disease. Judy Silberg, the Director of MATR, sent me data on heights of 1,486 pairs of twins in the MATR database to use for this example. The twins were full-grown, ranging in age from 20 to 72. The variances in height were 44.2 cm for females and 47.6 cm for males, similar to those for the Army personnel discussed in Chapter 3.

We don't know anything more about specific genes that influence height in this sample of twins than we do for the army personnel considered in Chapter 3, but we do know something that leads to a breakthrough in solving our problem. We know that monozygotic twins share 100% of their genes while dizygotic twins share 50% of their genes. The MATR sample included 754 monozygotic pairs and 732 dizygotic pairs, so we can progress by comparing these pairs. The first step in doing this is to see how heights of pairs of twins are related to each other. How many monozygotic twins are exactly the same height? How about dizygotic twins? How many twins in these two categories differ in height by 6 inches or more? We can ask these questions more systematically by making scatterplots of height for the two types of twins, as in Figure 7.1 for males and Figure 7.2 for females. Each point represents a pair of twins, with the height of one twin plotted on the horizontal axis and the height of the second twin plotted on the vertical axis.

It should be clear from these figures that variation in genes among people has something to do with variation in height. For both males and females and for both monozygotic and dizygotic twins, if one member of a pair of twins is taller than average the other is likely to be taller than average too. But this relationship is much tighter for monozygotic twins than for dizygotic twins. As I outlined in Chapter 5, we use the correlation coefficient, symbolized by r, to measure the closeness of relationships like these. Figures 7.1 and 7.2 show correlation coefficients for these monozygotic and dizygotic twins. In the next section, I'll use these data to develop a model for estimating genetic and environmental contributions to variation in phenotypic traits.

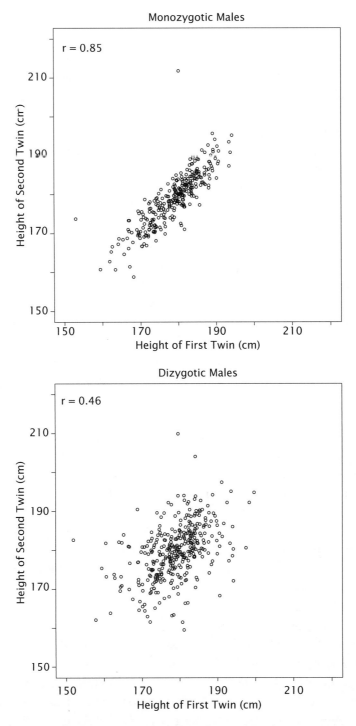

FIGURE 7.1 Heights of male twins from the Mid-Atlantic Twin Registry. Each point represents one pair of twins; monozygotic pairs are plotted in the top panel and dizygotic pairs are plotted in the bottom panel.

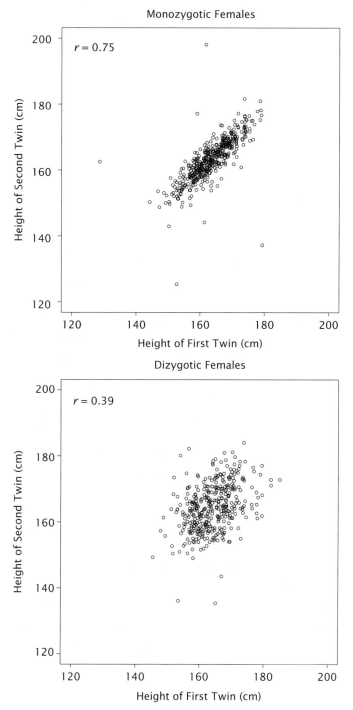

FIGURE 7.2 Heights of female twins from the Mid-Atlantic Twin Registry. Each point represents one pair of twins; monozygotic pairs are plotted in the top panel, and dizygotic pairs are plotted in the bottom panel.

A Model for Using Twins to Study Nature and Nurture

I need to introduce a new concept to take the next step in this process of converting a general question about nature and nurture to a specific question that we can answer with data on twins. This new concept is *heritability*, defined as the percentage of phenotypic variation in a population that can be attributed to genotypic variation. Heritability is defined numerically, so is something we should be able to estimate. Heritability of a trait in a population will be between 0%, if there is no genetic variation that affects the trait, and 100%, if there is no environmental variation that affects the trait. Like converting general questions to specific ones, quantification is another important part of doing science. In the case of heritability, however, this process of quantification has fostered a host of misunderstandings and misrepresentations, as we shall see.

We now need to think a little more deeply about how environmental conditions affect the development of twins. The environment includes a huge number of factors that influence the physical, psychological, and social growth of children. Each of us spends about nine months in our mother's uterus before birth; this is part of our environment. Our environment includes what we eat as we grow up, the kind of weather that we experience where we live, the books that we read at home and at school, our interactions with peers and teachers, the sports that we play, our hobbies—the list is virtually endless.

At this point, you may wonder why I bothered to simplify a complex general problem by stating it as a specific question leading to the concept of heritability, only to introduce the complex new problem of defining how the environment influences our phenotypes. Even if we could identify all the specific environmental factors affecting our growth and development, how could we measure all of them? The simple answer is that we can't. When thinking about twins, however, we can divide environmental conditions into two groups—those that are shared by a pair of twins and those that are unique to individual members of the pair. For example, suppose the twins eat the same quantities of the same foods every day, which may be approximately true up to a certain age. This would be a shared environmental factor for this pair. At age 5, however, one twin starts playing soccer and the other starts playing baseball. Their physical, psychological, and social experiences in these sports would be unique environmental conditions for each twin.

Our ultimate objective is to understand what contributes to variation in a phenotypic trait in a population. We use the variance to measure variation, and we can now write a model for three basic contributions to phenotypic variance:

phenotypic variance = genotypic variance + shared environmental variance
+ unique environmental variance.

In this example of height of adults in the MATR, we have four groups to consider—monozygotic male twins, dizygotic male twins, monozygotic female twins, and dizygotic female twins. Let's start with monozygotic male twins. There are 600 individual twins in 300 pairs in this group. Since monozygotic twins are genetically identical, the genotypic variance *within* pairs is zero, and the genotypic variance in the equation above only includes variation *between* members of different pairs of twins. Likewise, we defined shared environmental variance to be zero *within* pairs, so all of the shared environmental variance in the equation is that of different pairs. For example, one couple feeds their twin sons bacon and eggs every day and another feeds their twin sons oatmeal. To the extent that these differences in diets contribute to differences in adult height, these and other environmental factors that differ between pairs of twins but not within pairs contribute to the shared environmental variance.

We can now take a big step toward quantifying various contributions to phenotypic variation, even if we can't identify all the specific sources of these contributions; for example, differences in diets, education, social interactions, sports, and hobbies between and within pairs of twins. We can do this because the correlation coefficient for height of monozygotic twins, r_{MZ}, relates to these components of variation in our model as follows:

$$r_{MZ} = \frac{\text{genotypic variance} + \text{shared environmental variance}}{\text{phenotypic variance}}.$$

This is true because the genotypic variance and the shared environmental variance are the two factors that contribute to the similarity of monozygotic twins. The remaining factor that influences the height of individuals in our model is unique environmental variance.

I suggested above that analysis of monozygotic twins raised separately like those studied by Bouchard at the University of Minnesota is more straightforward than comparison of monozygotic and dizygotic twins raised together like those studied by Silberg in Virginia. Here's why. With monozygotic twins reared apart, researchers assume that there is no shared environmental variance because members of each pair are raised in different environments. Therefore our equation for the correlation in the value of a trait between these twins can be simplified to

$$r_{MZ} = \frac{\text{genotypic variance}}{\text{phenotypic variance}}.$$

Since the definition of heritability is the percentage of phenotypic variation that can be attributed to genotypic variation, the correlation coefficient for monozygotic twins reared apart is a direct estimate of heritability. Based on 56 such pairs of twins, Bouchard's team reported r_{MZ} for height of 0.86, which translates to heritability of 86%.

This analysis implies that 86% of the variation in height in a small group of monozygotic twins adopted by different families was due to genetic differences between pairs. What about the remaining 14%? Since Bouchard's team assumed that shared environmental variance was zero, this leaves unique environmental variation responsible for the remaining 14%. In the case of monozygotic twins reared apart, this means differences in diet and other environmental factors between families that adopt individual twins of a pair.

This analysis of monozygotic twins raised separately has two problems. First, the sample size is small, simply because few such pairs of twins exist and can be identified and recruited for research. Second, and more important, the analysis rests on the key assumption that shared environmental variance is zero. This assumption made it possible to simplify our equation for r_{MZ} to get a direct estimate of heritability. How valid is the assumption?

All twins share their mother's uterus for nine months, whether raised together or separately after birth. Suppose one mother of monozygotic twins is a smoker and another mother of monozygotic twins is not. This creates different environments during gestation for the two pairs of twins, contributing to shared environmental variance for members of each pair (recall that shared environmental variance results from environmental conditions that are the same for members of a pair of twins but different for different pairs). In addition, the monozygotic twins studied by Bouchard's team were together for an average of five months before being adopted by separate families. Therefore shared environmental variance wasn't literally zero for this set of twins. The question is whether this shared variation had a meaningful effect on their growth in height. We don't know the answer to this question (see Question 2).

The correlation coefficient for height of monozygotic twins reared apart gives a direct estimate of heritability *if* we assume that shared environmental variance is zero. We can take another approach to estimating heritability by comparing the correlation coefficient for monozygotic twins raised together to that for dizygotic twins raised together. Sample sizes are much larger for this analysis; 754 pairs of monozygotic twins and 732 pairs of dizygotic twins in the MATR, compared to 56 pairs of monozygotic twins reared apart for Bouchard's study. Box 7.2 describes this analysis using r_{MZ} and r_{DZ}. For the MATR data, I estimated heritabilities of 78% for height of males and 72% for height of females.

Like the analysis of monozygotic twins raised separately, this comparison of monozygotic and dizygotic twins depends on a critical assumption—that shared environmental variance is the same for the two types of twins. How reasonable is this assumption? It seems likely that parents provide the same diets to twins whether they are monozygotic or dizygotic, so this component of shared environmental variance should be similar for the two types, at least for the first few years of life. I gave an exaggerated example of different diets for different pairs of monozygotic male twins above—one set of parents feeds their sons bacon and eggs every day and the other feeds their sons oatmeal every day. To extend this example, suppose parents of monozygotic twins always feed their twins the same breakfast each day, either bacon and eggs or oatmeal, while parents of dizygotic twins always feed one twin bacon and eggs and the other oatmeal. This doesn't seem very likely, but if it were true shared environmental variance would be greater for monozygotic twins than for dizygotic twins.

BOX 7.2 Using Data for Monozygotic and Dizygotic Twins to Estimate Heritability

I developed the following equation in the text to relate the correlation coefficient for height to components of variation in our model of factors that contribute to total phenotypic variance of height for monozygotic twins:

$$r_{MZ} = \frac{\text{genotypic variance} + \text{shared environmental variance}}{\text{phenotypic variance}}.$$

This applies whether monozygotic twins are raised together or separately; we used a simplified version for the twins studied by Bouchard's group that were raised separately. As explained in the text, we have to include shared environmental variance together with genotypic variance for twins raised together. For monozygotic male twins in the MATR, $r_{MZ} = 0.85$ (Figure 7.1); we could substitute another value for a different trait or a different sample of twins.

$$r_{MZ} = \frac{\text{genotypic variance} + \text{shared environmental variance}}{\text{phenotypic variance}} = 0.85.$$

The corresponding equation for dizygotic male twins in the MATR database is

$$r_{DZ} = \frac{0.5 \times (\text{genotypic variance}) + \text{shared environmental variance}}{\text{phenotypic variance}} = 0.46.$$

This differs from the equation for monozygotic twins because monozygotic twins share all of their genes whereas dizygotic twins share only half of their genes on average. Therefore the genotypic variance is multiplied by 0.5 in the numerator of the equation for dizygotic twins.

If we assume that shared environmental variance is the same for monozygotic and dizygotic twins (see text), we can rearrange and simplify these two equations to get another estimate of heritability to complement the estimate based on monozygotic twins reared apart described in the text. This rearrangement and simplification produces:

$$\text{heritability for males} = \frac{\text{genotypic variance}}{\text{phenotypic variance}} = 2(r_{MZ} - r_{DZ})$$
$$= 2(0.85 - 0.46) = 0.78 = 78\%.$$

For females, $r_{MZ} = 0.75$ and $r_{DZ} = 0.39$ (Figure 7.2), so

$$\text{heritability for females} = \frac{\text{genotypic variance}}{\text{phenotypic variance}} = 2(r_{MZ} - r_{DZ})$$
$$= 2(0.75 - 0.39) = 0.72 = 72\%.$$

One component of the environment that might affect monozygotic and dizygotic twins differently involves social responses of parents, siblings, and others. Monozygotic twins look more alike than dizygotic twins; this may cause other people, from close relatives to strangers, to treat monozygotic twins more similarly, increasing their shared environmental variance compared to dizygotic twins. In fact, there is abundant evidence that the social environment is more similar for monozygotic twins than for dizygotic twins; for example, the former are more likely to have the same friends than the latter. However, "even though parents do appear to treat MZ [monozygotic] twins more similarly than DZ [dizygotic] twins, this similarity of treatment is uncorrelated with twin similarity for personality, vocational interests and cognitive abilities" (Evans and Martin, 2000). This is a key point, because the role of shared environmental variance in our model is not the environmental conditions per se (various diets or various social responses) but how these conditions affect the two twins in a pair. More generally, we can't say definitively whether shared environmental variance is the same for monozygotic and dizygotic twins, but we're stuck with assuming this because we can't separate the effects of shared environmental variance from genotypic variance otherwise. This illustrates an even more general point: everything we do in science—not just modeling but also correlational, comparative, and experimental research, depends on some assumptions. Isn't this as true for the decisions that set the arc of our lives as it is for the work of scientists?

In summary of this example, heritability for height was 86% for a small sample of monozygotic twins raised separately and studied by Thomas Bouchard and colleagues at the University of Minnesota, and 78% for men and 72% for women in a large sample of monozygotic and dizygotic twins from Virginia and nearby states. Twenty researchers published similar analyses for twins in seven European countries plus Australia in 2003. They used data for about 26,000 pairs of twins between 20 and 40 years old and estimated heritabilities of height ranging from 68% for women in the United Kingdom to 94% for men in Italy (see Question 3). These values of heritability are not precise measurements but estimates, so we shouldn't make too much of the differences between the various samples from the United States and other countries.

More Examples of Heritability

Height is an obvious characteristic that differs among people, but fairly mundane unless you're a basketball coach trying to recruit players for your team. I used height to show how heritability is calculated because I could use raw data from the MATR to make scatterplots for pairs of monozygotic and dizygotic twins that demonstrated the influence of genetic factors on human height. Height also provided a concrete numerical example for developing a model to estimate heritability. However, researchers have estimated heritability for many human traits, some of which may be more interesting to you than stature. In fact, you're bound to encounter references to heritability in the news media. I searched for the word "heritability" in the *New York Times* between January 1, 2000, and

TABLE 7.1 Recent estimates of heritability for some diverse human characteristics.

Trait	Estimated Heritability (%)
Schizophrenia, bipolar disorder, autism	>80
Exercise habits	60
Participation in politics	60
Confidence about academic ability at age 9	51
Popularity in high school	45

December 31, 2012, and found 123 occurrences, the most recent of which (on December 11, 2012) was in a story entitled "Understanding How Children Develop Empathy." Dr. Perri Klass, the author of this story, wrote that "twin studies have suggested that there is some genetic component to prosocial tendencies" (empathy and sympathy), although Klass did not report an estimate of heritability for prosocial behavior.

Table 7.1 gives a few examples of heritability estimates for psychiatric diagnoses, habits, and personality traits. It may surprise you that genetic differences between teenagers may even contribute to differences in their popularity in high school. James Fowler and colleagues analyzed interviews of high school students to study their social networks. These networks included 1,110 twins in various schools. The interviewers asked each student who their friends were; an individual identified by more schoolmates as a friend was considered more popular. Just as for height in Figures 7.1 and 7.2, the popularity of monozygotic pairs of twins tended to be more similar than that of dizygotic pairs, leading to the heritability estimate of 45% for popularity.

How Heritable is IQ?

General cognitive ability estimated by IQ tests has gotten more attention than any other human trait from people interested in the nature–nurture problem, and there is a long and contentious history of attempts to estimate heritability of IQ. In 1994, Richard Herrnstein and Charles Murray published *The Bell Curve: Inheritance and Class Structure in American Life*. Herrnstein and Murray argued that genetic factors have a major influence on IQ and that educational enrichment programs such as Head Start have little ability to improve IQ. They also claimed that differences in IQ between blacks and whites in the United States reflected genetic differences between races. They highlighted strong correlations between IQ and economic and social success of individuals and predicted increased class division in the United States between a "cognitive elite" and the rest of society with average or below average IQ. One of their major policy arguments was that welfare provides incentives for poor women to reproduce, exacerbating the division between the cognitive elite and the rest of the population, with the cognitive elite comprising a smaller and smaller fraction of the total population over time. This argument was based on the assumptions that intelligence is highly heritable and causally related to poverty and led to their recommendation to eliminate welfare programs that subsidize children and therefore, to their way of thinking, stimulate poor women to have more babies.

Not surprisingly, publication of *The Bell Curve* led to a huge amount of discussion, much of it critical, in the news media, scientific journals, and books such as *Intelligence, Genes, and Success: Scientists Respond to the Bell Curve* and *Inequality by Design: Cracking the Bell Curve Myth*. I won't directly address critiques and defenses of the policy aspects of Herrnstein and Murray's argument but rather focus on the heritability of intelligence as measured by IQ, which is one of the key foundations of their argument. In their study of monozygotic twins raised separately, Bouchard and colleagues reported that heritability of IQ was 75%, comparable to estimates from three smaller studies published before theirs. Herrnstein and Murray state that "half a century of work, now amounting to hundreds of . . . studies, permits a broad conclusion that the genetic component of IQ is unlikely to be smaller than 40% or higher than 80%" (Herrnstein and Murray 1994:105). They "adopt a middling estimate of 60% heritability, which, by extension, means that IQ is about 40% a matter of environment," and conclude by claiming that "the balance of the evidence suggests that 60% may err on the low side" (Herrnstein and Murray 1994:105).

Shortly after publication of *The Bell Curve* in 1994, Bernie Devlin and two colleagues reviewed 212 prior studies of the heritability of IQ. These studies reported correlations of IQ measurements for monozygotic and dizygotic twins raised together as well as correlations for both types of twins raised separately, correlations for siblings raised together and separately, correlations of children with their biological parents, and correlations of children with their adoptive parents. Devlin's group analyzed a total of 50,470 pairs of individuals in these various groups. Their summary estimate of heritability from all of these data was 48%, with a 95% confidence interval from 43% to 54%. Although the estimate of 48% by these researchers is less than the 75% estimated by Bouchard's team and the 60% suggested by Herrnstein and Murray, it indicates a substantial role for genetic factors in differences in IQ between people, as well as substantial scope for environmental factors to influence these differences.

In their comprehensive analysis including twins and non-twin siblings raised together and apart as well as children and their biological and adoptive parents, Devlin's group was able to test the key assumptions underlying estimates of heritability. They found strong evidence undermining the assumption of Bouchard's analysis that shared environmental variance equals zero for twins raised apart. The analysis by Devlin's team implied that early shared experience during gestation and in the first few months of life makes an important contribution to individual differences in IQ and that heritability of IQ is closer to 50% than 70%. In short, direct estimates of IQ using monozygotic twins raised apart are higher than estimates using both types of twins raised together because direct estimates are based on the faulty assumption that shared environmental variance equals zero for monozygotic twins raised apart.

As you've seen, the news media regularly report new studies of heritability of a host of human characteristics. For one more example, a reporter for the *U-T San Diego* newspaper wrote a story in May 2008 with the headline "Genetic Link Suggested in Voting Behavior." This story was based on studies of twins in which James Fowler and colleagues reported heritabilities of 53% to 72% for likelihood of political participation for two samples of twins. I hope that my explanation of how these estimates are made will give you a more realistic appreciation of these news stories. In summary, I want to highlight several specific misconceptions about heritability that can lead reporters, commentators, and innocent consumers of the news astray.

Some Limitations of Heritability

First, there is no direct relationship between heritability and genetic determination, with a heritability of 100% meaning complete genetic determination and a heritability of 0% meaning complete environmental determination. For example, most humans have 10 fingers. This is a legacy of our evolutionary history as vertebrates and happens because a set of genes function in a predictable sequence during normal embryonic development. Clearly, "number of fingers" is a genetically determined trait in humans. Yet if we looked at a large enough group of humans, we would find some who lost one or more fingers in an accident. In other words, accidents are environmental factors that can cause a person to have fewer than 10 fingers. What does this mean for heritability of finger number? Heritability is the proportion of phenotypic variance that can be attributed to genetic differences between individuals. In this case, all of the phenotypic variance is due to environmental differences, so heritability of finger number is zero. Although accidents cause some variation in number of fingers in humans, this doesn't negate the basic fact that this trait is genetically determined.

Second, heritability is not the same as inheritance. Forget-me-nots, mice, humans, and other organisms inherit copies of genes from their parents. In some cases, individual genes have major effects on phenotypes; genes for eye color and cystic fibrosis are two examples in humans. In these cases it's reasonable to say that someone has blue eyes or cystic fibrosis because of specific genes that she inherited from her mother and father. Many traits, however, are influenced by many different genes. Height is one example. Average heritability of height is greater than 70% in studies of both men and women in many countries. Based on this, you might assume that at least 70% of your own height is due to genes that you inherited and 30% is due to your environment growing up—how much nutritious food you ate, exposure to toxins or pollutants, and so on. This assumption would be invalid because the concept of heritability is defined for groups or populations, not individuals, and it doesn't work to directly transpose an idea from the population level to the individual level. If heritability of height is 75% in a particular population, this means that 75% of the variation in height between members of that population is due to genetic differences, which is quite different from saying that 75% of your height is due to genes that you inherited.

A third and related misconception is to assume that we can associate specific genes with phenotypic traits that have high heritability. In fact, researchers are now able to use molecular techniques to search for specific regions of DNA in the chromosomes of individuals that are associated with particular diseases or traits like height or aspects of personality. These searches sometimes involve hundreds of thousands of snippets of DNA in thousands of people, but they often turn up genes that account for only a small proportion of the total heritability of a trait (see Appendix 5 for details).

Suppose the average value of a trait differs between groups of people and the trait has moderate to high heritability in one group. For example, according to *The Bell Curve*, the average IQ of African Americans was 85 and that of Caucasians in the United States was 103, while Devlin and his colleagues estimated that heritability of IQ was about 48%. The fourth and most damaging misconception is to attribute group differences in a heritable trait to genetic differences between groups, and this was a major flaw of Herrnstein and Murray's argument. To restate the key point, heritability is not a measure of the contribution of genes to a trait but rather a measure of the contribution of genetic variation to variation of a trait in a population. Furthermore, it is not an absolute measure of this contribution but a relative measure. This means that it is specific to a particular population and depends on environmental contributions as well as genetic contributions to variation in the trait in this population.

For example, imagine that some researchers use data on 3-year-old twins to estimate heritability of height in a small country with limited ethnic diversity. They find that heritability is 75%. Then there is a severe famine that affects a large portion of the population for three years, although a substantial minority of the population has the financial resources to be minimally affected by local food shortages. The twins used in the initial study are now 6 years old, and the researchers decide to measure their heights and estimate heritability at age 6. Assuming the famine wasn't so severe that many children died, the genetic variance of this group of 6-year-olds will be the same as it was when they were 3 because they are the same individuals. But the environmental variance is now much greater—some kids didn't get enough nutritious food to eat in the intervening years, and others ate about as well as they had before. Genotypic variance remains the same while environmental variance increases, so total phenotypic variance, which is the sum of these two components, increases. Therefore heritability, which is genotypic variance divided by phenotypic variance, decreases, without any change in the frequencies of genes in the population that influence height.

Let's return to the contentious topic of IQ. Most twin studies used to estimate heritability of IQ have been done in Europe, the United States, Canada, and Australia, and most monozygotic and dizygotic twins used for these studies have been Caucasian, simply because Caucasian parents are more likely to enroll their children in twin registries. There is a difference in average IQ of African American and Caucasian children in the United States, although the gap has narrowed substantially since Herrnstein and Murray published *The Bell Curve* in 1994. Heritability of IQ estimated for a Caucasian population tells us something about the genetic contribution to total variation in IQ within the Caucasian population, but nothing about the difference in average IQ between this population and another, like African Americans. The reason is essentially the same as that illustrated by the example of height in the last paragraph—heritability is specific to the population in which it is estimated and depends on both genetic and environmental variation within that population. Even if there is substantial heritability of IQ in the Caucasian population of the United States, that doesn't tell us anything about heritability of IQ in the African American population. It also doesn't explain the average difference in IQ between Caucasians and African Americans, which might be due to differences in education or socioeconomic conditions between the groups rather than genetic differences.

Nature, Nurture, and the Complexity of Causation

These challenges of interpreting assertions about heritability of IQ and other human traits illustrate some aspects of the complexity of causation in understanding nature and nurture. One reason for these challenges is that we can't do experiments with humans like we can with other animals and plants. These experiments include selective breeding, cross-fostering, and common-garden experiments in which genetic and environmental effects are separated experimentally, allowing credible estimates of heritability for certain traits. They also allow researchers to study an important complexity of causation that we've ignored so far—interactions between genes and environments. I'll set the stage for discussing *genotype-environment interactions* by describing a recent study of small birds called zebra finches to illustrate one experimental approach to studying nature and nurture. Then I'll describe a different experimental approach with plants to illustrate a genotype-environment interaction, then two non-experimental studies of humans that suggest genotype-environment interactions.

Heritability of Size and Behavior in Zebra Finches

Zebra finches (Plate 12a) are small, colorful, Australian birds that feed on seeds. They breed readily in captivity and are used worldwide for laboratory studies of behavior. Wiebke Schuett and three colleagues from Germany and the United Kingdom have done several studies of consistent behavioral differences between different zebra finches, analogous to personality differences between people. In 2013, they reported results of a cross-fostering experiment to study exploratory behavior of these birds. They started by measuring the length of the heads, including bills, of members of a laboratory colony of adult males and females. Then they measured exploratory behavior by moving the birds temporarily from their home cages to a new cage containing 10 novel objects such as branches of various size and shapes placed at random locations in the cage. The researchers counted the number of novel objects visited by each bird in five minutes as an index of exploratory behavior. Each bird was tested twice, then males and females were paired randomly for mating. After eggs were laid, the researchers waited until just before they were ready to hatch and then transferred 53 clutches between parents, choosing foster parents at random. This created a cross-fostering design for their experiment.

There were 154 cross-fostered chicks that survived to adulthood, when their head-bill lengths and exploratory behavior were measured just as for their biological and adoptive parents. Figure 7.3 shows the correlations of offspring with their biological parents and their adoptive parents for head-bill length and exploratory behavior. Head-bill length of offspring was correlated with head-bill length of their biological parents but not their foster parents. By contrast, exploratory behavior of offspring was correlated with exploratory behavior of their foster parents but not their biological parents and more closely with exploratory behavior of foster mothers than fathers. The heritability of head-bill length was 69%, similar to that for height in humans, while the heritability of the exploration index was zero (see Question 4).

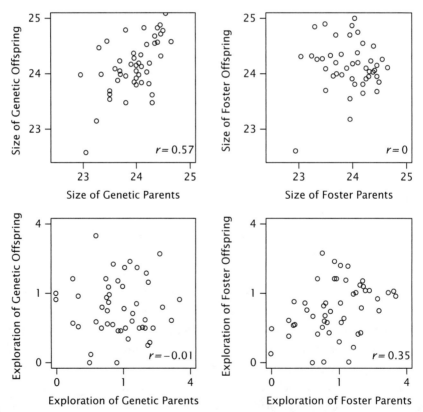

FIGURE 7.3 Size and behavior of cross-fostered zebra finches in relation to the size and behavior of their genetic and foster parents. Size is head plus bill length in millimeters and exploration is the number of novel objects investigated by a bird in a five-minute trial in a new cage.

An Experiment to Study Genotype-Environment Interactions in a Plant

Plant ecologists devised the first experiments to tease out genetic and environmental effects on traits in the late nineteenth and early twentieth centuries. For example, Jens Clausen, David Keck, and William Hiesey studied a wildflower called yarrow (Plate 12b) that is widely distributed in California. Clausen's group noticed that yarrow plants growing near peaks of the Sierra Nevada were much shorter than those growing in the foothills, and they wondered if this was due simply to the harsher growing conditions at high elevation or if height was also influenced by genetic differences between populations living at different elevations. To answer this question, they designed a common-garden experiment. First, they collected seeds from two populations living at Mather, in the foothills of the Sierra, and Timberline, near the top of the range. They germinated these seeds in a greenhouse and then transplanted seedlings to an outdoor garden at their research laboratory at Stanford University. After two years of growth, the researchers transplanted cuttings of plants from each of the original populations into two common gardens, one at Mather and one at Timberline. These gardens had very different environments since Mather was at 1,400 meters elevation and Timberline was at 3,000 meters.

When the researchers were ready to set up their common gardens at Mather and Timberline, they took two cuttings from each of 30 plants grown from seed collected in the source populations. They transplanted one member of each pair of cuttings to the common garden at Mather and the other to the common garden at Timberline. Since paired cuttings came from the same plant, they were genetically identical, that is, clones. This created four treatments for comparison (Table 7.2). If you found some differences in heights of yarrow plants between these treatments, which differences would indicate an influence of environmental conditions on height? Which differences would indicate an influence of genes on height?

TABLE 7.2 Design of common-garden experiment to determine genetic and environmental effects on height of yarrow plants growing in the Sierra Nevada of California (Clausen et al. 1948). Numbers identify the four treatments; for example, 1 represents seeds collected at Mather and planted at Stanford, with cuttings from these plants transplanted to the common garden at Mather.

		Site of Common Garden	
		Mather (1,400 m elevation)	Timberline (3,000 m elevation)
Source of Clones[a]	Mather	1	2
	Timberline	3	4

[a] Seeds from each source population were planted at Stanford; cuttings from the resulting plants were transplanted to the common gardens at Mather and Timberline. Cuttings were made in pairs with one member of each pair planted at Mather and one at Timberline. Therefore each pair comprised two clones of genetically identical plants.

Figure 7.4 shows the results of this experiment. The first thing that you might notice is that most of the lines on the graph slope down from left to right, indicating that clones tended to grow taller in the garden at Mather than in the garden at Timberline. This shows that different environmental conditions in the two gardens influenced growth. If you've ever done any gardening you won't be surprised by this result because you know that all sorts of environmental factors influence growth of plants such as temperature, availability of water and light, and nutrients in the soil.

Figure 7.4 also shows that genetic differences between individuals influenced growth—each line is a clone, and the lines are in different positions relative to the axis showing height of the tallest stem. This too shouldn't be very surprising if you think about all the genetic variation that exists in a population that may be more familiar to you, such as humans. Figure 7.4 shows not only genetic variation between individuals by the different positions of each clone but also genetic variation between populations. In the common garden at Mather, every single Mather clone was taller than all Timberline clones. In the common garden at Timberline there was some overlap in heights but about half of the Mather clones grew taller than all the Timberline clones. In other words, the average height of clones from Mather was greater than the average height of clones from Timberline in each environment.

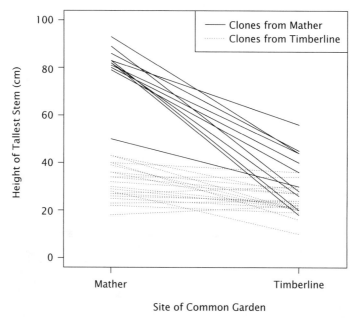

FIGURE 7.4 Size of yarrow plants originating from two different populations in the Sierra Nevada in California, Mather at 1,400 meters elevation and Timberline at 3,000 meters elevation, grown in common gardens at each site. Seeds collected at Mather and Timberline were first grown at Stanford and then cuttings of these plants were transplanted to gardens at Mather and Timberline. Each line is a separate clone, so the endpoints of any line represent genetically identical plants.

Figure 7.4 shows that genetic differences between populations and environmental differences between sites both influence stature of yarrow, but the most important thing illustrated by this example is a genotype-environment interaction. Specifically, the clones of yarrow from Mather were much more sensitive to the difference in environment between Mather and Timberline than the clones of yarrow from Timberline. What pattern in Figure 7.4 shows this greater sensitivity of clones from Mather? This is a subtle result that will be important later in the chapter when we discuss human examples, so it will be worthwhile to reread the last few paragraphs while studying Figure 7.4 to fix this idea in your mind (see Question 5).

While these experiments imply that genetic differences between populations of yarrow influence growth of yarrow, we don't know specific genes responsible for this effect. We also don't know specifically which environment factors influence growth, although there are several likely candidates. The growing season is shorter, and it's colder and windier at higher elevation; all of these factors might inhibit growth of yarrow at Timberline compared to Mather. But plants grown from Timberline seeds didn't respond to the more favorable environment at Mather, as shown by the fact that the lines for Timberline clones are parallel to the horizontal axis. This suggests that genetic variants promoting slow growth may have been favored by natural selection in the Timberline population, perhaps because short stature is adaptive for reducing exposure to high winds that can be very damaging to plants. If mutations favoring slow growth accumulated over many generations at Timberline, these mutations would still be present in the first generation of Timberline plants grown at Mather.

A Genotype-Environment Interaction Affecting Health in Humans

We can't do cross-fostering experiments with humans, much less common-garden experiments (why not? see Question 6). Recent comparative research, however, has produced intriguing evidence of genotype-environment interactions in human traits. Using modern methods of molecular biology, researchers have discovered specific genes associated with certain traits. Medical researchers have been particularly interested in finding genes that might be associated with risk of obesity because obesity increases the chance of dying from diseases like diabetes, heart disease, or cancer. In 2007, a group of researchers in the United Kingdom collected DNA from 38,759 Europeans and found that a mutated form of a gene called FTO was associated with higher body mass index (BMI). BMI is defined as body weight in kilograms divided by the square of height in meters; individuals with BMI greater than 30 are considered obese. Sixteen percent of the people whose DNA was studied by the researchers had two copies of the mutated FTO allele and weighed an average of 3 kilograms more than individuals without the mutated allele. This may not seem like much difference in weight, but it translated to a 67% greater risk of obesity for those with two copies of the mutated allele.

This study of variants of the FTO gene associated with obesity stimulated a huge amount of further research, but one of the most interesting follow-up projects was done by a group at the University of Maryland who studied an Old Order Amish community in Lancaster County, Pennsylvania. The Amish don't use cars or electricity; most work in physically demanding jobs such as farming or homemaking without benefit of modern appliances. The Maryland researchers found that even more of the Amish population than of the large European sample described in the last paragraph had two copies of the mutated allele of FTO associated with increased risk of obesity. In the Amish, however, physical activity also influenced BMI and risk of obesity. The Amish subjects wore recorders on their hips that measured physical activity for one week. When the researchers correlated BMI with the number of copies of the mutated FTO allele carried by an individual, they found different patterns for subjects with below average and above average activity. For the group with below average activity, BMI was greatest for individuals with two copies of the mutated allele and least for those with no copies of the mutated allele, just as for the general population. For the group with above average activity, BMI was the same for individuals with zero, 1, or 2 copies of the mutated allele. Above average activity for the Old Order Amish meant "3 to 4 hours of moderately intensive physical activity [daily], such as brisk walking, house cleaning, or gardening" (Rampersaud et al. 2008). This illustrates the influence of environmental conditions on the expression of genes, just as we saw for height of yarrow plants growing in common gardens in the Sierra Nevada and treatment of PKU by controlled diet in newborns. All of these cases exemplify genotype-environment interactions.

A Genotype-Environment Interaction and the Heritability of IQ

General cognitive ability of humans is a complex trait influenced by a large number of mostly unknown genes. As you saw above, psychologists often use IQ to index cognitive ability and study twins to estimate the heritability of IQ. I described some of the controversy surrounding the heritability of IQ, focusing on the many ways that heritability can be misinterpreted. However, I left out the most important problem with the concept of heritability in the discussion on pages 162–164. Our basic model for estimating heritability was

$$phenotypic\ variance = genotypic\ variance + shared\ environmental\ variance + unique\ environmental\ variance.$$

This model led to the specific equations explained in the text and Box 7.2 that we used to derive estimates of heritability from correlation coefficients for traits of monozygotic and dizygotic twins. The model, however, omits any consideration of genotype-environment interactions. The plus signs between the one genetic term and two environmental terms in the model assume that genetic and environmental effects on phenotypic variation are independent of each other. What if they aren't? Here's an example of how one group of researchers incorporated a genotype-environment interaction in an analysis of heritability of IQ.

Eric Turkheimer is a psychologist at the University of Virginia who heads a research program in behavioral genetics. Turkheimer and four colleagues analyzed data on IQ of 7-year-olds collected in the National Collaborative Perinatal Project. This was a large-scale effort to assess the physical and mental health of children in the United States between 1959 and 1974, although Turkheimer's group did their analysis much later and published their findings in 2003. The National Collaborative Perinatal Project involved about 48,000 mothers and 59,000 children in a wide range of socioeconomic circumstances. The children in the sample included 319 pairs of twins. Turkheimer's group was interested in these data because they differed from most data on twins in including a substantial number of families of low socioeconomic status (SES).

Turkheimer's research team discovered that heritability of IQ in this group of twins increased dramatically with SES, from close to zero for those growing up in families with the lowest SES to about 90% for those growing up in families with the highest SES. As SES increased, environmental contributions to variation in IQ decreased. This was especially true for the shared environmental component of variance in IQ, which accounted for about 60% of variance in IQ at the lowest level of SES and about 10% of variance in IQ at the highest level of SES. When they divided the sample in half at the median level of SES, the researchers found that heritability of IQ was 72% in the upper half of the SES distribution but only 10% in the lower half. This is a striking example of an interaction between genes and environments influencing variation in a human trait.

How does SES affect the heritability of IQ? I've emphasized that heritability is the *relative* proportion of variation in a trait in a population that can be attributed to genetic variation in the population. As such, heritability depends on both environmental conditions and genes—as environmental variation increases, heritability will decrease even with no changes in genetic variation because the two terms must sum to 100%. Environmental variation itself has two components, the shared environment and the unique environment, where the shared environment includes factors common to each pair of twins but differing between different pairs and the unique environment includes factors unique to each individual twin. In the study by Turkheimer's group, shared environmental variance was 58% of total IQ variance for twins growing up in families with SES below the median but only 15% of total IQ variance for twins growing up in families with SES above the median. In other words, there was more variance between families of low SES than between families of high SES in environmental conditions affecting IQ. Perhaps children are exposed to books earlier and more extensively in most families with higher incomes and better educated parents, causing relatively little environmental variance between these families in a factor that may influence IQ. In poorer families with parents who are less well educated, by contrast, some children may still be exposed to books early and extensively while others are not, leading to greater environmental variance between these families. Whatever the explanation for differences in genetic and environmental contributions to variation in IQ between children, this study by Turkheimer and colleagues clearly shows the complexity of interpreting heritability estimates for human populations. The same trait may show high heritability in one group but much lower heritability in another. If a trait is highly heritable in one group, that doesn't mean that genetic factors explain differences between groups or that the trait is resistant to environmental influences. Therefore discussions of heritability of IQ provide no basis for decisions about policies such as intensive early education for disadvantaged children.

Conclusions

This chapter was about genes, environments, and the complexity of causation. I used studies of zebra finches in the laboratory and yarrow plants in common gardens in the Sierra Nevada to illustrate how researchers use experiments to study effects of genetic and environmental factors on traits of organisms. These kinds of experiments aren't possible with humans, so researchers must instead use correlational and comparative approaches to answer questions about roles of nature and nurture in human traits. Therefore the examples in this chapter reinforce the contrast between experimental research discussed in detail in Chapter 4 and correlational and comparative research discussed in detail in Chapter 5.

My description of how researchers analyze data on twins to estimate heritability of traits also involved some modeling, like the modeling we discussed in Chapter 6. You saw several equations in this chapter, although they were expressed in words rather than symbols. These equations are a model of apportioning variability in a phenotypic trait between genetic and environmental causes. As you saw, this model depends on some critical assumptions (also see Appendix 5), and it's important to keep these assumptions in mind in applying the model. Like all models, this one should focus our attention on defining key terms, such as heritability, and understanding the implications of our assumptions.

Complexity of causation is an important general principle in science, and I'll give other examples in later chapters. In Chapter 1, I introduced the idea that scientists are ultimately interested in understanding causation—how things work and how and why they evolved. I've reiterated this point in other chapters, including this one. One problem with seeking an understanding of causation is that causation is rarely straightforward. In the case of genes and environments, the complexity is that these two types of factors are rarely independent causes of the characteristics of humans and other organisms. We discussed several cases of interactions between genes and environments that illustrate this key point. While heritability is a popular concept for both scientists and journalists, gene–environment interactions are often ignored in reports of heritability of human traits, even though this complexity can be the most interesting part of studying nature and nurture. Keep this in mind when you see a news story about heritability of creativity or sports ability or any interesting human trait in the future (see Question 7).

Questions to Ponder

1. In Box 7.1, I used an exponential model to extrapolate growth of the human population to 2100 based on estimated population sizes between 1930 and 2010. This is a standard model of population growth applied to many species and to other kinds of growth as well, like compound growth of money invested at a fixed interest rate. The exponential model is based on the assumption that growth rate is constant, so the percentage increase in population size is the same for each year from 1930 to 2100, although the absolute number of individuals added to the population is greater in each successive year because the growth rate is applied to a larger base each year.

   A simpler extrapolation based on a linear model instead of an exponential model would be to simply draw a straight line between population size in 1930 and population size in 2010 and extend this straight line to project population sizes in 2050 and 2100. There were about 2 billion people on Earth in 1930 and 6.9 billion in 2010. Using straight-line extrapolation, what would you project for the population in 2100? How does this compare to my projection of 32 billion using an exponential growth model and to the UN's projection of 10 billion?

   Figure 7.5a shows population growth from 1930 to 2010. The curved form of this graph is characteristic of exponential growth, whereas linear growth would appear as a straight line between 2 billion people in 1930 and 6.9 billion in 2010. If you imagine adding such a straight line to the graph, you can see why a linear model doesn't fit the data for this time period very well.

   When population size is plotted on a logarithmic scale (see Figure 3.8 and associated discussion), exponential growth appears as a straight line rather than the curved line in Figure 7.5a. Figure 7.5b shows population growth from 1930 to 2010 plotted on a logarithmic scale. How do you interpret this figure? Is it relatively straight, as I assumed in using an exponential model for extrapolation to 2100? How do you interpret the parts of the plot of population size in Figure 7.5b that aren't straight?

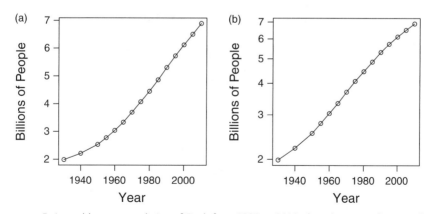

FIGURE 7.5 Estimated human population of Earth from 1930 to 2010 plotted on an ordinary arithmetic scale (a) and on a logarithmic scale (b).

2. We used the correlation coefficient between monozygotic twins raised by different adoptive families to make a direct estimate of heritability of height; that is, $r_{MZ} = 0.86 \rightarrow$ heritability = 86%, assuming that shared environmental variance is zero and genotypic variance is the only source of variation between different pairs of these types of twins. Another factor that might affect traits of twins adopted by different families is potential similarity between families that adopt members of a pair. Unlike doing experiments with animals, we can't arrange for monozygotic twins to be assigned randomly to adoptive families. Speculate on the likelihood that there is greater similarity between families that adopt members of a pair of twins than between families that adopt members of different pairs. How does this affect our assumption that shared environmental variance for monozygotic twins raised separately is zero?

3. Much variation in height in human populations can be attributed to genetic variation among individuals, as illustrated in Figures 7.1 and 7.2 and by the fact that estimated heritabilities of height are about 80% in most populations studied. Given this context, how do you interpret Figure 7.6, which shows how average height of three groups of adult males changed during the twentieth century? What are the similarities and differences in the patterns for the three groups? What do you infer from these results about how genetic factors, environmental factors, and genotype-environment interactions influence height?

4. Review the methods for the cross-fostering experiment with zebra finches done by Schuett's group. Is it legitimate to assume no shared environmental variation for

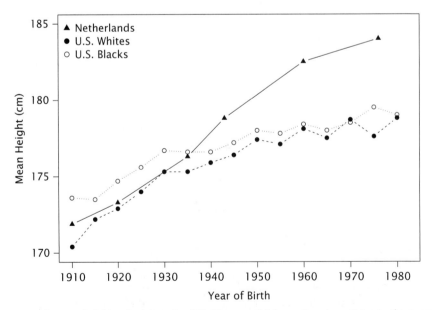

FIGURE 7.6 Average heights of adult males (23–47 years old) born at various times in the twentieth century in the United States and the Netherlands as reported by John Komlos and Benjamin Lauderdale. Births are aggregated over five-year periods centered at the location of each point; Caucasians and African Americans in the United States are plotted separately.

TABLE 7.3 Survival of yarrow clones from Mather and Timberline in common gardens at the two sites.

		Site of Common Garden	
		Mather (1400 m elevation)	Timberline (3000 m elevation)
Source of Clones	Mather	97%	41%
	Timberline	71%	96%

members of the same clutch transferred to foster parents just before eggs hatched? Why or why not?

5. As described in the text, Clausen and colleagues planted 30 clones of yarrow derived from seeds collected at Mather and Timberline in common gardens at Mather and Timberline. Figure 7.4 shows their results. You may wonder why there are fewer than 30 lines representing individual clones for each source population in Figure 7.4. The reason is because these lines represent only clones that survived long enough at both sites to be measured. Several clones did not survive over their first winter so could not be measured for plotting the height of the tallest stem in Figure 7.4. Table 7.3 shows the percentages of clones from each source population that survived in each garden. Do these results imply that genotypes and environments have independent effects on survival of yarrow or that survival is influenced by an interaction between genotypes and environments? Explain your answer.

6. Why can't we do cross-fostering experiments with people? This question has a moral dimension that may be obvious and a scientific dimension that is less obvious. To think about the scientific dimension, consider how Bouchard's study of monozygotic twins adopted by different families is different from Schuett's cross-fostering experiment with zebra finches.

7. As you saw in this chapter, heritability can be calculated for virtually any trait, including not just IQ but also popularity and political activism. What if genetic and environmental factors don't have independent influences on the trait but interact in their effects? Are estimates of heritability still meaningful? Why or why not? Use the results of Turkheimer's group on heritability of IQ in children of different SES to formulate your answer but think about how to generalize from these results to other traits.

Resources for Further Exploration

1. I introduced some basic principles of genetics in this chapter but didn't give a thorough introduction to genetics because my primary goal was different—to get you thinking about the complex problem of figuring out causes of things. A good vehicle for this was the story of nature and nurture because these two causes of human traits are inextricably intertwined, despite the almost irresistible temptation to assign

independent roles to genes and environments and use heritability to quantify these roles. There are lots of reasons why you might want to learn more genetics, however, ranging from the practical—genetic counseling might be important if you are thinking of becoming a parent—to the idealistic—genetics encompasses a fascinating set of topics. Learn.Genetics (http://learn.genetics.utah.edu/) is a website developed and maintained by the University of Utah and an excellent place to start if you want to broaden and deepen your understanding of genetics or find answers to specific questions.

2. Twins were an important subject of the research discussed in this chapter. There are many websites devoted to twins, although half of these sites in the first page of my initial Google search for "twins" were sites about the Minnesota Twins baseball team. If you'd like to learn more about biological twins, I suggest the Mid-Atlantic Twin Registry as a starting point (http://www.matr.vcu.edu/). This site includes general information about twins, descriptions of past results and current studies conducted by MATR researchers, news about twins, and links to books and websites for those interested in twins.

3. I devoted several pages in this chapter to describing a model for estimating heritability from data on correlations of values of a trait for monozygotic and dizygotic twins. If you'd like to learn more about this model, a website developed by Shaun Purcell called Behavioural Genetic Interactive Modules is the place to start (http://pngu.mgh.harvard.edu/~purcell/bgim/). This includes tutorials, web demonstrations, and downloadable computer programs for calculating statistics like variances, correlation, coefficients, and heritabilities (Purcell is now affiliated with a research unit at Massachusetts General Hospital but uses the British spelling of behavioral because he was educated at Oxford, University of London, and King's College).

From Causes to Consequences: Considering the Weight of Evidence

Scientists sometimes publish papers with intriguing titles. One of my favorites is "Epidemiology and the Web of Causation: Has Anyone Seen the Spider?" by Nancy Krieger. Epidemiology includes the study of epidemics, or the spread of diseases in populations, but has much broader scope as the scientific foundation of public health. Epidemiology is closely related to biology and medicine but also has strong ties to social sciences like psychology and sociology. For example, epidemiologists studied the recent outbreaks of measles described in Chapter 6 as well as the relationships between lead exposure and crime described in Chapter 5.

Krieger was near the beginning of her scientific career when she published her groundbreaking paper in 1994. She argued that epidemiologists use the "web of causation" as a metaphor for the idea that diseases result from a complex network of multiple interacting causes. This parallels my discussion of gene–environment interactions in the last chapter. Just as it's too simplistic to think that some traits are strictly determined by genes and others are strictly determined by the environment; it's too simplistic to consider only microbes as causes of infectious disease—spread of these diseases also depends on cleanliness of food and water, nutrition and general health of susceptible people, and the social, economic, and political factors that influence these things. Krieger didn't deny that diseases have multiple complex causes but argued that epidemiologists should focus on the *root* causes of webs of causation, that is, "spiders." For Krieger, spiders represent the socioeconomic factors that make different populations susceptible to different diseases.

Neither Nancy Krieger nor I would argue that we should give up trying to understand webs of causation just because they involve many factors interacting in complex ways. Webs of causation have important consequences for us as individuals and as members of communities. We may wish to break some webs and reinforce others to enhance individual and community well-being. To do so, we need to learn what links are most important and how to alter these links without inducing unintended consequences.

Hurricane Katrina, New Orleans, and the Web of Causation

Hurricane Katrina hit New Orleans on August 29, 2005, causing more than 1,800 deaths and at least $80 billion in property damage. It seems straightforward to designate the hurricane as the cause of this massive toll of death and destruction, but a closer look shows a more complex situation (Figure 8.1). Had the levees designed to protect New Orleans from storm surges not failed, damage in the city would have been much less. Perhaps the Corps of Engineers, which designed and built the levees, did a poor job, either because of incompetence or lack of funding. Development of the vast wetlands of the Mississippi Delta for housing and industry also contributed to the damage because these wetlands in their natural state would have absorbed much of the water brought ashore by the hurricane.

Politics played a role as well, as exemplified by poor planning and slow response of the Federal Emergency Management Agency (FEMA) as well as missteps by state and local governments. The tradition of presidents appointing political supporters to leadership positions in government agencies contributed to these problems. President George W. Bush appointed Michael D. Brown as head of FEMA in January 2003. Brown was a lawyer who had no prior experience in emergency services or management; before joining the Bush administration, he worked for the International Arabian Horse Association for 13 years. Although Bush congratulated Brown in early September 2005 for his management of the federal response to Katrina (Figure 8.1), Brown resigned on September 12 in reaction to widespread criticism of FEMA's performance.

Hurricane Katrina was the most obvious direct cause of damage to New Orleans in 2005, some of which remains eight years later. We can't eliminate or control hurricanes, however, so situating the hurricane in the web of interacting factors that influenced the extent of damage (levees, development of wetlands) and the recovery process (actions of governmental agencies) can help us figure out how to manipulate similar webs to reduce damage from future storms. But we must consider one additional component of the Katrina web if we hope to manage future storms more effectively. This is climate change, which has consequences for hurricanes and other severe weather events in the future and which requires more than single, specific actions to disrupt the web, like building stronger levees or appointing professionals to head FEMA (Figure 8.1).

At the time of Katrina, climatologists speculated that climate change might increase the frequency or intensity of hurricanes, but there was little direct evidence to test this idea. Since then, several researchers have shown that average intensity

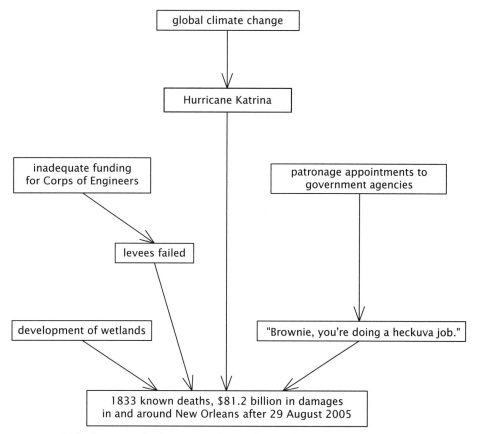

FIGURE 8.1 Web of causation for damage in and around New Orleans attributed to Hurricane Katrina. We could build a much more elaborate web by including additional direct and indirect causes and interactions between causes.

of hurricanes has increased as global temperatures have increased in recent decades, although there is no corresponding pattern of change in number of hurricanes. Hurricane intensity is measured on a scale from 1 to 5 based on wind speed, with speeds greater than 157 miles per hour for category 5. There were eight category 5 hurricanes in the Atlantic Ocean north of the equator between 2000 and 2010, including Katrina. This was an average of slightly less than one category 5 hurricane per year, compared to one every four years from 1970 to 2000. Climatologists have known for quite a while that warmer surface water in the ocean contributes to the energy of hurricanes by increasing evaporation, which ultimately increases wind speed as well as amount of rainfall. Since global climate change encompasses temperature increases in oceans as well as on land, climatologists were not surprised by recent analyses showing a correlation between temperature and average intensity of hurricanes in the past century. Combining these analyses with models of global climate change (Appendix 4), Aslak Grinsted and two colleagues predicted that hurricanes with storm surges as great as those of Katrina would occur at least twice as often in the twenty-first century as in the twentieth century.

Including climate change in the web of causes of large-scale death and damage in New Orleans following Hurricane Katrina (Figure 8.1) illustrates one important complexity of causation beyond simply the fact that events usually have multiple interacting causes, as shown by the web itself. Some of these causes of things are deterministic. If a person is bitten by a rabid dog and infected with the rabies virus, the virus will travel to the person's brain and cause acute encephalitis and death. Many infectious diseases are similar in that the disease invariably follows infection with a causal microorganism, although not all infectious diseases end in death. But other causes are probabilistic rather than deterministic. This means that a causal factor increases the likelihood of a particular result but doesn't invariably produce the result. The relationship of smoking to lung cancer is a good example—a long history of smoking increases the chance that a person will get lung cancer but doesn't make lung cancer a certainty. Global climate change has a parallel influence on hurricane intensity. Climate change increases the probability that any hurricane that forms will be stronger—category 4 or 5—but doesn't guarantee this result. With climate change, there will still be relatively weak hurricanes in categories 1 to 3 but a greater percentage will be in categories 4 and 5 than in the past. In addition, when Hurricane Alice first forms in the Atlantic in August 2018, climatologists can't say, "Because of climate change, we know this will be a category 5 storm." As Alice or any hurricane develops and moves toward land, climatologists can begin to predict its intensity, not based on general knowledge of climate change but on observations of the hurricane itself.

Challenges in Dissecting Webs of Causation

I've used several examples in previous chapters to show how and why scientists seek to understand causation, but you've also seen that this is a challenging problem. The challenges come from two main sources. One is the complexity of causation, illustrated by the web of causes associated with damage from Hurricane Katrina (Figure 8.1) and by the fact that many causal factors in this web are indirect and probabilistic rather than direct and deterministic. The second major challenge in understanding causation is limitation of our methods. Experiments offer the best opportunity for a clear test of an hypothesis about the cause of some phenomenon, but experiments aren't always possible. Even when experiments are possible, they can go astray for a host of reasons such as faulty design, unjustified assumptions, or inadequate sample sizes. In addition, experiments work best where potential causes are few and straightforward. This is one reason why microbiologists of the nineteenth century like Louis Pasteur were so successful in finding causes of diseases like anthrax, cholera, and rabies (another reason is that they were brilliantly creative scientists).

Other methods for learning about causation such as the comparative and correlational studies of cell phone use as a potential cause of brain cancer discussed in Chapter 5 may be better suited than experiments for phenomena that have multiple interacting causes. However, comparative and correlational studies usually don't generate results that are as definitive as those from experimental studies.

Crime stories, whether true or fictional, sometimes involve a "smoking gun" that leads to conviction of the person holding the gun. Stories in science are occasionally resolved by definitive evidence that constitutes a smoking gun. For example, the combination of Watson and Crick's model of DNA, Rosalind Franklin's X-rays, and Erwin Chargaff's discovery of consistent patterns in amounts of the four bases in different samples of DNA was a smoking gun that convinced everyone that DNA was a double helix (Chapter 6). Much more often, however, there is no smoking gun in scientific research and scientists must answer questions based on the weight of evidence for and against various hypotheses. This is especially true in ecology, where complex webs of causation are as pervasive as they are in epidemiology. It's also true for questions that impact decisions about policies that managers, legislators, and voters must make.

Wolves were extirpated from much of the contiguous United States during the nineteenth and twentieth centuries, even in national parks like Yellowstone. Wolves were protected under the US Endangered Species Act in 1973 and reintroduced in Yellowstone National Park in 1995 after an absence of 60 years. The population in Yellowstone grew rapidly and spread beyond the park borders, and wolves in other parts of the country have expanded their range as well. In June 2013 the US Fish and Wildlife Service proposed to remove wolves in the contiguous United States from protection under the Endangered Species Act, leaving it up to individual states to manage wolf populations and, if they wished, set hunting seasons for wolves. In response, 16 leading wolf researchers wrote a letter to the Secretary of the Interior outlining four major objections to this proposal. How can we evaluate the proposal of the US Fish and Wildlife Service to remove wolves from the endangered species list? Our evaluation won't be based solely on science but will involve economic and ethical considerations too. Even disregarding these added complexities, the scientific arguments for and against the proposal are complex enough. The only way to fairly evaluate the scientific foundation of policy proposals like these is to consider all the evidence, pro and con, and decide whether the weight of evidence favors one path or another. Our final decision may rest on economic or ethical arguments but should be consistent with scientific evidence. For example, if you were a rancher in Idaho who raised sheep and thought wolves should be killed to protect your income, the persuasiveness of your argument would depend on evidence about wolf predation on sheep. Is it incidental or pervasive? How does it depend on the distribution and abundance of wolves? Are there other methods for protecting sheep that are more effective or less expensive than killing wolves (see Question 1)? We'll return to the wolf story at the end of this chapter, but first I'll describe some other examples of ecological questions that can only be answered by thinking about weight of evidence rather than hoping for a smoking gun.

Keystone Species in Food Webs—Analyzing Complex Causation in Ecology

Ecologists study many features of the natural world, but many of us focus on interactions between individuals and populations that live together in ecological communities. Just as a sociologist might study the human community in the Haight-Ashbury District of San Francisco, an ecologist would study the plants and animals in a mountain meadow. This ecological community contains various species of grasses, sedges, forbs (non-woody flowering plants, commonly known as wildflowers), and shrubs. There are also grasshoppers, butterflies, ants, mice, snakes, ground squirrels, gophers, rabbits, foxes, and songbirds. Coyotes wander through from time to time, and hawks fly overhead. The mountain meadow community also includes invertebrates and microorganisms that live in the soil.

You've already seen examples of the fundamental ecological interactions of competition, predation, and mutualism. When two species use the same resources and these resources are in limited supply, the species are in competition. Each has a negative effect on the other, ultimately by suppressing its population size although this can happen through several different mechanisms. Yampah is a plant with delicate white flowers and a large starchy root that occurs in our mountain meadow. If a gopher traveling through its burrow under the meadow encounters a yampah root, it will probably eat it, killing the plant and making it unavailable to a ground squirrel that would have eaten the leaves and flowers. Predation is an interaction in which one species benefits at the expense of the other. If a coyote catches and kills a rabbit in our meadow, the coyote gets a meal that helps keep it alive and may contribute to its ability to reproduce, while the cost to the individual rabbit as well as the population of rabbits is obvious. The interaction between the gopher and the yampah plant or a ground squirrel and another yampah plant is like predation in that one party benefits at the expense of the other, although predation on plants is given the special name of herbivory. Mutualism is an interaction in which both species benefit, like the pollination of flowers by bees discussed in Appendix 3.

Feeding interactions in ecological communities are often illustrated by food webs, as in Figure 8.2 for a site called Wytham Woods that has been an environmental research site for Oxford University since 1942. Like other food webs, this one includes green plants such as oak trees, other trees, shrubs, and herbs; herbivorous insects, mammals, and birds; predators such as spiders and weasels; parasites; and parasites of parasites, or hyperparasites. It also includes organisms that live in the soil and feed on dead organic matter such as the leaves of plants that drop in fall and become part of the litter layer on the surface of the soil. This is an incomplete depiction of the food web at Wytham Woods because it doesn't show all individual species that live in the community. Except for very simple communities, it's difficult to draw complete food webs.

This discussion of basic ecological interactions and food webs sets the stage for our first example of an important question in ecology—are there certain species that have especially important roles in ecological communities—so important that if they weren't

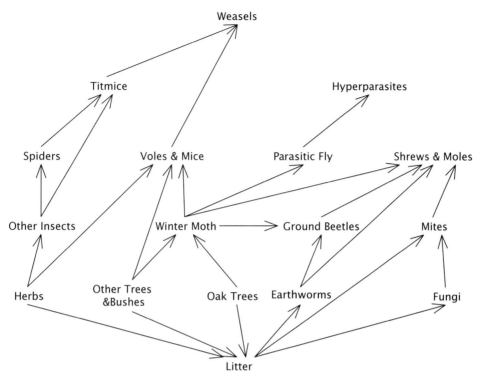

FIGURE 8.2 Food web for Wytham Woods, near Oxford in the United Kingdom. Like most diagrams of food webs, this is incomplete. For example, herbs, other insects, and spiders each comprise many different species.

present the structure and function of the community would be much different? The structure of a community is its species composition and general appearance; for example, an oak woodland looks quite different from a pine forest. The function of a community includes the flow of energy from the sun to green plants to herbivores to predators and the cycling of nutrients between plants, animals, and the soil. Suppose the food chains in one community include at most three species—a plant, an herbivore that eats the plant, and a carnivore that eats the herbivore—while some food chains in another community include as many as five species. This would be a functional difference between the communities (see Question 2).

Species that have especially important roles in their communities are called *keystone species*. More specifically, Mary Power and several colleagues defined a keystone species as "one whose impact on its community or ecosystem is large, and disproportionately large relative to its abundance." These researchers offered their definition in a 1996 paper entitled "Challenges in the Quest for Keystones," indicating that the identities of keystone species in various communities aren't obvious. In some cases, ecologists disagree about whether any species should be singled out as a keystone.

A keystone species is analogous to the keystone in an arched, stone bridge (Figure 8.3). The keystone in a bridge is just a single stone under the highest point at the center of the

Keystone

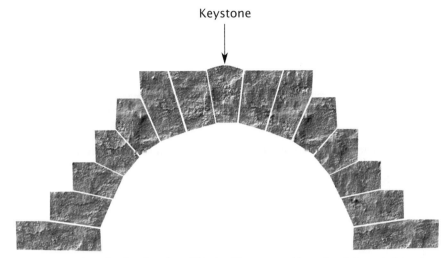

FIGURE 8.3 An arch made of wedge-shaped blocks. The structural integrity of the arch depends on the keystone block.

bridge. It holds the least weight of all the stones, but the bridge collapses without the keystone. This analogy to keystones in bridges illustrates why Power's group defined a keystone species as one with a *disproportionate* influence on its community. For example, distemper is caused by a virus that infects dogs, cats, and their wild relatives. The Serengeti is a savannah ecosystem in east Africa with huge populations of large mammals such as wildebeest, zebras, gazelles, lions, cheetahs, wild dogs, hyenas, and many other species (Plate 1d). If distemper became established in the lion population of the Serengeti and killed a substantial number of lions, there would be ramifying effects on the prey species that lions eat, on the other predators that compete with lions, on the vegetation, and even on things like termites and songbirds. As a tiny virus, the biomass of distemper would be minuscule, but its impact could be huge, so we would consider it a keystone species.

In introducing keystone species, I suggested that removal of a keystone would cause major changes in the structure and function of a community. This definition complements that of Power and her colleagues, but it also implies that we can identify keystones by doing experiments. To test whether or not a particular species is a keystone in its community, remove it and see what happens. If there are unexpectedly big changes relative to the abundance of the species, it is considered a keystone.

Robert Paine introduced the concept of keystone species in 1969 to describe the results of such an experiment that he did at Mukkaw Bay on the coast of Washington state. He worked on a rocky ledge in the intertidal zone, which is covered by water at high tide and exposed to the air at low tide. The intertidal community includes organisms like barnacles, mussels, sponges, and anemones that attach themselves to the rocks and filter nutrients from the water as the tide rises. There are also seaweeds and various mobile herbivores that feed on seaweeds such as snails, limpets, and chitons. The food web at Mukkaw Bay included two main predators, whelks that drill holes in the shells of barnacles and mussels and extract the meat within and starfish that feed on chitons and limpets but also can detach barnacles and mussels from the rock and turn them over to access the edible parts of these animals. Not counting microscopic organisms called plankton in the water that covered the intertidal at high tide, there were 15 species in the food web at Mukkaw Bay.

Paine marked out two large plots on the rocks in July 1963. He visited his study site once or twice a month and removed all of the starfish from one of the plots, throwing them into the ocean. At the end of one year, the food web on the removal plot was much simpler, with only eight species, while the food web on the control plot still had the original 15 species. The removal plot was now dominated by mussels, which apparently had outcompeted many of the other species of herbivores. Starfish eat most of the species attached to rocks in this intertidal community but prefer mussels (Plate 13a), keeping the population of mussels low enough under natural conditions that other herbivores can coexist with them. When mussels were released from control by predators, their population grew rapidly at the expense of other herbivores.

Paine continued removing starfish from his experimental plot through 1968 and monitored the study site periodically until 1973. At that time, the control plot still had 15 species but the experimental plot had just two, mussels and starfish. As top predators, starfish weren't especially abundant, but their presence had a noticeable effect on the species composition of the rocky intertidal community at Mukkaw Bay, so Paine coined the term keystone species to describe the role of starfish in this community (see Question 3).

Some researchers argue that experiments are the only way to test whether a species is a keystone. For some species in some communities, however, manipulative experiments aren't feasible, so we have to rely on observations, comparisons, correlations, and what we might call natural experiments (Chapter 6). Sea otters are one of the most famous examples of an animal considered by many researchers to be a keystone species, but this is based on the weight of various kinds of evidence, most of which is from observational and comparative studies rather than manipulative experiments. Therefore I'll tell the story of sea otters as a keystone species to introduce how we can evaluate hypotheses by considering the weight of evidence both for and against the hypotheses.

Is the Sea Otter a Keystone Species? What Does the Weight of Evidence Suggest?

Sea otters (Plate 13b) are the second smallest marine mammals and spend their entire lives close to shore, unlike seals, sea lions, dolphins, and whales. Sea otters have extremely thick and dense fur, with up to 1 million individual hairs per square inch, so they were highly sought by fur traders. Before exploitation by humans, sea otters were widely distributed along coasts in the North Pacific Ocean from Japan to the Aleutian Islands to Baja California. Rough estimates of their total number ranged from 150,000 to 300,000, but by 1900 there were only about 2,000 left in widely scattered populations in Russia, Alaska, and California. The International Fur Seal Treaty signed in 1911 banned hunting of sea otters and the US.

Marine Mammal Protection Act of 1972 provided further protection. As a result of these actions and because their habitat was largely intact, sea otters recolonized much of their original range and the population rebounded to about 106,000 in 2007, although the species is still classified as endangered by the International Union for Conservation of Nature.

Sea otters eat a wide variety of foods, including fish and various marine invertebrates, but they especially favor sea urchins, which are mobile grazers that feed on large seaweeds called kelp. In the coastal habitats where they live, sea otters dive 20 to 50 meters to the ocean floor to collect sea urchins, abalones, and other invertebrate prey. For a prey item with a thick hard shell like a mussel, a sea otter may bring a rock to the surface of the water, roll over on its back, set the rock on its chest, and pound the mussel against the rock to break the shell. When this use of tools was first studied by Hall and Schaller in 1964, it was thought to be the only example of tool use in a mammal outside the primates (although a handful of other examples have been reported since).

Jim Estes was a graduate student at the University of Arizona in 1970 when the Atomic Energy Commission was planning underground nuclear tests on Amchitka Island, in the middle of the Aleutian Chain between the Alaska Peninsula and the Kamchatka Peninsula of Russia. The Atomic Energy Commission wanted to fund research on sea otters, and Estes was hired to work on the project. In his first summer, he mostly got oriented to the study site—tagging the otters, observing their behavior from shore, and scuba diving to get a closer look at the sea floor where they foraged. In 1971, his coworker John Palmisano suggested that Estes visit Amchitka when Robert Paine was going to be on the island. By this time, Estes knew that an important food chain went from kelp to sea urchins to sea otters, and he started to wonder how the abundance of kelp might influence the population of otters, thinking that more kelp would support more sea urchins and therefore a higher population of otters since sea urchins were a preferred food for otters. This was a "bottom–up" hypothesis about the control of an ecological community—traditional thinking was that green plants produce the food that all the animals in a food web rely on directly (herbivores like sea urchins) or indirectly (carnivores like sea otters). But Paine told Estes about his experiments with starfish at Mukkaw Bay and suggested that Estes give some thought to top–down effects instead.

If Estes and Palmisano wanted to study the effects of sea otters on nearshore ecological communities in the ocean, they needed a study site where otters were absent to compare to Amchitka Island where they had survived through the fur-harvesting period of the 1800s and early 1900s. Russians were the main hunters of sea otters in the western Aleutians, and a few otters survived at Amchitka because it was relatively far from Russia and from the eastern Aleutians where Americans hunted. Shemya Island is much farther west in the Aleutians, thus closer to Russia, and otters had been completely eliminated at Shemya and were still absent in the 1970s. Therefore Estes and Palmisano decided to compare the abundance of kelp and sea urchins at Amchitka and Shemya.

Their basic methods were straightforward although "their data did not come cheaply" (Stolzenburg 2008). Palmisano drove the boat; Estes dove. They worked in summer when the water averaged 45°F and winter when it was 35°F. Stolzenburg reported,

> Estes would dive in the morning for an hour, collecting samples of kelp and urchins, staring into square meter quadrates, filling data sheets with names and numbers. He would write underwater on slates until his chilled hands were shaking too violently to continue. After a hot shower to thaw the blood, he would head out again in the afternoon for another hour, until he was too numb to write anything more. The next day he would repeat the process.
>
> (STOLZENBURG 2008:60)

Estes and another researcher named David Duggins extended this research to three more sites in the Aleutians and three in southeast Alaska in 1987 and 1988. Sea otters had been eliminated by hunting throughout southeast Alaska in the nineteenth and early twentieth centuries but were reestablished at Surge Bay and Torch Bay by 1970. They were still absent at Sitka Sound, however. Figure 8.4 shows the quantitative results. In both

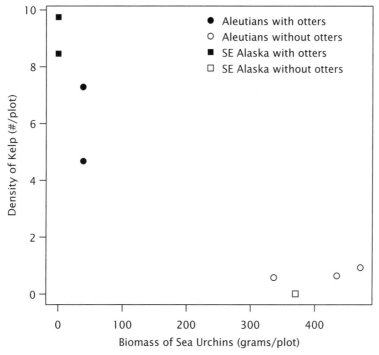

FIGURE 8.4 Average density of kelp and average biomass of sea urchins in 0.25 m² plots on the ocean floor at five sites in the Aleutian Islands and three sites in southeast Alaska. Density of kelp is the average number of individual plants in each plot; biomass (total weight) of sea urchins is an estimate based on counts of individuals and their estimated sizes. Each site is shown by a different symbol: filled symbols for sites with sea otters present and open symbols for sites lacking sea otters.

regions, sites with sea otters were dominated by kelp while sites without sea otters were dominated by sea urchins. Plate 14 illustrates these differences visually.

Although these studies by Estes and his colleagues gave very clear results, they weren't true experiments because sites were not randomly selected to have sea otters present or absent. Amchitka and Adak Islands in the Aleutians had otters because they were a long distance from the home base of the Russians who hunted them to extinction at Shemya and other islands closer to Russia. Without random assignment of treatments (presence or absence of sea otters) to sites, we can't exclude the possibility that other differences between sites might have accounted for the results. For example, the weather might be even harsher at islands like Shemya lacking otters than at islands closer to the Alaskan mainland with otters, and this might explain the low density of kelp at islands lacking otters. In southeast Alaska, the Department of Fish and Game didn't select random sites to release otters but used other criteria—perhaps where they thought the otters might be most successful, most accessible to tourists for observation, or in least conflict with anglers, or simply for convenience. This makes these studies a natural experiment rather than a true experiment. As we discussed in Chapter 6, the possibility of confounding factors makes the interpretation of natural experiments shaky.

Estes and his colleagues did their work in the far north. What about California? Sea otters were almost completely eliminated from California by 1900, although the population increased from about 50 in 1938 to 2,800 in 2012. Michael Foster and David Schiel reviewed survey data for 224 sites lacking sea otters along the California coast to examine the relationship between sea urchins and kelp in the absence of otters. They found some sites with 100% kelp cover and no large sea urchins, like those shown by the filled symbols in Figure 8.4, and some sites with no kelp but abundant sea urchins, like those shown by the open symbols in Figure 8.4. But these sites were only a small percentage of the total number that they considered, while 89% of the sites had both sea urchins and moderate cover of kelp. Foster and Schiel used this result to argue that sea otters are not a keystone species, at least in the southern part of their range in California.

The last remaining population of sea otters in California after hunting was prohibited in 1911 lived at Big Sur, about 150 miles south of San Francisco. As the population increased, animals dispersed north and south along the coast, so that the current range extends from Half Moon Bay, just south of San Francisco, to Point Conception, near Santa Barbara, a distance of 300 miles. Glenn VanBlaricom compared the amount of kelp along two stretches of coastline before and after the reestablishment of sea otters. There were no otters and little kelp south of Point Lobos in 1911–1912 but abundant otters and widely distributed kelp in 1983. There were no otters and only scattered kelp north of Point Estero as late as 1973, but otters had colonized this area by 1983 and kelp was now widely distributed (Figure 8.5).

These are three independent lines of evidence about the potential role of sea otters as a keystone species. In the Aleutian Islands and southeastern Alaska, Estes's team measured the abundance of sea urchins and kelp at sites with and without otters. In California, VanBlaricom plotted the distribution of kelp before and after the arrival of otters. Foster

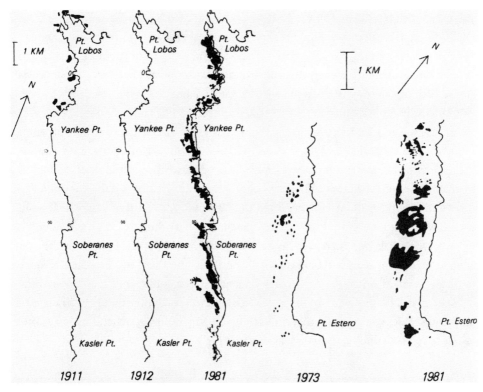

FIGURE 8.5 Distribution of kelp along the California coast south of Point Lobos (left three maps) and north of Point Estero (right two maps) before and after recolonization by sea otters. Sea otters were absent from the area south of Point Lobos in 1911–1912 but present in 1981. Sea otters were absent from the area north of Point Estero as late as 1973 but present in 1981.

and Schiel examined sites without otters in California but found that most such sites contained both kelp and sea urchins. At these sites, however, they only found sea urchins living in small patches or in crevices, suggesting that something other than otters limited the abundance of sea urchins at these sites. They didn't speculate about what this might be, but damage by strong waves or currents might be a possibility. What are the strengths and limitations of these three lines of evidence? How relevant are the observations of Foster and Schiel, since they didn't examine any sites *with* sea otters? The research by Estes and his colleagues in the Aleutians and southeastern Alaska produced clear and dramatic quantitative results (Figure 8.4), but since this was only a natural experiment, it may be that other differences between sites besides presence or absence of otters might have accounted for the results. The research by VanBlaricom in California showed increases in kelp distribution after colonization by sea otters, but he didn't provide quantitative data on the abundance of kelp because he did broad-scale surveys rather than intensive sampling. However, these two studies by Estes's group and VanBlaricom provide complementary evidence consistent with the keystone species hypothesis for sea otters (see Question 4).

Keystone species have large effects on the structure and function of their communities even though they aren't the most abundant members of these communities. The starfish studied by Robert Paine and the sea otters studied by Jim Estes meet the second qualification because they are top predators and not as abundant as the plants and herbivores in their communities. This characteristic of food webs occurs because top predators depend on herbivores for energy, herbivores depend on plants, and less energy is available at each higher level in the web (see Chapter 3). Paine showed that starfish had a big effect on the composition of a rocky intertidal community because the number of species of animals was much higher when starfish were present than when they were absent. By influencing the abundance of sea urchins and kelp, sea otters can change the appearance of a nearshore oceanic community from a site covered with sea urchins (Plate 14a) to a kelp forest (Plate 14b). This change has effects that ramify through the community. In their initial work in the Aleutians, Estes and Palmisano found more kelp at Amchitka Island, where otters were present, than at Shemya, where otters were absent. Kelp provides cover for fish, and there were more rock greenling and other fish at Amchitka than at Shemya. Harbor seals and bald eagles prey on fish, and these two predators were also more abundant at Amchitka. So there is evidence that sea otters influence several different species in nearshore environments in addition to sea urchins and kelp. The appearance and species composition of a community are aspects of its structure, but sea otters also influence community function, as described in Box 8.1.

BOX 8.1 Sea Otters, Kelp, and Climate Change

In 2013, Christopher Wilmers, Jim Estes, and three colleagues extended their analysis of nearshore communities to ask a functional question. Their ultimate goal was to estimate the potential contribution of kelp forests to ameliorating climate change. Since green plants use carbon in photosynthesis, plants that are large and long-lived may sequester substantial amounts of carbon in their tissues. Individual kelp plants don't live for centuries, like some trees, but when kelp dies and sinks to the bottom of the ocean, it is removed from an environment in which it would decompose and release its stored carbon to the atmosphere. Since predation by sea otters on sea urchins favors the development and maintenance of kelp forests, Wilmers's group wondered if sea otters might help ameliorate climate change.

The researchers collected all the kelp rooted in 10 quadrats at each of four sites in the Aleutian Islands, two with and two without otters. They dried these samples and then used an elemental analyzer to measure the percentage of carbon in the samples. In previous work, Estes, Duggins, and others had counted kelp rooted in quadrats at other sites with and without otters (Figure 8.4) but had not collected samples to measure carbon content. To extend the scope of their analysis, Wilmers's team used these earlier data as well. This introduced potential error because they had to estimate carbon storage from counts of individual plants, but the authors described their approach as conservative, that is, minimizing estimates of potential

(Continued)

BOX 8.1 Continued

effects of sea otters on carbon storage. By including the earlier data, the authors added 3,215 quadrats at 153 sites in six regions from Vancouver Island to the western end of the Aleutian Chain.

Box Figure 8.1 summarizes the results of Wilmers and colleagues. The figure shows estimates of rates of transfer of carbon to and from the nearshore ecological community and amounts of carbon stored in kelp—the kelp carbon pool. Note that each of these estimates is a range of values, not a single value, because the researchers collected multiple samples and there was variation in the biomass and carbon content of kelp in these samples. Despite this variation, areas with sea otters and abundant kelp had much more carbon stored in the kelp and greater potential for transfer of this carbon to the ocean depths than areas without otters and with little kelp.

What are the broader implications of these results? Suppose sea otters recolonized their original range in the North Pacific, either on their own or with human help.

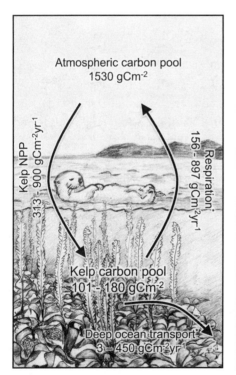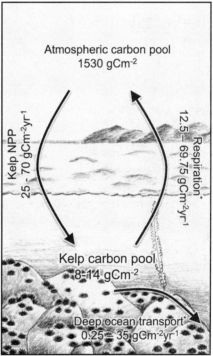

BOX FIGURE 8.1 Estimated effects of sea otters on carbon transport and storage in nearshore environments of the North Pacific. The left panel illustrates a site with sea otters and a healthy population of kelp; the right panel shows a site without sea otters and with little kelp. C. C. Wilmers and colleagues reported amounts of carbon in the atmosphere and in the kelp (pools) in grams of carbon per square meter. The arrows represent transfer of carbon from the atmosphere to kelp in photosynthesis (NPP = net primary production), from the biological community to the atmosphere in respiration, and to the bottom of the ocean when kelp die and sink. The units for these transfers are grams of carbon per square meter per year.

(Continued)

BOX 8.1 Continued

Wilmers's group extrapolated from the kelp carbon pool shown in Box Figure 8.1a to the area from the western Aleutians to British Columbia to estimate that 4.4 trillion to 8.7 trillion grams of carbon could be stored in the kelp that might exist along these rocky coasts if otters were present to control sea urchins that would otherwise eat the kelp. This is 21% to 42% of the total amount of carbon our burning of fossil fuels has added to the atmosphere above the sea otter's potential range in the North Pacific since the start of the Industrial Revolution. Since this range is only 0.01% of the Earth's surface, restoring otters won't have a measurable effect on total carbon in the atmosphere of Earth as a whole. The next step will be to extend this analysis to other ecosystems, especially on land, with different top predators that may increase plant production by keeping herbivores under control, thereby contributing to additional carbon sequestration.

The Plot Thickens for Sea Otters

Following protection of sea otters from hunting in the early 1900s, the population increased rapidly, especially in the Aleutian Islands, where it reached a peak of about 74,000 between 1965 and 1990. In the 1990s, however, the population in the Aleutians dropped sharply to fewer than 10,000, and in 2005 the US Fish and Wildlife Service classified sea otters in the Aleutians as threatened based on the Endangered Species Act. Just as for testing the idea that sea otters are a keystone species, searching for the cause of this dramatic decline of their population involved weighing various kinds of evidence, much of which was circumstantial and none that could be considered a smoking gun. This detective story also illustrates the benefits of long-term, intensive research. The Aleutian Islands are a difficult place to do biological research, especially on sea otters. The archipelago is remote, some of the islands are widely separated from each other, the weather is often harsh, and the subjects live in and under the water, although they generally can be observed from shore. But Estes began working in the Aleutians in 1970 and has continued studies of sea otters in this area with colleagues and students to the present. Consequently the researchers accumulated a substantial amount of data in the 1970s and 1980s that provided context for understanding the dramatic changes that occurred in the 1990s.

Estes's team was working intensively on Adak Island in the 1990s when they noticed fewer and fewer sea otters each year. They soon saw similar patterns at Little Kiska, Amchitka, and Kagalaska Islands, so that by 1997 the numbers of otters on all four islands were at most 10% of what they had been in the early 1990s. Since changes in population size are determined by the relative values of birth rate, death rate, and migration rate, Estes's team knew that declines of the magnitude that they saw had to be due either to an increase in death rate, a decrease in birth rate, or large-scale emigration of the otters—or some combination of these factors. They discounted emigration as a cause for two reasons. First, the decline occurred over a large geographic area—most of the length of the Aleutian Chain. Second, they radio-tagged sea otters on Amchitka and Adak Islands in the central Aleutians but never found them in aerial surveys of other islands. They also found that radio-tagged females on Amchitka and Adak were just as likely to have pups during the population decline of the 1990s as were females before the beginning of the decline, suggesting that a change in birth rate wasn't responsible for the decline.

These observations led Estes's team to consider possible reasons for increased mortality. Perhaps their food supply had declined for some reason, causing starvation of otters; the population had been hit by infectious disease; or a new pollutant in the Aleutians had killed animals. In the past, when they found occasional animals who died from starvation or disease or pollution, the animals came ashore to die, but in the 1990s die-off they found virtually no carcasses as they walked along the shores of various islands.

There is one more possible source of increased mortality of sea otters in the 1990s: predation. In a summary of the biology of sea otters published by the American Society of Mammalogists in 1980, Jim Estes reported observations of bald eagles preying on young otters at Amchitka Island and observations by earlier researchers of predation by sharks in California. Otherwise, the main predators are humans, and there was no evidence of otter pelts showing up in the black market, indicating increased poaching of otters in the 1990s. The mystery deepens, but before describing a possible solution let me generalize from the last three paragraphs.

Estes's approach to solving the mystery of missing sea otters illustrates an important element of reasoning in science—it's helpful to consider a wide range of alternative explanations for something that you are trying to explain. If you can marshal evidence to eliminate most of the alternatives, that gets you closer to a solution by limiting your search. By taking this approach, Estes and his coworkers were able to discount migration, decreased birth rates, and several possible sources of increased mortality as explanations for the decline of sea otters in the Aleutians. The only remaining possibility seemed to be an increase in predation. In this example, it was relatively straightforward to identify a complete set of potential hypotheses to explain a phenomenon, although this isn't always the case.

Because of their intensive studies, the researchers had a handful of observations supporting the predation hypothesis—for a new predator of sea otters. During their two decades of work prior to 1990, they sometimes saw killer whales (Plate 13c) in the same general areas as sea otters, but the frequency of sightings of these predators increased substantially in the 1990s. In addition, their first observation of an attack by a killer whale on an otter occurred in 1991 and they saw five more attacks in the next five years.

These observations showed that killer whales attacked and sometimes killed sea otters but certainly didn't prove that killer whales were responsible for the steep decline in number of otters in the Aleutians in the 1990s. The researchers estimated that the population of otters in the west-central Aleutians dropped from about 53,000 in 1991 to 12,000 in 1997. If predation by killer whales accounted for all of this loss of 41,000 animals, you might think that many more attacks would have been seen. Despite intensive sampling, however, the study area is very large, and it was impossible to watch more than a handful of the otters at any one time. The group estimated the chance of seeing an attack based on the number of person-hours that they spent in the field and the distance that they could observe with binoculars or spotting scopes relative to the total length of the coast of the islands in the study area. Although they spent a total of 21,677 person-hours observing and could see an average distance of 1 kilometer, this distance was a tiny fraction of the total coastline of 3,327 kilometers. These numbers implied that the proportion of attacks that they might have seen if killer whales were entirely responsible for the decline of sea otters was 0.0001, which translates to 5 of the 41,000 otters that died. This is similar to the six attacks that they actually saw. In short, even if killer whales were responsible for most of the estimated mortality of otters in the Aleutians in the 1990s, we wouldn't expect Estes's team to see very many of the attacks in this huge area.

Killer whales are fascinating animals and quintessential predators. Unlike the filter-feeding baleen whales discussed in Chapter 3, killer whales have teeth like dolphins and porpoises. Killer whales are much larger than dolphins and porpoises but smaller than many of the baleen whales as well as the largest toothed whale, the sperm whale, made famous by Herman Melville in *Moby Dick*. Killer whales have incredibly diverse diets, with some populations eating other marine mammals and others eating exclusively fish. Like wolves, killer whales often hunt cooperatively and kill prey much larger than themselves, including blue whales and sperm whales. They are extremely intelligent and use complex acoustic communication to mediate social behavior within and between groups. Numerous observers have reported cases of killer whales inventing new methods to capture prey. For example, William Stolzenburg describes how a young killer whale at Marineland in Canada learned to catch seagulls by eating its daily meal of fish and then spitting out one of the fish and sinking out of sight below water. When a nearby gull landed to scavenge the bait, the whale lunged to the surface and caught the gull. The whale's brother soon imitated this tactic, then his mother, and finally the whole group of killer whales at Marineland.

Despite the diverse diets of killer whales, there are very few reports of them eating sea otters. A Russian biologist reported one case in 1965. Karl Kenyon studied sea otters in the Aleutians before Estes and saw killer whales swimming near otters but never saw an attack. Estes's team saw six attacks in the 1990s, but it's not clear how many of these resulted in death of the otter. Yet killer whales readily learn to exploit new foods, so it's conceivable that a group of killer whales started feeding on sea otters in the 1990s if other food sources had become scarce. Even granting this possibility, it may strain credibility to think that killer whales could account for the death of 41,000 otters in six years.

We can get a sense of the plausibility of this scenario by comparing the energy requirements of killer whales to the energy content of sea otters, as Terrie Williams, Jim Estes, Dan Doak, and Alan Springer did in 2004. Box 8.2 summarizes their calculations, which implied that only four killer whales could account for 41,000 deaths of sea otters in six years. Is this a surprising result? These calculations don't mean that killer whales that started eating otters in the Aleutians relied exclusively on otters, nor that four killer whales killed all 41,000 otters. The calculations do suggest that a relatively small number of killer whales might have been responsible for the dramatic decline in the number of sea otters in the Aleutians during the 1990s. Sea otters are much smaller than other mammalian prey of killer whales like seals, sea lions, and other whales, so killer whales would have to eat many more otters to meet their energy requirements. In fact, otters have been likened to popcorn for killer whales (Figure 8.6).

The most concrete evidence of the influence of killer whales on sea otters in this region in the 1990s came from a natural experiment at Adak Island. Otters lived in Clam Lagoon and Kuluk Bay at Adak. Clam Lagoon had a shallow entrance that prevented access by killer whales, while Kuluk Bay was freely accessible to killer whales.

BOX 8.2 Energetic Calculations for Killer Whales and Sea Otters

Williams and her colleagues compared the energy requirements of killer whales to the energy content of sea otters to estimate how many sea otters a killer whale would need to eat to stay alive if sea otters were its only source of food. Male killer whales weigh more than females, but the average of the two sexes is 4,500 kilograms. Individuals require 55 kilocalories (kcal) of energy per kilogram per day to sustain activity, for a total of 247,500 kcal/day (1 kcal = 1,000 Calories = 1 Calorie in human nutrition; we use 2,000 to 3,000 Calories [2,000 to 3,000 kcal] per day). Sea otters weigh 28.5 kilograms, with an energy content of 1,810 kcal per kilogram, for a total of 51,585 kcal. Dividing 247,500 by 51,585 gives 4.8, which is the number of sea otters a killer whale would need to eat daily if it ate nothing else. There are 2,190 days in six years, so a single killer whale feeding only on sea otters would need to eat 10,512 to survive for six years. Since the sea otter population in the Aleutians declined by about 41,000 from 1991 to 1997, four killer whales could have accounted for the entire decline. Although not as complex as the models discussed in Chapter 6, these calculations are a simple model of the interaction between killer whales and sea otters. The results of the model don't prove that killer whales accounted for the crash of the sea otter population. Instead, they support the plausibility of the hypothesis and imply that it shouldn't be rejected out of hand (see Question 5).

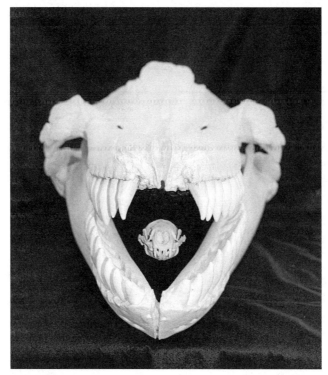

FIGURE 8.6 A sea otter (interior skull) is like a piece of popcorn for a killer whale (exterior skull).

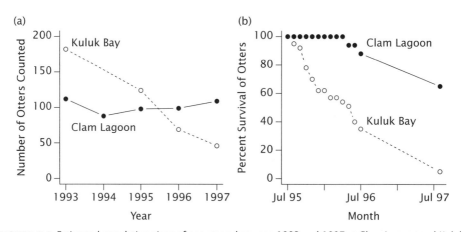

FIGURE 8.7 Estimated population sizes of sea otters between 1993 and 1997 at Clam Lagoon and Kuluk Bay on Adak Island in the Aleutians (a); survival of marked sea otters from July 1995 to August 1997 at Clam Lagoon and Kuluk Bay (b).

Estes's team of researchers counted the number of otters in these two sites every year from 1993 to 1997 and found a marked decline at Kuluk Bay but a relatively constant number at Clam Lagoon (Figure 8.7a). In 1995, they marked 37 sea otters at Kuluk Bay and 17 at Clam Lagoon with flipper tags and radio transmitters and then monitored the survival of these individuals for two years. Sixty-five percent of the marked animals at Clam Lagoon survived, but only 5% of those at Kuluk Bay survived (Figure 8.7b).

As you know from other examples of natural experiments discussed in this book, it's always possible to imagine an alternative explanation for the results. Two researchers at the Cetacean Research Lab of the Vancouver Aquarium Marine Science Center used this approach to challenge Estes's interpretation of the results at Adak Island. Specifically, Katie Kuker and Lance Barrett-Leonard noted the existence of extensive military activity with attendant pollution on several islands in the Aleutians, including Adak. Levels of polychlorinated biphenyl (PCB) and other pollutants were higher in blue mussels and bald eagles at Kuluk Bay than elsewhere, and in 1989 the military discharged 2 million liters of JP-5 jet fuel into Kuluk Bay. Kuker and Barrett-Leonard argue that one or more of these pollutants may have killed sea otters in Kuluk Bay, while exposure of sea otters in Clam Lagoon was much less.

As noted by Kuker and Barrett-Leonard in their 2010 article, this story of killer whales and sea otters has become a "textbook case of top-down predator control," commonly known as a *trophic cascade*. However, Kuker and Barrett-Leonard went on to criticize the argument implicating killer whales on several grounds, including the possibility that differential pollution in Kuluk Bay and Clam Lagoon rather than differential access to killer whales may have been responsible for differences in survival of otters in the two sites. Estes and nine other researchers wrote a rebuttal of Kuker and Barrett-Leonard's article that hadn't been published as of June 2014 (in a footnote, these 10 authors state that they "are all scientists who believe their data or analyses have been misrepresented by Kuker and Barrett-Leonard"). Estes's group defended their interpretation of the Kuluk Bay–Clam Lagoon comparison in several ways, most notably by pointing out that "the sea otter decline was not limited to Kuluk Bay but rather occurred broadly across the Aleutian archipelago and western Alaskan peninsula and included areas far removed from established military bases or other human settlements [where otters might be exposed to pollutants]." They also reported that pollutant levels in blood samples of sea otters in the Aleutians were less than those in central California, even though the California population was increasing in size unlike the Aleutian population. Finally, they suggested that "it is difficult to reconcile how contaminants could have led to widespread decline across the Aleutians while leaving animals unaffected in Clam Lagoon, which is subject to constant flushing and tidal exchange with Kuluk Bay."

The main theme of the critique by Kuker and Barrett-Leonard is that Estes's team didn't provide definitive evidence against alternative hypotheses for the sea otter decline. One more example will give the flavor of Kuker and Barrett-Leonard's approach. They thought predation by sharks rather than killer whales might have caused the otter decline, based on the fact that white sharks are known predators of otters in California. White sharks are uncommon in the Aleutians, although Pacific sleeper sharks occur there. Sleeper sharks swim in deep water during the day but shallower water at night where they might encounter resting sea otters. No one has ever seen a Pacific sleeper shark attack a sea otter, but Kuker and Barrett-Leonard argued that people wouldn't be observing otters at night, so there might be no recorded attacks even if these sharks prey on otters at night. This all seems pretty hypothetical, but the main thrust of the argument by Kuker and Barrett-Leonard was that the number of sleeper sharks increased in the North Pacific in the 1990s, so they may have been just as likely as killer whales to start attacking sea otters and even less likely to have been seen doing so. The 10 defenders of the killer whale hypothesis pointed out, however, that the region in which shark populations increased was mainly south and east of the region where sea otters declined (see Appendix 6).

Perhaps the killer whale–sea otter story deserves its iconic status as an example of top–down predator control of an ecological community, at least until more substantive objections to the evidence favoring this hypothesis are proposed. In fact, there was one other interesting result of the studies by Estes's group at Adak Island. Estes and Duggins had surveyed the sea floor for kelp and sea urchins in 1987, before the beginning of the sea otter decline. Estes, Tinker, Williams, and Doak repeated these surveys in 1997, when the otter population had dropped by 80%. In that 10-year interval, the abundance of sea urchins increased by 800% while kelp density decreased by 92% (Figure 8.8). These results have two implications. First, one of the alternatives to the killer whale hypothesis that the

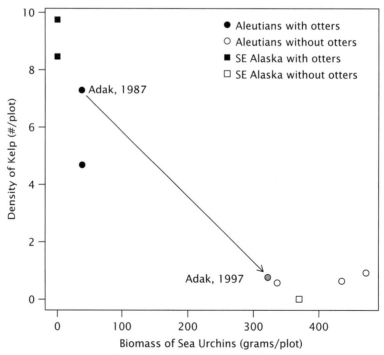

FIGURE 8.8 Average density of kelp and average biomass of sea urchins in 0.25 m² plots on the ocean floor at five sites in the Aleutian Islands and three sites in southeast Alaska. This is the same as Figure 8.4 except that data for Adak Island in the Aleutians are shown for both 1987, as in Figure 8.4, and 1997, when the population size of sea otters had declined by about 80% from its peak, indicated by the gray shading of the symbol for Adak in 1997.

researchers considered and rejected was a shortage of food leading to starvation of otters. Kuker and Barrett-Leonard thought this food-shortage hypothesis was still viable. However, sea urchins are a favorite food of otters, and urchins were abundant by the mid-1990s, making this alternative seem highly unlikely. Second, the results shown in Figure 8.8 illustrate another test of the keystone species hypothesis for sea otters. The decline of the sea otter population was followed by an increase in sea urchins and a decrease in kelp, just as the keystone species hypothesis predicts. These results contribute to the weight of evidence that sea otters are a keystone species in nearshore marine environments.

The Stage Expands for Killer Whales

I've left one big question unanswered in telling this story about killer whales and sea otters—why did killer whales start preying on otters in the 1990s? Researchers don't know the answer to this question, but they do know that sea otters weren't the only marine mammal whose population crashed in the Aleutian Islands. For example, Steller sea lions (Plate 13d) lost 80% of their population in Alaskan waters between 1970 and 2000, and the population west of the Gulf of Alaska was classified as endangered by the National Marine Fisheries Service in 1997. Although the decline of Steller sea lions started in about 1970, it was especially steep between 1985 and 1990, just before sea otters started to decline.

Steller sea lions are much larger than sea otters but much smaller than killer whales. Sea lions and their relatives among the pinnipeds (fin-footed carnivores, including seals, the walrus, and sea lions) mate and give birth on land but spend most of their lives at sea and range much farther from shore than sea otters. Steller sea lions eat fish, and the first hypothesis proposed for their population crash was a decline in the quantity or quality of their food. This, in turn, might have been due to overharvesting by humans in commercial fishing operations or climate change causing changes in the abundance or distribution of fish eaten by Steller sea lions.

The Endangered Species Act requires that species classified as endangered be protected from human activities that would further depress their populations. If the population of Steller sea lions crashed because of lack of food due to overfishing by humans, then managers would be legally justified, indeed obligated, to limit fishing in the North Pacific where sea lions live. In fact, the National Marine Fisheries Service has instituted such limits.

When Jim Estes and his colleagues were documenting the decline of sea otters in the Aleutians in the 1990s, they didn't connect it to the earlier decline of Steller sea lions and they had no clear ideas about why killer whales might have started hunting sea otters. In 2003, however, they published an hypothesis in a paper entitled "Sequential Megafaunal Collapse in the North Pacific Ocean: An Ongoing Legacy of Industrial Whaling?" The researchers suggested that killer whales in this region initially fed on other whales, some much larger than themselves. These great whales were depleted by industrial whaling after World War II, causing killer whales to switch to smaller prey: first harbor seals, then fur seals, then sea lions, and finally sea otters. According to these researchers, the killer whales caused marked population declines of each of these species in sequence (as illustrated in Figure 8.9), hence the name of their hypothesis.

The sequential megafaunal collapse hypothesis offers an explanation for why killer whales began feeding on sea otters in the 1990s, but it also entails an alternative hypothesis for the decline of Steller sea lions and other pinnipeds before the 1990s. These declines had been attributed to depletion of their food supply, a bottom–up process, but the new hypothesis illustrated in Figure 8.9 suggested that the declines were caused by predation, a top–down process. This new hypothesis about why killer whales started hunting sea otters was much more contentious than the basic idea that killer whales caused sea otters to decline (see Chapter 9). There were three critical articles and two rebuttals totaling 99 pages in the journal *Marine Mammal Science*, plus exchanges in other journals. The critical articles and rebuttals addressed scientific issues, such as the question of whether killer whales even attack great whales, but one reason why the argument was so ferocious may have been the implications for conservation and management. For example, during this time Ted Stevens was a US senator from Alaska who strongly supported commercial fishing in Alaskan waters. Shortly after Springer's group published their idea of sequential

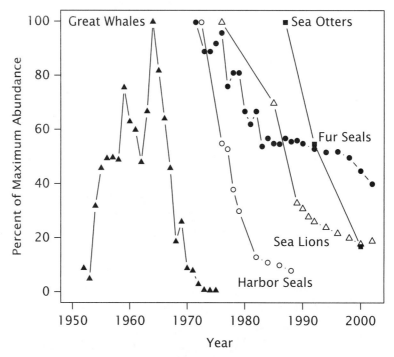

FIGURE 8.9 Population estimates for great whales, harbor seals, sea lions, fur seals, and sea otters in the North Pacific Ocean and southern Bering Sea (just north of the Aleutian Islands) from 1952 to 2002. The great whales include 12 species of baleen whales (see Chapter 3) plus the sperm whale; these are large species that are primary targets of industrial whaling. Estimates are shown as percentages of maximum abundance. For whales, the index of abundance was harvests for the region reported by the International Whaling Commission; for the other species, abundance was indexed by counts or population estimates for intensively studied sites within the region.

megafaunal collapse due to killer whales, Stevens suggested that "the government begin looking into this new evidence 'that rogue packs of killer whales' were to blame for the North Pacific's endangered sea lion" (Stolzenburg 2008:75–76). At the same time, killer whale biologists worried that Alaskans would use the report by Springer's group as a justification for attacking killer whales.

Springer, Estes, and their colleagues ended their main rebuttal of the scientific criticisms of their hypothesis by acknowledging the difficulty of definitively showing that killer whales were sequentially responsible for crashes of harbor seals, fur seals, sea lions, and sea otters after depletion of their primary food source by industrial whaling after World War II. The main source of difficulty is that their hypothesis attempts to explain historical events that weren't tracked in detail at the time that they happened. No one did systematic studies of the foraging behavior of killer whales before, during, or after the industrial whaling that occurred between 1949 and 1969. Population estimates for pinnipeds in the North Pacific were also imprecise. There's no way to recreate these key data retrospectively. The researchers suggested, however, that future events may bring an opportunity for an indirect test of the hypothesis. Great whales are now protected by the International Whaling Convention and the Convention on International Trade in Endangered Species of Wild Fauna and Flora. If these regulations are effective, populations of great whales should recover, albeit slowly because their reproductive rates are low. The sequential megafaunal collapse hypothesis predicts that killer whales may eventually switch back to feeding on great whales, allowing recovery of alternate prey such as Steller sea lions. In other words, this will be a natural experiment that can provide evidence bearing on the hypothesis—*if* the behavior and population dynamics of great whales, killer whales, and pinnipeds are closely monitored.

It may also be possible to test the alternative bottom–up hypothesis for the decline of Steller sea lions by a carefully designed manipulative experiment, even at the large scale of the North Pacific range of these animals. Steller sea lions come ashore to mate and give birth at numerous rookeries scattered throughout the Aleutians. This region could be divided into several sampling areas, each centered on a rookery. Some of these areas could be selected at random and opened to commercial fishing, while other areas are closed to fishing. If sea lions increase in the protected areas but not in the fished areas, this supports the bottom–up hypothesis that food abundance drives sea lion numbers, at least in part.

Wolves and Elk in Yellowstone—Another Trophic Cascade?

Ecological processes like interactions between kelp, sea urchins, sea otters, killer whales, and other plants and animals of the North Pacific are complex scientifically because they occur over large areas and long spans of time, because they can't be studied with manipulative experiments, and because much of the evidence about their causes and consequences is circumstantial. They are also complex because they don't happen in a realm of nature that is isolated from human society. The bottom–up and top–down hypotheses for the decline of Steller sea lions have very different implications for commercial fishing in Alaskan waters, so anglers, owners of fishing companies, politicians, conservationists, and the general public have deep but diverse interests in the resolution of the scientific issues. As we saw earlier, the reintroduction of wolves to Yellowstone National Park has similar complexity involving science, economics, and ethics.

When I gave a brief history of wolves in the Yellowstone region on page 180, I focused on the dilemmas that arose when wolves spread beyond the boundaries of the national park after being reintroduced in the park in 1995. Wolves outside the park are killed during hunting seasons established by the states of Wyoming, Montana, and Idaho and outside hunting seasons if they prey on livestock; while wolves within the park are protected, along with all other animals and plants. Wolf biologists and other researchers have been fascinated by ecological changes in the park since 1995. Wolves eat a variety of prey species, but their main food source in Yellowstone has been elk. As the wolf population increased to about 100 between 2003 and 2007, elk declined from about 20,000 to fewer than 10,000. Elk feed heavily on aspen and willow, two deciduous woody plants that are especially abundant in riparian habitats along streams. With a large population of elk in the decades before wolf reintroduction in 1995, Yellowstone had little regeneration of aspen and willow because elk favored tender, young shoots of these plants. After 1995, as the elk population declined researchers began to see the first evidence of regeneration of aspen and willow stands (Plate 15). William Ripple and Robert Beschta documented these changes, which show the same pattern as the effect of sea otters on kelp in the ocean. Just as sea otters eat sea urchins, allowing kelp to flourish, wolves eat elk, benefitting willow and aspen.

This is another example of a top–down effect, or trophic cascade. Like the sea otter story, there were apparently ramifying effects of the return of wolves to Yellowstone. With the resulting growth of aspen and willow, the abundance of beavers and bison increased. Also like the sea otter story, there is disagreement among researchers about what's happening. In the case of wolves and elk, Ripple and Beschta proposed that two mechanisms are at work—a reduction in population size of elk due to predation by wolves and a change in the behavior of elk to avoid wolves. They suggested that elk are more vulnerable in riparian areas because these don't have expanses of open ground where the elk can outrun wolves. Therefore, with wolves present, elk avoid riparian areas, allowing even more regeneration of aspen and willow. This "behaviorally mediated trophic cascade" has been more controversial than direct predation by wolves on elk as a cause of increased regeneration of aspen and willow. As with sea otters, most of the scientific controversy has been due to the lack of evidence from rigorous manipulative experiments and reliance instead on observations, comparisons, and correlations associated with the natural experiment of reintroducing wolves to Yellowstone. This is a system in which manipulative experiments certainly are possible, although their relevance may be questionable. For example, researchers could use fencing to build a set of exclosures to prevent elk from foraging in certain plots and compare regeneration of aspen and willow in these plots with regeneration in unfenced control plots. These would be small-scale experiments, so one question would be whether the results would be meaningful at the scale of many square miles that constitute the home ranges of elk. These experiments also don't manipulate wolves, so don't directly assess the top–down effects of wolves on the system.

Aldo Leopold was an influential early conservation biologist, best known for his book *A Sand County Almanac* published in 1949. He began his career as a US Forest Service employee in Arizona and New Mexico in 1909 where he trapped and shot mountain lions and wolves to favor deer and elk populations, consistent with Forest Service policy at that time. By 1949, he reflected on this part of his early career with some regret: "I was young then, and full of trigger-itch; I thought that because fewer wolves meant more deer, that no wolves would mean hunter's paradise. But after seeing the green fire die, I sensed that neither the wolf nor the mountain agreed with such a view" (Leopold 1968:130). With this poetic language, Leopold foreshadowed current scientific interest in the top–down effects of predators on ecosystems.

Conclusions

The title of this chapter is "From Causes to Consequences: Considering the Weight of Evidence." The two main examples focused on sea otters, first as a keystone species and second as part of a trophic cascade from killer whales through sea otters, sea urchins, and eventually kelp. We discussed various kinds of evidence about keystone species and trophic cascades, but most was indirect or circumstantial. Therefore we couldn't rely on a critical observation or experiment—a smoking gun—to evaluate the hypotheses that sea otters are a keystone species in nearshore marine environments and that killer whales caused a crash in the population of sea otters in the Aleutian Islands in the 1990s. Instead, we had to make our decisions based on the weight of evidence for and against these hypotheses. Furthermore, these decisions have consequences beyond science; for instance, they will impact how commercial fisheries are managed in Alaskan waters. This process of considering the weight of evidence to evaluate scientific hypotheses about causation that have practical consequences applies to many decisions that we make as individuals and as members of society. Going through this process doesn't mean considering only scientific evidence and ignoring factors like economics and ethics in making our decisions. But it does mean evaluating the science fully and carefully, as we'll discuss further in Chapter 10.

Has anyone seen the spider? This is what Nancy Krieger asked in thinking about complex webs of causation in her field of epidemiology. Ecologists ask the same question when they think about the control of complex food webs. For many years, ecologists focused on bottom–up control of food webs—through photosynthesis, plants are the source of energy for everything in food webs, starting with herbivores and ending with top predators. But we've learned from the work of Bob Paine, Jim Estes, and many others that keystone predators (including actual spiders in some habitats) can exert control on both the structure and function of food webs from the top down. In other words, killer whales and sea otters, and perhaps wolves in Yellowstone, are metaphorical spiders in complex webs of causation.

Questions to Ponder

1. What if you were a conservationist or animal rights advocate? What scientific issues would you be obligated to consider in thinking about wolf management (see page 180)?

2. The food web for Wytham Woods shown in Figure 8.2 includes several different food chains. How many can you identify that include three species? Four? How about five?

3. Although Paine's experiment that led to the keystone species concept was certainly planned and is considered a classic field experiment by ecologists, it had some basic deficiencies in experimental design. What were some of these deficiencies? You might want to review Chapter 4 in thinking about your answer to this question.

4. Only one manipulative experiment has been published that is directly relevant to the keystone species hypothesis for sea otters. David Duggins did this experiment from 1976 to 1978 at Torch Bay in southeast Alaska. No sea otters occurred in Torch Bay at this time, although they had recolonized this site by 1985 (Figure 8.4). Duggins laid out five underwater plots at Torch Bay, each about 50 meters square. He removed all sea urchins from in and near three of these plots and used the other two plots as controls; he repeated his removals of sea urchins from experimental plots every two months. Various species of kelp colonized the experimental plots, resulting in an average density of 150 individuals per square meter after two years. By contrast, there were no kelp in the control plots at any time during the experiment.

   Compare this experiment to the three lines of evidence about the keystone species hypothesis for sea otters discussed in the text. How is the experiment similar to and different from each of these lines of evidence? Based on these comparisons, what are the strengths and limitations of this experiment as evidence relating to the hypothesis?

5. Some killer whales rely exclusively on fish. How is this possible, considering that fish are much smaller than sea otters?

6. Estes and colleagues argued that killer whales were responsible for the dramatic decline in the population of sea otters in the Aleutian Islands in the 1990s. Plate 22 summarizes some of the evidence consistent with this hypothesis in the form of an argument map (see Appendix 6). Which line of evidence related to the killer whale predation hypothesis is not included in Plate 22? Can you extend the argument map to include this additional line of evidence?

Resources for Further Exploration

1. Tim van Gelder started Austhink Software and has written valuable articles about critical thinking as described in Appendix 6. He has an interesting blog called "Bringing Visual Clarity to Complex Issues" at http://timvangelder.com/.

2. Parks Collins of Mitchell Community College has developed a case study about the reintroduction of wolves to Yellowstone National Park. It is available at the National Center for Case Study Teaching in Science: The Return of Canis Lupus? (http://sciencecases.lib.buffalo.edu/cs/collection/detail.asp?case_id=691&id=691).

Science as a Social Process

Science is often portrayed as a linear process: a new observation leads a scientist to ask a question, she devises a tentative answer in the form of an hypothesis, plans further observations or experiments to test the hypothesis, collects data, and finally concludes that the hypothesis is supported or not supported based on the results. Many observers of science including historians, philosophers, journalists, and scientists themselves have shown that science is much messier than this idealized portrait of "the" scientific method. Frederick Grinnell is a cell biologist at the University of Texas Southwestern Medical Center who described this messiness as the "everyday practice of science" in his 2009 book by this name. I've used diverse biological examples to make the same point. The sources of our questions are diverse. We may be motivated by a very specific question (Where did the wolverine that got its picture taken by a camera in California come from?—Chapter 2) or by much more general questions (Can mothers influence the sexes of their offspring? If so, how and why?—Chapter 4). We may be motivated by practical concerns or simple curiosity. We may formulate one or more hypotheses to guide our research (*Migration* or *food shortage* or *increased predation* accounted for the decline in number of sea otters in the Aleutian Islands in the 1990s—Chapter 8) or we may simply want to discover something new, like the winter home of Monarch butterflies (Chapter 1). Methods for discovery and hypothesis testing are also diverse and sometimes not definitive.

There isn't a single recipe for doing science but lots of different pathways for posing and answering scientific questions. What about the working styles of scientists? What is your image of a scientist, and where did it come from? Many people become acquainted with scientists through movies. I recently watched *The Gods Must Be Crazy*, having remembered how hilarious it seemed when it first came out in 1980. Andrew Steyn is a wildlife researcher in the remote and forbidding landscape of the Kalahari Desert in Africa. He has one assistant whose main responsibility is to keep Steyn's ancient Land Rover working. Steyn is portrayed as a knowledgeable and competent scientist who is extremely awkward in social interactions, especially with women such as Kate Thompson, a new teacher in a local village. For example, when they first meet and Kate asks Andrew what he does, he says that he collects elephant manure. It gets worse from there, although the slapstick elements of the story of Andrew and Kate are interspersed with a delightful tale of a Bushman named Xi who sets out on a journey to get rid of a coke bottle that is thrown out of a small plane, lands in the camp of Xi's family, and causes bad luck for the family (see Question 1).

This is one portrayal of a scientist as someone who works alone, perhaps because he is uncomfortable in social settings. But the personalities and working styles of scientists are as varied as those of people in general. Let me give two examples from my own career to illustrate the everyday practice of science. These examples will show how one individual scientist worked differently in different circumstances, that is, how the practice of science varies not only between different scientists but also within individuals over the course of their careers.

One Researcher, Two Examples of the Everyday Practice of Science

My first research project was a study of food selection by beavers (Plate 16a) in central Massachusetts. Many of my fellow graduate students in ecology at Harvard were working in exotic locations like the West Indies or Africa, and part of my motivation for working on beavers near home was simply a desire to do something different. My work was also motivated by new ideas about food selection by animals that ecologists were talking and writing about in the early 1970s. This general area of research is called optimal foraging ecology and my advisor, Tom Schoener, was one of the leaders of this new area. Although I thought a lot about optimal foraging in planning my field work on beavers, I did not design it to test specific a priori hypotheses but rather to collect more detailed data about patterns of tree cutting by beavers than previous researchers had collected. Most people who studied optimal foraging in those days focused on animals eating other animals, especially birds and lizards eating insects. I wanted to apply ideas about optimal foraging to an herbivore but had no clear idea how to do this when I began.

I need to provide a little background about optimal foraging ecology and about beavers before describing my own work. Optimal foraging ecology is an evolutionary approach to food selection because it asks what choices an animal should make to maximize its ability to survive and reproduce. Imagine a lizard sitting on a branch waiting for a potential prey item to happen by. A small insect lands within striking distance. Will the lizard be better off trying to capture this prey item or waiting so it doesn't miss a bigger insect that lands soon after the first has flown away? "Better off" means gaining energy at a faster rate, which will depend on the success rate for the lizard capturing large and small prey, its handling time for large and small prey, and the abundance of prey in the its territory.

Beavers eat many different kinds of plants but usually rely on the inner bark of trees and shrubs in fall, winter, and spring when green vegetation is not available. They use their strong, sharp incisors to fell trees around the ponds where they live, and then cut branches off these trees and carry them to a food cache in the pond close to their lodge. The lodge has a platform above water level where the beaver family rests and an underwater passageway where they enter and leave the lodge (Plate 16b). Beaver ponds freeze over in winter in northern latitudes, and the beavers swim out under the ice to harvest food from the food cache, which is buried in the mud and frozen in by the ice.

I mostly worked alone collecting data at several field sites in the Quabbin Reservation. This is the site of a large reservoir in central Massachusetts that supplies most of the water for metropolitan Boston. The reservoir was built in the 1930s and required abandonment and flooding of four towns. A tall fence surrounds the entire reservoir and adjacent lands, although some of the original roads from the 1930s remain in areas that weren't flooded. I got permission to access the reservation through a couple of locked gates; although the area is open to hikers, there were many days when I was probably the only person in this "accidental wilderness" of 38 square miles. It is certainly not as remote as the Kalahari but more remote than many people imagine is possible in the densely populated northeastern United States.

I also worked alone at my desk analyzing data, making graphs, and writing up my results. My wife and I had an apartment in Athol, near the Quabbin, and I commuted once or twice a week to Cambridge where I taught a class, met with my advisor, and used the library. In short, I was largely on my own in designing my project, doing the research, and preparing the results for publication. In fact, unlike most of the examples you've seen in previous chapters, I was the sole author of all four papers that reported the results of this research, although my advisor and members of my graduate committee reviewed the papers before I submitted them to journals, and other scientists reviewed them before publication. I'll describe this process of peer review in more detail later, but I want to conclude this example of the working style of one scientist early in his career with one more part of the story.

My advisor, Tom Schoener, is both a theoretical and empirical ecologist. He builds models, including some of the early models of optimal foraging, and uses observations and experiments in the field to test these models. He works mostly on birds and lizards, especially in the Bahamas, so didn't get involved in my field work on beavers in Massachusetts. While I was doing this work, however, Tom was developing a model of food selection by animals that repeatedly travel out from a nest or burrow to harvest food. They may eat food items where they collect them or bring them back to the nest or burrow, like a pair of robins bringing worms to their nest to feed their hungry chicks. This is called central-place foraging because the nests or burrows are often centrally located in an animal's territory.

Tom's initial model predicted that central-place foragers would be more selective about the sizes of prey that they harvest at greater distances from the nest or burrow and, specifically, that they would favor *large* prey items over small ones at greater distances. We talked about this model on my periodic trips to Cambridge, and I read a draft of Tom's paper. In the meantime, I was collecting data on the trees being cut by beavers at their ponds in the Quabbin and found that they cut all sizes of trees close to shore but only *small* trees at greater distances from shore. This was just the opposite of what the model predicted, so we tried to figure out why. After several weeks of discussion, we realized that Tom had made an assumption that was not explicitly stated in the description of the model. This assumption was that predators are larger than their prey; it was based on Tom's experience working with birds and lizards feeding on insects and other invertebrates. Beavers, however, are much smaller than the trees that they cut for food. When Tom revised his model to allow for the possibility that some central-place foragers are smaller than food items that they use, the new version of the model predicted different patterns for foragers based on the relative sizes of their prey. These predictions were consistent with my data for beavers felling trees as well as data for birds and lizards feeding on insects and other invertebrates. Tom published his model in *The American Naturalist* in 1979, and I published my study of a size–distance relationship in tree selection by beavers in *Ecology* in 1980.

Twenty years after working mostly on my own in the woods of central Massachusetts, I was a faculty member at the University of Nevada, Reno. I had worked on various research projects in the Sierra Nevada west of Reno and the Great Basin Desert east of Reno, but I will use a study of seed-eating animals in the desert as a second example to contrast with my beaver study because it illustrates some big differences in working style. I did this study of granivory with two faculty colleagues, Bill Longland and Steve Vander Wall. We shared interests in foraging behavior and food selection of animals and in the impacts of herbivores on plant populations, but we brought different kinds of expertise to the project. Bill worked for the Agricultural Research Service of the US Department of Agriculture. Agricultural Research Service scientists in most of the country focus on growing crops like corn and wheat in the Midwest or wine grapes and almonds in California, but the Agricultural Research Service lab in Reno does a lot of research on native plants that are important food sources for wildlife as well as livestock. One of these plants is Indian ricegrass (Plate 16c). Large, grazing animals forage on Indian ricegrass, while birds, ants, and desert rodents (Plate 16d) harvest its nutritious seeds. Birds just eat the seeds, but ants and rodents also store seeds for future use when fresh seeds aren't available. Both ants and rodents store seeds deep in their burrows, but some rodents also fill their cheek pouches with seeds, carry these seeds away from their burrows, and bury them just below the surface of the soil. These buried caches of seeds are called scatterhoards and are a potential source of regeneration of the Indian ricegrass population if the rodent that makes a scatterhoard doesn't go back and dig it up later.

We were interested in the effects of birds, ants, and rodents on the population dynamics of Indian ricegrass. By eating seeds, do these granivores cause population sizes of Indian ricegrass to be lower than they would otherwise be? This was likely to be the case for birds that simply eat seeds and don't store them and for ants that store seeds too deeply for germination. The situation is more complex for rodents, which eat some seeds but scatterhoard others in conditions that may aid germination and establishment. If rodents fail to recover all of the seeds that they scatterhoard and if shallow burial enhances germination success compared to seeds that simply fall to the ground, then there might be a mutually beneficial relationship between Indian ricegrass and these rodents, like that between bees and the flowers that they pollinate discussed in Appendix 3.

We designed an experiment to study the impacts of birds, ants, and rodents on Indian ricegrass at a field site about 40 miles east of Reno. My beaver research in the 1970s had been strictly observational, producing evidence for testing models like that discussed in Chapters 5 and 6. I had done a laboratory experiment on food hoarding by kangaroo rats in the 1980s, but large-scale field experiments like this study of Indian ricegrass pose different kinds of challenges. Bill, Steve, and I wrote a grant proposal to the US Department of Agriculture (USDA) Competitive Grants Program that was funded on our second try, so we went to work. We first established 36 large plots at our field site in which we applied 11 different treatments. The treatments involved various combinations of access to and exclusion of birds, ants, and rodents. Setting up the experiment involved building fences that were sunk below ground (to discourage rodents from digging under them) and had metal flashing attached at the top (to discourage rodents from climbing over). Fortunately we secured enough money from the USDA to hire two graduate students and several undergraduates to help with this hard work as well as running the experiments and collecting data for three years in the field.

This research was quite different from my earlier beaver study in being very much a team effort. All three of the principal investigators—Bill, Steve, and I—made different contributions to the project based on our interests and expertise, and we had good help from graduate students and undergraduates. My major contributions were designing the experiment, writing key parts of the proposal that funded the study, analyzing the results, and writing parts of the main paper reporting the results. We published these results in *Ecology* in 2001 with the title "Seedling Recruitment in *Oryzopsis hymenoides* [Indian Ricegrass]: Are Desert Granivores Mutualists or Predators?" Our answer to this question was that desert rodents, especially kangaroo rats (Plate 16d), do have a mutualistic relationship with Indian ricegrass by scatterhoarding seeds in sites favorable for germination and not recovering all of these scatterhoards.

Science Is a Social Process

In 1989, the US National Academy of Sciences published a handbook for students starting their scientific careers. The handbook begins with this story from 1937:

> Tracy Sonneborn, a 32-year-old biologist at Johns Hopkins University, was
> working late into the night on an experiment involving the single-celled organism

Paramecium. For years biologists had been trying to induce conjugation between paramecia, a process in which two paramecia exchange genetic material. . . . Looking through the eyepiece [of his microscope], he witnessed for the first time what he would later call a "spectacular" reaction: The paramecia had clustered into large clumps and were conjugating. In a state of delirious excitement, Sonneborn raced through the halls of the deserted building looking for someone with whom he could share his joy. Finally he dragged a puzzled custodian back to the laboratory to peer through the microscope and witness this marvelous phenomenon.

(COMMITTEE ON THE CONDUCT OF SCIENCE, National Academy of Sciences 1989)

This story paints a picture of science as a solitary process of discovery, with the successful scientist being a heroic figure who works harder and longer than everyone else. This is a popular image that is reinforced in literature, film, and the news media. For example, in 1993, the *Reader's Digest* told how an Australian physician named Barry Marshall discovered that bacteria contribute to ulcers, partly by purposely infecting himself: "As the hands of the wall clock edged towards midnight, Barry Marshall rubbed his eyes and placed a slide under the microscope. . . ." (Oreskes 1996:111). Yet not all scientists work alone or through the night. As you saw, our study of kangaroo rats and Indian ricegrass was a group project. Most of the research that I described in previous chapters was done by teams of scientists, from pairs to groups of 40 or more. Despite these stories of how Sonneborn discovered genetic exchange by conjugation of paramecia and Marshall discovered a role for bacteria in stomach ulcers, science is a highly social process. Indeed, the National Academy of Sciences handbook for young scientists emphasizes the importance of the social aspects of science despite the initial story about Tracy Sonneborn (see Question 2).

Science is a social process not only because many research questions are pursued by groups of scientists, especially today, but, even more important, because the credibility of science depends on how scientists share their work with the scientific community and how the community responds to new discoveries. In *Everyday Practice of Science*, Frederick Grinnell divides science into two parts, discovery and credibility. So far in this book, we've discussed the role of simple observations, experiments, comparisons and correlations, models, and the weight of evidence in discovering new facts and new relationships of causation. Now I will describe how scientific discoveries gain credibility and why this makes science such a powerful force in society.

Peer Review

Teachers often caution students against relying too much on Web resources for writing papers. They may require students to use a certain number of *peer-reviewed* sources for their papers. What are peer-reviewed sources? Why do teachers consider these to be more credible than other information students may find with a Google search?

The World Wide Web is a wonderful source of information for students, teachers, researchers, and the general public. When I started my scientific career as an undergraduate in the 1960s, I found material for writing assignments in large, heavy bibliographic reference books with very small print and then looked in rows upon rows of library stacks for the printed journals containing this material. This continued through my graduate work and the first 20 years of my teaching career. These resources have been totally supplanted by electronic databases and journals and now by electronic books, at least in the sciences. Doing library research is *much* easier these days and often doesn't even involve physical libraries. But this new world of interconnectedness poses new challenges for students and researchers, especially the challenge of separating the wheat from the chaff in all of the information on the Web. The fundamental problem is that anyone can post anything on the Web, from a well-documented observation or well-reasoned argument to a pure rant. Both the good and the bad pass for information and may show up when we do a Web search, with no indication of the quality of the information. For example, in May 2009 Anna Kata did searches of the US and Canadian Google sites for the terms "vaccination," "vaccine," and "immunization." She restricted her analysis to the first 10 hits for each of these terms because previous research had shown that most people seeking health information on the Web only look at these 10 sites. Using the US version of Google, Kata found that 24% of the first 10 sites opposed childhood vaccination. Kata entitled her paper on this research "A Postmodern Pandora's Box: Anti-Vaccination Misinformation on the Internet" (see Question 3).

Peer review is a filter that can help separate the wheat from the chaff in information about scientific topics that may be available on the Web or from other sources. It isn't 100% successful—letting all reliable information through the filter and blocking all misinformation. In particular, using peer-reviewed sources doesn't relieve you of the responsibility to examine critically the arguments and evidence in these sources. However, you can increase the efficiency of your research by using peer review to help filter wheat from chaff in sources of information.

Peer review in science has informal and formal components. Let's return to our research on Indian ricegrass to illustrate these components. Conversations between Steve Vander Wall, Bill Longland, and me were the beginning of the project. These conversations and later conversations with the students who worked with us were not aspects of peer review although they were critical social interactions that improved the design and execution of the research. The first peer reviews came when we submitted our proposal to the USDA Competitive Grants Program. The program officer selected several scientists whom he thought would have some knowledge about the type of work that we wanted to do and asked them to evaluate the proposal. They provided written comments and summary ratings that the program officer used in making a decision about funding the proposal. As I mentioned earlier, we didn't get funding until the second try. As a standard part of the review process, the program officer sent us the anonymous reviews of the unsuccessful proposal, and we revised the proposal to address the problems identified by the first set of reviewers. We also received reviews of the second, successful version of the proposal, which gave further advice for strengthening the project. This process of grant submission and evaluation is part of formal peer review.

Informal peer review includes discussion of the project with colleagues who aren't part of the research team. These discussions happen in offices and hallways at our home institution, by phone or email with colleagues elsewhere, and in person at professional meetings or over coffee, lunch, or beer. Colleagues include fellow scientists, from students to senior researchers, as well as spouses and others who sometimes provide fresh insight because they see problems from a different perspective. Informal peer review also happens when a member of the research team gives a seminar at another institution or a presentation at a professional meeting and gets questions or comments from listeners.

We completed our field experiment on granivores and Indian ricegrass in 1996. The next phase of the project was collating, organizing, and analyzing the data so that we could write a paper explaining our rationale, methods, results, and interpretations. Once the team members were satisfied with the written paper, we asked friends for comments in another phase of informal peer review. We revised the paper and submitted it to the editor of the journal *Ecology* for consideration. The editor sent the paper to two anonymous reviewers for their advice. Based on the reviewers' comments and the editor's own evaluation, the paper was accepted for publication pending further revision, which we did. The review and revision process took about a year, and our paper was finally published in 2001. Like the handling of our grant proposal several years earlier, our dealings with *Ecology* were part of the formal peer review of our work.

Does this seem like a long and arduous process? Bill, Steve, and I were motivated by curiosity about whether enough scatterhoards made by kangaroo rats "escaped" being harvested and eaten by their makers or other rodents that Indian ricegrass got a net benefit from scatterhoarding. Why didn't we just satisfy our curiosity and leave it at that? One reason is that we needed money to do the research and one of the conditions of our grant from USDA was to make the results publicly available. Then why not simply create a blog and post the results on the Web? The short answer to these questions is that our motivation included more than just satisfying our curiosity but also a desire to get credit for making a contribution to knowledge. Like all scientists, we wanted recognition for our work, especially from other scientists. By tradition, the source of this recognition is the peer-reviewed, published literature, which in turn is cited by other researchers in their own work. Our paper in *Ecology* became part of an ongoing conversation between those who preceded us in studying Indian ricegrass, desert rodents, and mutualisms between plants and animals and those who followed in studying these and related issues and used our work as part of the foundation for theirs.

Formal peer review as described above has a long history that began in 1665 when Henry Oldenburg became the publisher and editor of the *Philosophical Transactions of the Royal Society of London*, the oldest scientific journal that is still published. Oldenburg sent submitted papers to experts for comments and criticisms before publication, so the origin of formal peer review coincided with the origin of scientific journals as we know them today. Informal peer review must go back much further than this, although details are obscure and individual early scientists probably used informal peer review in different ways.

I briefly described Gregor Mendel's experiments with pea plants between 1856 and 1863 that were the foundation of our understanding of genes, alleles, and basic inheritance (Chapter 7) but didn't describe the social context of this groundbreaking work. Mendel lived and worked in the Augustinian abbey in Brünn (now Brno, in the Czech Republic). He was the epitome of the solitary researcher who worked entirely alone. He published only two papers in his life, the most important of which was called "Experiments on Plant Hybridization" and described his crossing experiments with peas. This paper is a classic in biology not only because of its contribution to understanding of genetics but also because of its rigorous form and style, including statement of an explicit hypothesis, description of the logic of Mendel's methods for testing the hypothesis, and detailed quantitative analyses that were not typical of biological papers at the time. When he completed his pea experiments, Mendel described his results in two talks to the Natural History Society of Brünn, and these talks were published in 1866 as one paper in the *Proceedings of the Society*. These results were largely ignored until 1900. It's easy to imagine that Mendel's isolation at Brno, that is, his apparent lack of involvement in informal peer review, was responsible for this delay in appreciation of his work. But this explanation is inconsistent with three facts. First, his paper was sent to at least 115 other libraries, including major ones in Great Britain where scientists who were intensely interested in the questions addressed by Mendel worked. Second, Mendel arranged for 40 copies of his paper to be printed and he sent these to other scientists. Third, Mendel initiated correspondence about his research with Carl Nägeli, one of the foremost botanists of the nineteenth century. In other words, Mendel was not indifferent to the importance of informal peer review in science, nor was he immune to the desire for recognition that motivates most scientists.

The basic principles of inheritance that Mendel deduced from his experiments with peas were rediscovered by two botanists in their own breeding experiments in the late 1890s. Both published their results in 1900 and credited Mendel as the original source of these principles. Unfortunately, Mendel didn't have an opportunity to appreciate his ultimate success because he died in 1884. In his book *The Growth of Biological Thought: Diversity, Evolution, and Inheritance*, Ernst Mayr suggests several reasons why Mendel's work was ignored for 34 years after publication. The *Proceedings of the Natural History Society of Brünn* where Mendel's paper was published was a regional natural history journal; although it was sent to many libraries in Europe, it probably wasn't read as widely as more prominent journals. Mendel exchanged letters about his work with Carl Nägeli, but Mendel's work was inconsistent with Nägeli's ideas about inheritance, so Nägeli rejected Mendel's conclusions. In fact, Nägeli didn't even mention Mendel's research in the chapter on hybridization experiments in his magnum opus on evolution and inheritance. Finally, Mendel did not promote his work as aggressively as many scientists might: "After having been snubbed by Nägeli, he apparently made no effort to contact other botanists or hybridizers or to lecture at national or international meetings" (Mayr 1982:723).

Consequences of Peer Review

The contents of peer-reviewed journals constitute the primary record of scientific discovery (books also report new discoveries in science, although their role was more important in the past; for example, Darwin's *Origin of Species*, first published in 1859). Thousands of journals publish biological research, and these journals vary in breadth of content (*Journal of Insect Physiology* and *Foot and Ankle Quarterly* to *Bioscience* and *Journal of the American Medical Association*), geographic scope (local or regional to national and international), and selectivity and prestige. For authors of research papers, there are three main outcomes of submission to a peer-reviewed journal: (i) the paper is accepted contingent on satisfactory revisions in response to reviewer comments; (ii) the paper is rejected, and the authors give up trying to get it published; or (iii) the paper is rejected, and the authors revise it for submission to another journal, often one that is less prestigious than the original journal. (Very rarely, a journal editor will accept a paper without requiring revision.)

In my experience and that of my colleagues, revision usually improves papers. Reviewers identify points that aren't clear, arguments that are muddy, and interpretations that aren't justified by the results. Fixing these problems strengthens papers before they are seen by other scientists and the general public.

Giving up after rejection of a paper by one journal is often an unwise move. It's possible that the editor or reviewer identified a fatal flaw that can only be fixed by collecting more data or redoing an experiment with a different design, but the researchers no longer have the resources to do this. In this case, giving up may be the only option. Alternatively the editor may have rejected the paper because the conclusions weren't justified by the results, even though the results were of some interest, or because the authors didn't make a convincing case for the broader significance of the results. For example, an ecological journal with a diverse readership might publish a paper about sea otters that provided new insight about keystone species in general but not a paper that simply contained new information about the natural history of sea otters in a particular location. If this is the reason for rejection by one journal, authors should revise and resubmit the paper to another journal.

These consequences of peer review for authors have implications for the published record of scientific discovery that are important for you as a consumer of research. I've already mentioned that peer review is imperfect. Here are some reasons why: editors and reviewers have biases that can influence their decisions; they may lack knowledge that would be helpful in evaluating a paper; most reviewers and many editors are volunteers, limiting the time that they can spend doing reviews; and most papers are read by only one editor and one to three reviewers who may lack the breadth of knowledge for a comprehensive evaluation. These imperfections in the review process have two main outcomes: flawed research may be published, sometimes in high profile journals like *Science* or *Nature*, and interesting research may be relegated to lower profile journals that aren't as widely read as these, perhaps even to obscure and difficult-to-find journals. I'll use several examples to illustrate these outcomes, starting with interesting research that was treated skeptically by initial editors and reviewers.

Peer Review and Controversial Ideas

You learned a lot about sea otters in Chapter 8, especially about the long-term research of Jim Estes and his colleagues and students in the Aleutian Islands. In 1974 Estes and Palmisano presented the first evidence that sea otters were a keystone species in coastal marine environments, and in 1998 Estes, Tinker, Williams, and Doak presented evidence that killer whales started eating sea otters in the Aleutians, causing the sea otter population to crash. Both of these papers were published in *Science*. This led to the question of why killer whales switched to eating sea otters in the early 1990s, when there was no evidence of this before 1990. Alan Springer and several colleagues, including Estes, proposed the hypothesis of sequential megafaunal collapse (Figure 8.9) to answer this question. The researchers published their paper explaining this hypothesis in the *Proceedings of the National Academy of Sciences* in 2003, but only after it had been rejected by *Science*. As explained by William Stolzenburg, "The *Science* paper came back twice—first with a few encouraging suggestions for revision, the second time with a fatal rejection, appended with a caustic dismissal from one of the reviewers: 'Everybody knows that Steller sea lions starved to death. That's a fact'" (Stolzenburg 2008:74).

Proceedings of the National Academy of Sciences is as prestigious as *Science*, so Estes's group didn't have to settle for an obscure publication in which to present their explanation for why killer whales sequentially depleted populations of harbor seals, fur seals, Steller sea lions, and eventually sea otters in the North Pacific. In fact, Bob Paine, also featured in Chapter 8 as the source of the keystone species concept, was a member of the National Academy of Sciences and promoted publication of the paper by Springer and colleagues in the *Proceedings*. The main point of this example is that peer review may not deal effectively with controversial ideas. As discussed in Chapter 8, the hypothesis of sequential megafaunal collapse of large marine mammals in the North Pacific was controversial because the evidence for it was circumstantial and because it had major implications for management of commercial fishing. With Paine's help, Springer and colleagues were able to publish their hypothesis in a widely read source. Critics wrote their own papers denouncing the hypothesis, Springer's group defended their arguments, and the controversy continues.

Peer Review and the Origin of AIDS

Research on the origin of AIDS offers another fascinating example of the role of peer review in science. AIDS is caused by human immunodeficiency virus (HIV). There are two main types of HIV, 1 and 2, with HIV-1 being the type that has spread throughout the world from its origin in west-central Africa and caused most deaths from AIDS. HIV-1 is related to a similar virus that infects chimpanzees. AIDS was first identified in the United States in 1981, but much later researchers used stored tissue samples to show that two individuals had been infected with HIV-1 in what is now the Democratic Republic of Congo in 1959 and 1960. These are the earliest verified cases of infection with HIV-1, but they don't show how or when HIV became established in humans.

Since HIV-1 is closely related to a form of simian immunodeficiency virus (SIV) found in chimpanzees, scientists generally agree that HIV is a zoonosis, a disease of humans that is acquired when a virus or other parasite that normally infects a non-human animal is transferred to a person. Eventually, the disease organism may evolve to spread directly from person to person, as with HIV. There are two main hypotheses for how humans acquired SIV from chimps that ultimately became HIV-1 in humans. The most widely accepted is the cut-hunter hypothesis. People who live in the forests of west-central Africa kill and eat many different kinds of wild animals, including chimpanzees. A hunter might cut himself while butchering a chimp infected with SIV, resulting in contact between the blood of the chimp and the hunter's wound.

The main alternative to the cut-hunter hypothesis is that SIV from chimpanzees was accidentally transferred to people through contaminated polio vaccine. Hilary Koprowski was a virologist who invented the first live polio vaccine in 1948–1950. In the late 1950s, Koprowski tested his vaccine, which was a weakened form of the virus given orally, in the Congo. Koprowski's team grew the polio vaccine in kidney cells from monkeys at the Wistar Institute in Philadelphia, and advocates of the tainted polio vaccine hypothesis initially suggested that these cells were contaminated with SIV. When research later showed that HIV-1 was derived from SIV specific to chimps, not monkeys, the hypothesis was modified to suggest that Koprowski's team in the Congo had a chimpanzee colony as well as monkeys and used some material from chimps when they made polio vaccine.

The tainted polio vaccine hypothesis for the origin of AIDS was initially developed by Louis Pascal, described by Brian Martin at the University of Wollongong in Australia as "an independent scholar . . . based in New York City." Pascal came up with this hypothesis in 1987 and sent it to 13 scientists for informal peer review, receiving only one brief response. He then sent his manuscript to *Nature*, then to a medical journal called *The Lancet*, and then to the *New Scientist*. All rejected it. Pascal then wrote another manuscript explaining his ideas for the *Journal of Medical Ethics*, which rejected the manuscript because it was much longer than their guidelines. Brian Martin learned about Pascal from a colleague at Australian National University, wrote Pascal, and eventually arranged to publish Pascal's article in 1991 as Working Paper No. 9 in the Science and Technology Studies series of the University of Wollongong.

It would be difficult to find a more obscure outlet for Pascal's paper than this publication, with no disrespect intended to the University of Wollongong. Others, however, were thinking about similar hypotheses at about the same time. A journalist named Tom Curtis proposed the idea that AIDS came from contaminated polio vaccine used by Koprowski in Africa in a popular article in *Rolling Stone* in 1992, an attorney named Walter Kyle published a similar hypothesis in *The Lancet* in 1992, and a writer named Edward Hooper developed the idea at great length in his 1999 book *The River: A Journey to the Source of HIV and AIDS*. In the meantime, Brian Martin has published a series of papers claiming that biologists and physicians have conspired to discredit the tainted polio vaccine hypothesis because it implicates careless researchers as the cause of the worldwide scourge of AIDS.

Although Hooper and Martin are still promoting the tainted polio vaccine hypothesis, recent genetic work has convincingly disproven it. The genetic material of the HIV virus is ribonucleic acid (RNA) rather than DNA, and there's quite a bit of genetic variation in different samples of HIV because mutations accumulate rapidly. Michael Worobey and coworkers extracted HIV from two human tissue samples collected in the Congo in 1959 and 1960 and found that their RNA differed by about 12%. Based on the rate of genetic change of HIV RNA, which can be calculated from comparisons of many samples collected after 1960, Worobey's team estimated that the 1959–1960 samples from the Congo diverged from a common ancestor about 1908. This implies that HIV was present in humans about 50 years *before* the polio vaccinations that Koprowski did in the Congo in the late 1950s; that is, contamination of polio vaccines couldn't have been the origin of HIV in humans.

Worobey's team published their genetic comparison of HIV samples in 2008, but Brian Martin continued to promote the tainted polio vaccine hypothesis for the origin of AIDS as late as 2010 in a paper called "How to Attack a Scientific Theory and Get Away with It (Usually): The Attempt to Destroy an Origin-of-Aids Hypothesis." In this paper, Martin dismissed Worobey's work by stating "Scientists like Michael Worobey . . . have been trying to sink the polio vaccine theory for over a decade. Their arguments are theoretical but they have managed to establish the bushmeat theory [cut-hunter hypothesis] as the dominant view" (Martin 2010:216). This fails a basic criterion of effective argument, that a critic must rebut the specific evidence and logic presented by the other side. Worobey's group had presented empirical data for rejecting the tainted polio vaccine hypothesis, not simply a theoretical argument. For Martin to make a convincing rebuttal, he would have to explain why the data and analyses of Worobey's group were wrong. Instead, Martin ignored these data and analyses and simply mischaracterized Worobey's argument as theoretical. This sentence near the beginning of Martin's paper caused me to be highly skeptical of the remaining 20 pages (see Question 4).

The results of Worobey's group that were described in a paper in *Nature* in 2008 were a smoking gun that caused all but a few diehards to reject the tainted polio vaccine hypothesis for the origin of AIDS, but these results didn't provide similar certainty that the cut-hunter hypothesis was correct. One of the key unanswered questions about the cut-hunter hypothesis is why HIV infection didn't become established in humans until the twentieth century, since Africans probably had been killing and eating chimpanzees for millennia. One explanation involves a constellation of changes that occurred in west-central Africa in the first half of the twentieth century. There was rapid population growth, urbanization, and large-scale movement from small towns in the forest to cities such as Léopoldville (now Kinshasa), which grew from fewer than 10,000 people in 1908 to 49,000 in 1940 to more than 400,000 in 1960. Before World War II when the Belgians controlled this part of Africa, they used Léopoldville as a labor camp where the sex ratio was 10 males for each female and prostitution thrived. Perhaps a hunter living in a small village on one of the tributaries of the Congo River got infected by butchering a chimpanzee and then traveled downriver to Léopoldville to sell smoked fish, ivory, or other forest goods to people in town. He may have visited a prostitute, passing his infection to her, and she may have passed it on to others. David Quammen explains this hypothesis in detail in his 2012 book *Spillover*.

A complementary explanation for establishment of HIV in humans involves vaccination campaigns against tropical diseases like sleeping sickness, syphilis, yaws, and leprosy between 1921 and 1959. These days, doctors and nurses use disposable syringes for injections, but this wasn't the case in Africa in the middle decades of the twentieth century. Glass syringes that were used until the 1920s were expensive and facilities to sterilize them between patients weren't always available to medical workers in Africa. For example, a French doctor treated 5,347 individuals for sleeping sickness between 1917 and 1919 with six syringes. Unlike the tainted polio vaccine hypothesis, this explanation doesn't suggest that medical practices were responsible for transfer of HIV from chimpanzees to humans but that these practices may have aided the spread of HIV among humans once it had been transferred.

Correction of Faulty Results after Publication

Chronic fatigue syndrome (CFS) is a mysterious disease that was first reported in the 1980s. It has a variety of symptoms, most commonly extreme and long-lasting tiredness, but it can also be associated with depressed immune function, increased likelihood of cancer, and early onset of dementia. Extreme cases of CFS are similar to end-stage AIDS. Between 1 million and 4 million people in the United States and possibly 17 million people worldwide have CFS.

One of the biggest mysteries about chronic fatigue syndrome is its cause. Two main hypotheses have been proposed—that an unknown virus causes CFS or that CFS is a psychosomatic condition in which stress or other psychological factors cause the physical symptoms of the disease. Until recently, the Centers for Disease Control and Prevention has emphasized the psychosomatic hypothesis in their research on CFS, although several scientists have looked for associations between CFS and infection with various viruses. Many CFS patients would be relieved if researchers could show that CFS is caused by a virus because it would reduce the stigma attached to being diagnosed with a psychosomatic illness and because it might eventually lead to a treatment.

In October 2009, *Science* published a dramatic article that promised to overturn conventional thinking about the cause of chronic fatigue syndrome. Judy Mikovits of the Whittemore-Peterson Institute in Reno, Nevada, led the team of 13 researchers who did this study. Annette and Harvey Whittemore and Daniel Peterson had started this Institute in 2005, motivated in part by the fact that the Whittemore's daughter was diagnosed with CFS at age 12 and was a patient of Peterson's, who eventually prescribed an experimental antiviral drug that improved the daughter's condition. The Whittemores, other private donors, and the State of Nevada provided much of the funding for the Institute. Annette Whittemore was particularly interested in testing the idea that CFS was an infectious disease. As president of the Institute, she hired the immunologist and AIDS researcher Judy Mikovits as research director in 2006.

Two clues suggested that CFS might be an infectious disease: patients often experience acute, flu-like symptoms before full-scale CFS sets in and there are occasional apparent outbreaks of disease like one near Lake Tahoe between 1984 and 1988 documented by Daniel Peterson. Therefore Mikovits's team initially tested tissue samples from Peterson's patients for several viruses that had been suggested as possible causes of CFS but found none of these viruses in any of the samples.

Mikovits then turned to a virus called XMRV that she learned about from Robert Silverman of the Cleveland Clinic at a meeting near Lake Tahoe in October 2007. XMRV normally infects mice, but Silverman found it in some samples from tumors of prostate glands in humans. XMRV is an RNA virus, like HIV. Also like HIV, XMRV can insert a copy of its genetic material in a chromosome of its host, allowing it to live for a long time in cells of the host. As these cells divide from generation to generation, the genetic material of the virus is carried along like a hitchhiker on the host chromosome. Eventually, the host cells produce new virus particles that exit the cells and can infect new hosts.

Mikovits and her coworkers tested for XMRV in tissue samples from 101 patients with chronic fatigue syndrome and 218 healthy controls. After about a year of lab work, they got their results in November 2008—evidence of the virus in 67% of samples from CFS patients but in only 3.7% of samples from healthy controls. The researchers prepared a paper describing these results and submitted it to *Science* in May 2009. The editors rejected the paper but offered to reconsider a revised version "if the authors could both retain the 'novelty of its main message' and 'address the referees' concerns with new data rather than with counterarguments'" (Cohen and Enserink 2011:1697). Mikovits and her coworkers revised and resubmitted their paper, which was accepted and published by

Science in October 2009. The paper became an instant sensation, and Mikovits became a heroine to many CFS patients, although the president of a patient support group, the CFIDS Association of America, adopted a wait-and-see attitude because of her experience with previous false alarms in the form of other infectious agents initially thought to be related to CFS but later disproven as causes of the disease.

Before describing attempts to replicate Mikovits's results, I want to highlight two aspects of her paper. First, the researchers found XMRV in a much greater percentage of CFS patients than of healthy controls, but a few healthy controls did have evidence of the virus. If the results for healthy controls are representative of the population as a whole, 10 million people in the United States might be infected with XMRV without currently suffering from CFS. This in turn raised the concern that the virus could spread to more vulnerable people through blood donations or organ transplants. Second, Mikovits and her coworkers acknowledged that they had only found a correlation of XMRV with CFS, which didn't necessarily prove that XMRV was the cause of CFS. Another infectious agent or even something like stress might have suppressed people's immune systems, resulting in secondary infection by XMRV.

The discovery of an association between XMRV and chronic fatigue syndrome by Mikovits's team was important for several reasons, fully justifying its publication in a top-tier journal like *Science*. Although not proving that XMRV caused CFS, the discovery implied that this was a strong possibility and set the stage for further research to test causation. The discovery also had significant practical implications—that antiviral medications might help alleviate symptoms of CFS and that public health personnel should act quickly to limit the spread of XMRV through the nation's blood supply. I remember first hearing about Mikovits's work on National Public Radio while driving to my office in the Biology Department at the University of Nevada, at the opposite end of campus from the Whittemore-Peterson Institute at the Medical School. I was proud of the fact the researchers at my institution had made such an important discovery. Like all important research in science, however, the first report is never the last word. Other researchers are eager to replicate the initial results, partly to see if they are really true and partly as a basis for extending the results in their own laboratories, for example, by designing experiments to test whether XMRV is the cause of CFS or just a byproduct of the disease. Unfortunately, the connection between this virus and CFS quickly began to unravel.

The first critiques of the study by Mikovits's group appeared within a few weeks of publication of their paper. These critiques didn't present new data but questioned the assumptions and methods of the Whittemore-Peterson researchers. For example, two cancer virologists at the University of Pittsburgh pointed out that Mikovits had not insured that the members of her team who tested tissue samples for XMRV were blind to the source of those samples—CFS patients or controls. By knowing the source of the samples, the researchers who collected the data may have been biased, that is, more likely to record a questionable result as positive for the virus if it came from a CFS patient than if it came from a control. At about the same time, Simon Wessely at King's College in

London argued that finding the virus in two-thirds of the CFS patients tested was inconsistent with the great variability in symptoms reported for CFS patients.

Wessely is a psychiatrist who has studied CFS for many years, and he teamed up with a virologist to test for XMRV in 186 British patients with CFS. They reported finding no virus in these patients in January 2010. Within a few months, researchers in Germany and the Netherlands also reported negative results. In December 2010, four separate sets of researchers showed that the method used by Mikovits and others to test for XMRV could be compromised by minuscule amounts of contamination by mouse DNA or RNA. Judy Mikovits reportedly described these papers as "Christmas garbage" (Callaway 2012).

In March 2011, Vinay Pathak and colleagues at the National Cancer Institute reported an important discovery about the evolutionary history of XMRV that supported the idea that contamination might have accounted for the dramatic findings at the Whittemore-Peterson Institute 18 months earlier. Although this virus can infect mice, it doesn't naturally occur in mice. However, Pathak's group found two viruses naturally present in mice, each genetically similar to a component of XMRV. The researchers then studied prostate cancer cells that had been maintained in mice since the early 1990s to keep the line of cancer cells alive. By tracing the history of passage of these cells through successive generations of mice, Pathak's team found that XMRV eventually showed up in the cell line. As described by Carl Zimmer:

> The scientists concluded that as prostate cancer cells were injected in mice, viruses migrated from the mice to the cancer cells. Then the viruses were carried to the next mouse inside the cancer cells. At some point during these passages, two different mouse viruses recombined in the human prostate cells to produce XMRV. Thus, rather than being a virus that naturally circulates in humans or mice (or both), XMRV is a virus that emerged in human cell cultures in labs and can contaminate samples that scientists bring into their labs.
>
> (ZIMMER 2011, *see Question 5*)

Based on this research showing how XMRV could have arisen as a laboratory contaminant of prostate cell cultures in mice and on several unsuccessful attempts to replicate the original findings of XMRV in patients with chronic fatigue syndrome, the editors of *Science* retracted the 2009 paper by Mikovits's group. The editors first asked the authors of the 2009 paper to retract it, but only some of these authors were willing to do so. Editor-in-chief Bruce Alberts wrote on behalf of *Science*: "It is *Science*'s opinion that a retraction signed by all the authors is unlikely to be forthcoming. We are therefore editorially retracting the Report. We regret the time and resources that the scientific community has devoted to unsuccessful attempts to replicate these results" (Alberts 2011).

Judy Mikovits was one of the authors unwilling to sign this retraction. She dismissed the contamination hypothesis and argued that no one had exactly replicated their methods, so the results of these replication attempts weren't reliable. In the meantime, Annette

Whittemore fired Mikovits as research director of the Whittemore-Peterson Institute in September 2011, and Mikovits was arrested in November 2011 and charged with stealing notebooks from her lab at the Institute (Mikovits spent four nights in jail before posting bond; eventually the Washoe County District Attorney dropped criminal charges against her). Things looked pretty bad at this point, not just for the infectious disease hypothesis for the cause of CFS but also for the scientific credibility of Judy Mikovits. Amazingly enough, however, there's one more chapter in the story.

Ian Lipkin is an epidemiologist at Columbia University who has found viral causes for several diseases but also has a reputation for "de-discovering" viruses (his terminology), that is, using rigorous methods to disprove new claims for causation of disease by viruses. Based on this experience, Lipkin organized a careful and thorough attempt to test the association between XMRV and chronic fatigue syndrome. He sent blood samples from 147 patients with CFS and 146 healthy controls to researchers at three laboratories and asked each group to test for XMRV by whatever method that they chose. One of these groups included Mikovits and her colleagues, and they used the same methodology as in their initial study. None of the researchers who tested the samples knew the sources of the samples, so their results couldn't be biased by knowing which samples came from CFS patients and which came from healthy controls.

When the researchers reported their results to Lipkin, all three groups found no evidence of XMRV in any of the samples of CFS patients or controls. Mikovits was the second author of the paper reporting these results in the online journal *mBio* in September 2012. She also participated in a press conference on September 18, 2012 announcing the new results, and said, "It's simply not there," referring to XMRV (Enserink 2012). I don't know Judy Mikovits personally, but I'm impressed by her willingness to change her mind about an hypothesis that she had been strongly committed to in response to new evidence. This story illustrates the self-correcting nature of science in a particularly dramatic way because it happened so rapidly and with such intense public interest.

Conclusions

Mikovits and her colleagues published their evidence for a relationship between chronic fatigue syndrome and infection with the XMRV virus in October 2009 and the multi-lab project organized by Ian Lipkin definitively refuted this relationship about three years later. This outcome doesn't mean absolute refutation of the hypothesis that CFS is caused by an infectious agent—another virus or even another type of infectious agent could be responsible for the disease. In addition, although demonstration that XMRV isn't the cause of CFS has disappointed some patients, the research that followed Mikovits's study and culminated in Lipkin's project has spurred greater scientific attention to CFS and may stimulate more research on various hypotheses about causes of the disease.

Not all stories about correction of faulty results in science play out as this one did. The idea that XMRV causes CFS was corrected relatively quickly and definitively. In other cases, resolution of controversial ideas may be delayed or ambiguous or may never happen at all. Unlike this example, in which the main proponent of the XMRV hypothesis eventually conceded, "It's simply not there," in other cases scientists may only accept correction of their faulty results after a long delay or not at all. There are also cases in which results aren't deemed interesting or important enough to try to replicate or in which the rewards to scientists for attempting to replicate previous research aren't as great as the rewards for doing novel research, so we design a new experiment to test a new hypothesis instead of repeating a previous experiment to test the robustness of an old hypothesis (see Appendix 7 for further discussion of replication in science).

Most of this book has been about the various ways that we do science. These include specialized techniques for field studies and laboratory experiments—from radio telemetry to follow the movements of animals in nature, to DNA microarrays to rapidly determine the genotype of an individual from a small sample of its DNA, to sophisticated statistical methods for analyzing huge amounts of data generated in the field or lab. Doing science also involves tools for thinking critically, like being wary of anecdotal evidence and remembering that correlation doesn't necessarily mean causation. I've illustrated these tools in previous chapters and argued that they can be useful to you even if you aren't a scientist. This use of specialized techniques and critical thinking is only half of doing science, however. As explained by Frederick Grinnell, this *discovery process* must be complemented by a *credibility process* to count as a contribution to science. As outlined in this chapter, the credibility process is mediated by social interactions in which scientists discuss and debate their plans and ideas. It also includes formal and informal peer review as well as the responses of other scientists to our work once it is published, as illustrated by the story of XMRV and CFS.

The credibility process is essential to science because individual scientists are subjective and our tools for discovery can't overcome this subjectivity. T. C. Chamberlin was an eminent geologist who published a paper in 1890 called "The Method of Multiple Working Hypotheses" that is still influential today. Chamberlin recognized our inherent subjectivity: "The moment one has offered an original explanation for a phenomenon which seems satisfactory, that moment affection for his intellectual child springs into existence; and as the explanation grows into a definite theory, . . . his intellectual offspring . . . grows more and more dear to him" (Chamberlin 1965:755). As a way to counter this favoritism for a particular hypothesis, Chamberlin suggested that scientists try to overcome their subjectivity by deliberately considering alternative hypotheses. Many scientists in the 120 years since Chamberlin gave his advice have realized that this is a valuable suggestion for enhancing our objectivity but is often difficult to realize in practice. No matter how objective that we try to be, each of us has biases, some of which are unconscious so especially difficult to root out.

Postmodernism is an approach to scholarship developed primarily in the arts and humanities that denies the possibility of objectivity. Various writers have produced postmodernist critiques of science that say essentially that science, like all other human endeavors, is irredeemably subjective. This implies that scientific knowledge is culturally specific. This means, for example, that there is no basis for favoring evidence-based science over astrology. For postmodernists, there is no absolute truth.

In the *Everyday Practice of Science*, Grinnell builds an extended metaphor around a traditional classification of baseball umpires: those who "call balls and strikes as they are," those who "call them as [they] see them," and those who say "what I call them is what they become." Grinnell argues that the first type of baseball umpire represents the traditional view of how science works—by using a linear scientific method, researchers discover truth. He argues that the third type of umpire represents the postmodernist view of science—different scientific theories hold sway at different times because their scientific proponents hold the most power, not because they provide better explanations of how the world works. Grinnell believes the second type of umpire best represents how science really works. Individual scientists, like all humans, are inherently subjective—they call them as they see them. When individual scientists or teams of like-minded scientists present their ideas in seminars or conferences or publish them in the peer-reviewed literature, these ideas are exposed to critical evaluation by other scientists who have different biases. If the ideas hold up under challenge from this community of scientists, they gain credibility. High credibility doesn't produce absolute truth, but it does produce better and better understanding of the natural world because ideas and evidence originally proposed by individuals have met the test of critical review and replication by other scientists. Thus the social processes of science discussed in this chapter are one key to the success of science: "Objectivity of science does not depend on the individual. Rather, objectivity is a function of the community" (Grinnell 2009:17).

Questions to Ponder

1. What is your image of a scientist, and where did it come from? Do you remember a scientist in a movie you've seen recently? How was he or she portrayed?

2. In the third edition of their handbook for students, *On Being a Scientist*, published in 2009, the National Academy of Sciences changed their introductory story from a description of Tracy Sonneborn's discovery of conjugation of paramecia to the following: "Climatologist Inez Fung's appreciation for the beauty of science brought her to the Massachusetts Institute of Technology where she received her doctoral degree in meteorology. 'I used to think that clouds were just clouds,' she says. 'I never dreamed you could write equations to explain them—and I loved it'" (National Research Council 2009). Why do you think the authors of the handbook made this change? This question doesn't have a "right answer" (at least I don't know the real reasons), so use your imagination.

3. Kata used both the US and Canadian versions of Google for her study of Web sites discussing vaccination. I reported the US results, but Kata found that fewer searches of Canadian Google yielded sites opposing childhood vaccination among the top

hits: 13% compared to 24% for searches of US Google. When she expanded her scope to the top 50 hits each for vaccine, vaccination, or immunization for the Canadian search, only 6% of sites opposed childhood vaccination. Speculate on possible reasons for this difference between using US Google and Canadian Google for these searches.

4. In earlier papers, Martin focused mainly on what he considered to be weaknesses of peer review, based largely on the experiences of Louis Pascal. I described Pascal's experience as related by Martin on pages 216–217. Compare this argument of Martin with his argument following publication of Worobey's results.

5. When was CFS first described? Why is this important in relation to the work of Pathak's team on the history of XMRV in mice?

Resources for Further Exploration

1. Understanding Science is a website developed by the Museum of Paleontology of the University of California at Berkeley that is an extremely valuable resource for learning about the processes of science, including the social aspects of science discussed in this chapter. Understanding Science is available at http://undsci.berkeley.edu/. If you search for "peer review" and "replication," you can find definitions, explanations, and interesting examples of these tools for achieving credibility for scientific discoveries.

2. Mendel Web, available at http://www.mendelweb.org/MWtoc.html, is a collection of resources about Gregor Mendel's life and work, including his paper "Experiments on Plant Hybridization" in the original German and in English translation and extensive commentary on this paper and on Mendel's significance in the history of biology.

3. "Patient Zero: The Origins, Risks, and Prevention of Emerging Diseases" is a case study about the emergence of HIV. It features several audio clips of interviews with scientists such as Michael Worobey who studied DNA extracted from tissue samples from the earliest known people infected with HIV in the Congo in 1959. Patient Zero is available at http://sciencecases.lib.buffalo.edu/cs/collection/detail.asp?case_id=697&id=697.

10

Critical Thinking about Climate Change

As you begin this last chapter of *Tools for Critical Thinking in Biology*, I hope that you might be reflecting on what you've learned so far from the book. One of my goals was to use a variety of biological examples to introduce you to the process of science without dodging its complexity. You probably found some of the concepts more challenging than others and some of the examples more interesting than others. Perhaps you were intrigued by experiments on the medicinal value of marijuana but not so much by our mathematical model of the effects of vaccination on the spread of disease (see Question 1).

I had a more important goal in writing this book, however, than simply introducing how science works. I wanted to present some tools for critical thinking that would be useful to you in the future. I used various examples to demonstrate some of these tools in earlier chapters. For instance, critical thinkers need to be wary of confirmation bias, whether in evaluating eyewitness identifications as a juror in a criminal trial or deciding whether to use an anti-inflammatory drug in preparation for the Western States 100-mile endurance run (Chapter 2). In this chapter, I'll reinforce my primary goal for this book by describing several tools for critical thinking and applying them to the common example of global climate change.

I've alluded to climate change in several places in previous chapters, most extensively in Appendix 4 about modeling climate change. Therefore you may have guessed that I think climate change is an important issue. Indeed I do—*the* most important issue at the intersection of science and society today. This is why critical thinking about climate change is critically important—individuals, companies, governments, and other organizations face decisions that will have profound impacts on future generations. These decisions involve not just science but also economics, politics, and especially ethics, so I'll also use this chapter to explore the connections between science and these other perspectives on climate change. Ultimately your own decisions in life will influence how future generations of people will live and whether many kinds of wild plants and animals will persist at all. Because these decisions are so important, I will use this final chapter to highlight six key tools for critical thinking and three key elements of the science of climate change to help you make the best decisions possible.

FiLCHeRS—An Introduction to Six Tools for Critical Thinking

In 1990, James Lett published an article in *Skeptical Inquirer* called "A Field Guide to Critical Thinking." Lett used the acronym FiLCHeRS to stand for six principles of critical thinking: falsifiability, logic, comprehensiveness, honesty, replicability, and sufficiency. "Apply these six rules to the evidence offered for any claim, . . . and no one will ever be able to sneak up on you and steal your belief. You'll be filch-proof" (Lett 1990).

Lett teaches a course on anthropology and the paranormal at Indian River State College in Fort Pierce, Florida. Paranormal beliefs encompass a diverse array of phenomena, from astrology and extrasensory perception (ESP) to ghosts and telepathy. Lett uses FiLCHeRS to show how paranormal beliefs don't satisfy any of the principles of critical thinking. Anthropologists study human cultures; since paranormal beliefs are widespread in US society, such beliefs are a suitable topic for investigation by budding anthropologists like Lett's students. However, Lett's ulterior motive is the same as mine in this book: to explain some critical thinking skills that are valuable for everyone.

I'm going to use mainly examples relating to climate change to illustrate the FiLCHeRS principles because climate change is a more consequential issue than paranormal beliefs, despite the prevalence of the latter. Keeping track of your horoscope is relatively harmless, while promoting illogical claims about climate change contributes to confusion about scientific questions that are well settled, which in turn contributes to paralysis of our political institutions that should deal with climate change.

There is a strong consensus among climate scientists about the reality of climate change caused in part by greenhouse gases added to the atmosphere by human use of fossil fuels. One kind of evidence for this consensus comes from analyses of published, peer-reviewed articles about climate change. In the first such analysis published in 2004, Naomi Oreskes considered 928 articles and found that none disputed the main conclusion of the Intergovernmental Panel on Climate Change (IPCC) that most global warming in the preceding 50 years was caused by increased atmospheric concentrations of greenhouse gases. Most of my examples of flaws in critical thinking about climate change come from those who deny this scientific consensus. The reason for this is not that the existence of a consensus in itself is evidence of sharp critical thinking but that the consensus arose from a long history of research by hundreds of climate scientists and casual dismissal of this research history is unwarranted. As you learned in Chapter 9, the credibility of scientific discoveries depends on critical review and replication by the community of scientists. These scientists are skeptical of new ideas, new arguments, and new evidence about topics in their areas of interest and expertise. As illustrated by the story of chronic fatigue syndrome in Chapter 9, scientists may be especially skeptical about new research by members of their "tribe"—other scientists working on the same research problem, and this skepticism can lead to rapid progress in solving this problem. Since the general consensus among climate scientists arose from a long history of critical review and replication of new results, it exemplifies the process of critical thinking (see Question 2 for a follow-up to Oreskes' study and Appendix 8 for more evidence about the scientific consensus on climate change).

Those who deny the scientific consensus about climate change consider themselves skeptics but aren't truly skeptical when they don't adhere to the principles of critical thinking. As explained by Michael Mann, "Scientists should in fact strive to be skeptics. . . . That is to say, they should always apply healthy scrutiny to any new claim or finding. True skepticism, however, demands that one subject all sides of a scientific contention or dispute to equal scrutiny and weigh the totality of evidence without prejudice" (Mann 2012:26). Unfortunately, many who deny the consensus on climate change are not really skeptics but rather contrarians who practice "a kind of one-sided skepticism that entails simply rejecting evidence that challenges one's preconceptions" (Mann 2012:26).

Falsifiability

The first principle of critical thinking in Lett's field guide is *falsifiability*. This means that it must be possible to *disprove* an hypothesis if the hypothesis is to be a useful basis for research. This may seem counterintuitive—isn't our goal to discover hypotheses that explain how things work in the natural world? If so, shouldn't we focus on evidence that is consistent with the hypothesis, not evidence that disproves the hypothesis? The problem with constructing hypotheses that can't be disproven is that such hypotheses can't be proven either. As Lett explains: "If nothing conceivable could ever disprove the claim, then the evidence that does exist would not matter . . . because the conclusion is already known. . . . This would not mean, however, that the claim is true; instead it would mean that the claim is meaningless" (Lett 1990).

One way that hypotheses can fail to meet the test of falsifiability is to be too general or too vague. I introduced this idea in a different context in Chapter 7. We were interested in the roles of nature and nurture in the development of human traits, but we had to think about how we could measure the contributions of nature and nurture to variation in human traits before we could study this problem. More specifically, we couldn't make progress on the nature–nurture problem until we defined our question more precisely, just as we can't test and potentially falsify hypotheses that are too general or too vague. Astrologers and psychics are fond of hypotheses that are too general or vague to be falsifiable. For example, Jeane Dixon predicted that the "greatest problem" for Whitney Houston in 1986 would be "balancing her personal life against her career" (Lett 1990). How would you test this hypothesis? Is it falsifiable? If not, an acolyte of Dixon could claim that anything that happened to Whitney Houston in 1986 proved Dixon's astrological skill.

Hypotheses can also be nonfalsifiable if their proponents repeatedly respond to evidence against the hypothesis by an excuse for why the hypothesis didn't apply under the circumstances in which the evidence was gathered. This is a favorite tactic of those who believe in ESP. ESP is the idea that some people have a "sixth sense"—an ability to detect things about the world without relying on one of the known sensory systems of seeing, hearing, touching, smelling, or tasting. ESP encompasses phenomena such as dreaming about an unusual event that ultimately happens or thinking about someone with whom you rarely speak who calls you on the phone at that moment. Researchers have studied ESP for at least 150 years, sometimes using rigorous experimental protocols. In one common protocol, a test subject is seated in one room while a person in a different room is given a well-shuffled deck of 25 cards, each with one of five symbols: a circle, cross, star, square, or set of wavy lines. The person with the cards is asked to turn them over one at a time, concentrating intently on the symbol showing on the card. The test subject then attempts to name the symbol on each card. Since there are five choices, the test subject should name about 20% of the symbols correctly just by random guessing. The ESP hypothesis predicts that a person with ESP ability should do much better than this.

Many psychologists in many different laboratories have done this experiment over a period of several decades. Sometimes results are impressive—a test subject might name 15 or more of the 25 cards correctly; many other times, however, performance is no better than expected by chance. In Box 4.1, I used coin tosses as a model to understand sex ratios in litters of common opossums in Venezuela. If I repeatedly flip a fair coin eight times, I'll occasionally get six, seven, or even eight heads, just as occasional litters of opossums will have more than six females. The same is true of experiments with so-called Zener cards to test ESP ability. In the card studies, just as in the opossum experiments, we have to consider the collective results of many trials to see if there is a general pattern of nonrandomness.

Here is where advocates of ESP violate the principle of falsifiability. If a test subject scores no better than expected by chance, believers claim that not all people have ESP ability and this subject is one who doesn't. Unfortunately, there's no way to determine in advance whether a person has ESP ability, so no way to predict his performance on the card test. Here's another example of an excuse for poor performance that believers in ESP might use. Suppose a subject gets 20 out of 25 cards right when tested on Monday. This looks like someone who has powerful ability at ESP, so the subject is invited back for a second test on Tuesday. He only gets four cards right in this test, about as expected by chance. Believers might say the subject did poorly because he was under more stress on Tuesday, because the card turner in the other room wasn't a believer so wasn't sending a strong mental signal, because the room was too hot or too cold on Tuesday, or . . . . By embracing positive evidence but excusing negative evidence about ESP, believers make ESP into an hypothesis that can't be tested because it can never be disproven.

Garrett Hardin explained the importance of falsifiability as follows: "*To win you must be willing to lose:* this is the central principle of the art of disputation. Our chances of persuading others are greatest when we convince the audience that we are willing to lose if our arguments are without merit" (Hardin 1976:465 [emphasis in original]). He continues "By contrast, the antiscientific mind dares not risk refutation: it builds its world-systems out of nonfalsifiable statements" (Hardin 1976:465). In his brief essay, Hardin is mainly concerned with arguing that the fundamental flaw of creationism is that it deals in non-falsifiable hypotheses, in contrast to evolution, which is the central unifying principle of biology (see Appendix 2). Hardin sees the vulnerability of proposing hypotheses that can be falsified as "the strength of science," and he illuminates this strength through vulnerability with a sonnet called "Paradox" by Clarence R. Wylie Jr. (1948).

> Not truth, nor certainty. These I foreswore
> In my novitiate, as young men called
> To holy orders must abjure the world.
> 'If . . . , then . . . ,' this only I assert;
> And my successes are but pretty chains
> Linking twin doubts, for it is vain to ask
> If what I postulate be justified,
> Or what I prove possess the stamp of fact.
>
> Yet bridges stand, and men no longer crawl
> In two dimensions. And such triumphs stem
> In no small measure from the power this game,
> Played with the thrice-attenuated shades
> Of things, has over their originals.
> How frail the wand, but how profound the spell!

Wylie's sonnet links the process of science to advances in technology that result from this process (see Question 2). Although this process is a "frail . . . wand," it has enabled engineers to design and build bridges and airplanes and all the other technological marvels of our modern world. Of course modern technology depends on abundant energy, much of which is derived from burning fossil fuels. How about the hypothesis that our use of fossil fuels contributes to global warming and other changes in climate? Is this hypothesis falsifiable?

Anthropogenic global climate change is a complex hypothesis with many components that can be tested (and potentially falsified) in various ways. As you saw in Appendix 4, global climate models play an important role in climate science. These models use equations to represent key chemical and physical processes in the atmosphere, oceans, and terrestrial environments. Climate scientists run computer experiments to test different versions of these models. Plate 17 shows results of a test in which they started model runs in 1860 and examined the match between measured temperatures from 1860 to

2000 and average global temperatures predicted by three versions of the model. Only a model that included both natural and anthropogenic factors matched the observed temperature record.

This approach illustrated in Plate 17 uses *hindcasts*—"predictions" of the past rather than forecasts of the future. Skeptics argue with some justification that hindcasts matching observed data aren't as credible as forecasts, because modelers may tweak their models to match known past results when they do a hindcast. However, climate modelers have also tested forecasts of average global temperature. Figure 10.1 shows such a test in which several climate models developed by different groups of researchers were run in 2000 and the predictions of these models were compared to actual temperature data through 2012. By running each model multiple times, researchers generated a 95% confidence interval of predicted temperature, and the observed temperature record was well within this range for the 2001–2012 forecast.

There are several signals of contemporary global climate change in addition to increasing average global temperature. One of the most dramatic of these signals is the minimum size of the ice pack in the Arctic Ocean, which occurs in August or September of each year (Figure 10.2). Models that predicted average global temperature from 2001 to 2012 accurately (Figure 10.1) didn't do so well for minimum Arctic ice, which fell *below* predicted values in 2009–2012. In other words, the model was too conservative: with

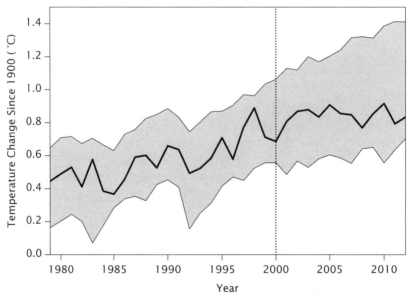

FIGURE 10.1 Gavin Schmidt's comparison of modeled and measured global average temperature from 1979 to 2012. Schmidt compared global temperature reported by the US National Climatic Data Center (heavy black line) to the 95% confidence interval of predictions of a set of global climate models run in 2000 (gray area). Temperature is measured as the deviation in degrees Celsius from the 1880–1920 average. For example, average global temperature in 1998 was about 0.9°C (1.6°F) above that at the beginning of the twentieth century. Model predictions for 2001–2012 (right of dotted line at 2000) are forecasts; "predictions" for 1979–1999 (left of dotted line) are hindcasts.

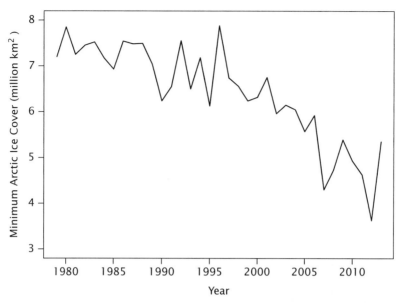

FIGURE 10.2 Minimum extent of ice cover in the Arctic Ocean from 1979 to 2013. The minimum occurred in September of each of these years. Ice cover is estimated from satellite measurements.

respect to Arctic sea ice; modelers were missing an element of climate change (see Question 3).

These are not ideal tests of models of anthropogenic climate change because the models project long-term trends and 12 years is too short a time period for a rigorous test. As described in Appendix 4, the models make projections for the next century under various scenarios, ranging from business as usual (with a growing human population and carbon-intensive economy) to rigorous and effective worldwide programs to limit emissions of greenhouse gases. If modeling global climate were purely an intellectual exercise, we might test our models by asking our children and grandchildren to collect climate data for the next 90 years and compare these results with model projections for 2100. Unfortunately, understanding climate change is not just an intellectual exercise but fundamentally important for the kind of lives our children and grandchildren will lead. So we can't wait to make decisions about dealing with ongoing climate change. Instead, we have to rely on less than ideal tests of models, like the hindcasts in Plate 17, the short-term forecast in Figure 10.1, and other evidence about the causes and consequences of climate change.

Models are just one component of the hypothesis of anthropogenic climate change. They satisfy the principle of falsifiability, although testing these models is challenging. Global climate models are also difficult for nonexperts to appreciate, much less understand. There are other ways, however, that anthropogenic climate change might be tested and potentially falsified. Here are two examples.

As described in Appendix 4, researchers have documented increasing concentrations of carbon dioxide (CO_2) in the atmosphere since 1958 by direct measurements at almost 14,000 feet elevation at the top of Mauna Loa, Hawaii. These measurements can be extended back in time by studying ice cores drilled from the ice packs on Greenland and Antarctica. Because of seasonal deposition patterns of ice, these cores can be dated. The ice also contains bubbles of air that were trapped when the ice formed, and researchers can measure CO_2 in these bubbles. Amount of CO_2 in the atmosphere was relatively constant from 1000 to 1850, and then it began to increase at an accelerating rate coincident with human use of fossil fuels (Figure 10.3). As you know, correlation doesn't necessarily mean causation, but there's an extra piece of evidence that confirms the fact that use of fossil fuels caused the increase in CO_2 in the atmosphere since the beginning of the

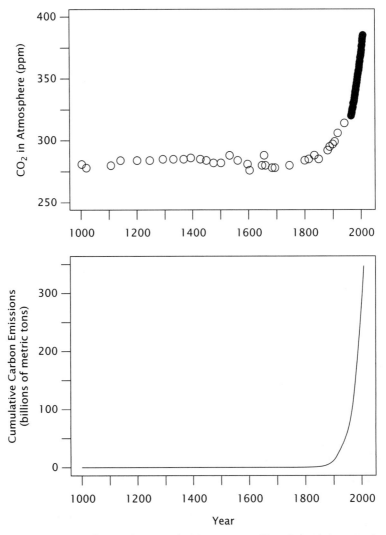

FIGURE 10.3 Concentration of CO_2 in the atmosphere in parts per million. Open circles are estimates from air bubbles trapped in ice cores from Antarctica, and closed circles are direct measurements at Mauna Loa, Hawaii (a); cumulative emissions of carbon from human use of fossil fuels (b). Ice core data from Figure SPM-10a of IPCC Third Assessment Report, Summary for Policymakers; Mauna Loa data and emissions data from Carbon Dioxide Information Analysis Center of the US Department of Energy.

Industrial Revolution. There are two naturally occurring, stable isotopes of carbon on Earth. These isotopes are forms of carbon atoms that both contain 12 protons in their nuclei but have different numbers of neutrons: 12 for the most common isotope and 13 for the less common one. Green plants use CO_2 from the atmosphere to grow, but they selectively use the isotope with 12 neutrons. Since fossil fuels are derived from plants, the CO_2 added to the atmosphere when humans burn these fossil fuels has more of the carbon isotope with 12 neutrons and less of the isotope with 13 neutrons than the atmosphere from which the plants extracted CO_2. Therefore the hypothesis of anthropogenic climate change predicts that the proportion of the 12-neutron isotope of carbon in the atmosphere increases as emissions of CO_2 from burning fossil fuel increases. Researchers have collected these data since 1980 and verified this prediction.

Second, the hypothesis of anthropogenic global warming by our enhancement of the greenhouse effect predicts increased temperature in the lower portion of the atmosphere, called the troposphere, but decreased temperature in the stratosphere, above the troposphere. If warming of Earth were due solely to increased solar radiation, both troposphere and stratosphere would get warmer (see Question 4). Measurements using satellites and weather balloons show warming of the troposphere and cooling of the stratosphere between 1958 and 2005, consistent with anthropogenic global warming but not with the alternative hypothesis that recent warming of the Earth was due just to increased solar radiation.

In the last few paragraphs, I've outlined several tests of the hypothesis of anthropogenic climate change. These tests involved comparing changes in average global temperature to predictions of climate models, documenting human contributions to increased greenhouse gases in the atmosphere, and showing that recent temperatures increased in the lower atmosphere but not the upper atmosphere, consistent with the greenhouse effect. The hypothesis of anthropogenic global climate change is clearly falsifiable, meeting Lett's first criterion of critical thinking. Furthermore, the hypothesis passed the tests described in the last few paragraphs, that is, it was not falsified by these tests.

Logic

Scientists are motivated by a combination of curiosity and a desire to solve practical problems, but their raison d'être is to find successful explanations of natural phenomena. The success of an hypothesized explanation in science depends on the weight of evidence supporting the hypothesis (or refuting alternatives) and on the soundness of arguments connecting the evidence to the hypothesis. How is evidence connected to an hypothesis? Once we have a clear and concrete statement of our hypothesis, we deduce one or more predictions that we test by observations or experiments that provide evidence. The structure of our argument is this: *IF* our hypothesis and supporting assumptions are true, *THEN* we predict that our observational or experimental test should produce specified results, that is, evidence. If we do the test and don't get the predicted results, then we should reject the hypothesis (or reconsider our assumptions). If we do get the predicted results, that should increase our confidence in the truth of the hypothesis, although doesn't prove the hypothesis because an alternative hypothesis might make the same predictions.

This depends on having a *valid* and *sound* argument in which the conclusion follows logically from the premises (making the argument valid) and the premises are true (making the argument sound). Lett gives a simple illustration of an invalid argument: *IF* all dogs have fleas and Xavier has fleas, *THEN* Xavier is a dog. But suppose Xavier is a cat. Then the two premises *might* be true—all dogs have fleas and Xavier has fleas, but the conclusion is clearly false.

We can make this a valid argument by changing it slightly: *IF* all dogs have fleas and Butch is a dog, *THEN* Butch has fleas. This is a logically valid argument because the conclusion follows from the two premises. This doesn't make it a *sound* argument, however, because at least one of the premises is still false. Butch may well be a dog, but some dogs don't have fleas.

Here's a more meaningful example (unless you're a veterinarian). The average annual temperature at the surface of the Earth has increased by about 0.4°C (0.7°F) since 1979 (Figure 10.1). This rate of increase is unique for the period since 1850, when meteorologists started recording direct measurements with thermometers over a large portion of Earth's surface. Written reports of things like growing conditions for crops provide evidence for earlier times when thermometers weren't widely used. These reports and other evidence show that Europe and parts of North America were unusually warm from about 950 to 1250. Unlike now, the southern and western coasts of Greenland were free of ice during summer in this Medieval Warm Period, and Norwegian explorers who had colonized Iceland moved north to Greenland as well, which became a stepping stone on their voyages of discovery to North America.

One common argument of climate change contrarians is that our current warm temperatures might simply be due to natural variability rather than human modification of climate by adding CO_2 to the atmosphere. There is natural variability in the amount of solar radiation hitting Earth due to cyclic changes in Earth's orbit as well as changes in sunspot activity. We can construct a valid argument that natural variability in solar radiation caused elevated temperatures during the Medieval Warm Period if we assume that the only two causes of long-term variation in climate are human use of fossil fuels and natural factors such as variation in solar radiation This is a valid argument because the Medieval Warm Period occurred long before the Industrial Revolution when we started burning fossil fuels (Figure 10.4a). It's not a sound argument, however, because burning fossil fuels is not the only way humans influence climate. We also do so by cutting and burning forests to clear areas for growing crops, and this started long before the Industrial Revolution.

Climate change contrarians go further. They claim that it was just as warm during the Medieval Warm Period as today, therefore contemporary warming might be a natural process as well (Figure 10.4b). This is an invalid argument because it simply

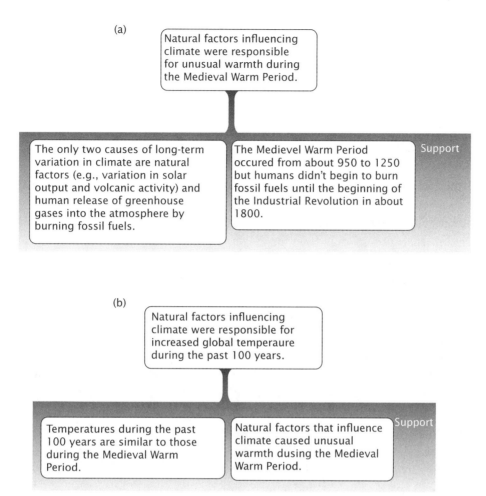

(a)

Natural factors influencing climate were responsible for unusual warmth during the Medieval Warm Period.

The only two causes of long-term variation in climate are natural factors (e.g., variation in solar output and volcanic activity) and human release of greenhouse gases into the atmosphere by burning fossil fuels.

The Medievel Warm Period occured from about 950 to 1250 but humans didn't begin to burn fossil fuels until the beginning of the Industrial Revolution in about 1800.

Support

(b)

Natural factors influencing climate were responsible for increased global temperaure during the past 100 years.

Temperatures during the past 100 years are similar to those during the Medieval Warm Period.

Natural factors that influence climate caused unusual warmth dusing the Medieval Warm Period.

Support

FIGURE 10.4 Two simple arguments about causes of elevated temperatures during the Medieval Warm Period and today. Argument (a) is *valid* because, if both premises are true, the conclusion is true, but not *sound* because the premise that the only two causes of long-term variation in climate are natural factors and human use of fossil fuels is false. Argument (b) is neither valid nor sound (see text for details).

assumes that the same result—elevated temperatures during the Medieval Warm Period and today—must have the same cause. But there are many situations in which different causes produce the same result. For example, one forest fire in the Sierra Nevada might have been caused by lightning while another was human caused, either accidentally or on purpose.

Even if the argument that contemporary warming is caused by natural processes based on comparison with the Medieval Warm Period could be made logically valid, it wouldn't be a sound argument because one of its premises is false. Humans didn't use thermometers 1,000 years ago, but temperature can be estimated from data such as the widths of annual growth rings of trees on land and corals in the oceans, the types of pollen found in lake and ocean sediments, and the chemical composition of air bubbles in ice cores extracted from Greenland and Antarctica. Several research teams have used these proxy data to extend estimates of average global temperature back 1,000 years, contradicting the assumption that current global average temperatures are comparable to those during the Medieval Warm Period. In fact, the past 50 years have been warmer than any time during the last millennium, including the Medieval Warm Period (Plate 18). I'll describe this evidence in more detail when I discuss replicability on pages 239–240.

Comprehensiveness

Comprehensiveness means considering the full array of evidence bearing on a set of hypotheses, as we did in Chapter 8 when we discussed reasons for the crash of the sea otter population in the Aleutian Islands in the 1990s. Critical thinking depends on comprehensiveness because evaluating the weight of evidence helps us avoid confirmation bias (see Chapters 2 and 9).

One of the favorite tactics of climate change contrarians is selective use of evidence, which violates the principle of comprehensiveness. During bouts of unusually cold weather, politicians in the contrarian camp provide some of the most egregious examples of this abdication of critical thinking. For example, during a heavy snowstorm in Washington, DC, in February 2010, the family of James Inhofe, a contrarian US senator from Oklahoma, built an igloo on the National Mall with a sign on one side saying "Honk if you ♥ global warming." This approach has superficial plausibility, since some hypotheses can be refuted by a single counterexample. For instance, if I hypothesize that "all swans are white" and you show me a single black swan, then I must abandon my hypothesis. The hypothesis that human activities contribute to a warming climate is a very different kind of hypothesis, however. Climate is influenced by many factors, both natural and human caused, and climate is inherently variable, so one or more bouts of cold weather don't refute a general pattern of increasing average temperature.

Furthermore, Inhofe's igloo doesn't just illustrate selective use of evidence but also failure to understand the difference between weather and climate. Weather is short-term, local variation in temperature, rainfall, and other atmospheric conditions from hour to hour and day to day—what weather forecasters predict on all the news media. Climate, by contrast, is long-term variation in average daily conditions at regional or global scales. One week of cold weather in one location like Washington does not refute the hypothesis of anthropogenic global climate change, which makes predictions about long-term and large-scale patterns of change.

Climate change contrarians use evidence selectively in another way. They focus on trying to refute the evidence that global air temperatures are increasing but often ignore other evidence of climate change such as melting of mountain glaciers, reduced sea ice in the Arctic Ocean, melting of permanently frozen soil at high latitudes, increased temperature of ocean waters, increased frequency and intensity of droughts in some parts of the world, and more and bigger forest fires.

Honesty

Honesty is the fourth and most fundamental of Lett's principles of critical thinking, because it is the basis for all the others. Lett's definition of honesty as an element of critical thinking is that "the evidence offered in support of any claim must be evaluated without self-deception."

We've seen a series of record-breaking temperatures in the past 30 years. According to the National Climatic Data Center, the 10 warmest years since 1880 occurred between 1998 and 2012, and the hottest decade overall was 2000–2009. However, a graph of global average temperature shows an increasing trend from about 1979 to 1998 followed by an apparent plateau since then (Figure 10.1). Climate change contrarians have interpreted this plateau as evidence that global warming has stopped. This illustrates dishonesty as well as selective use of evidence. Why do you suppose contrarians focus on 1998 as a starting point? It's convenient that 1998 was an unusually warm year—the third warmest on record after 2010 and 2005. Since global average temperature doesn't increase steadily year after year but fluctuates around an increasing trend, starting at a high level like 1998 leaves less scope for the next several years to show higher temperatures (see Question 5). In fact, there's a good reason in addition to climate change why average temperature was especially high in 1998—it was a strong El Niño year, as were 2005 and 2010. El Niños occur periodically when surface waters of the Pacific Ocean off the coast of South America get warmer. These changes, in turn, have effects on global climate, causing flooding in some parts of the world, droughts in other regions, and an overall increase in global temperatures. El Niños occur naturally, although their frequency and intensity may be affected by climate change. By highlighting an apparent plateau in global average temperature after 1998, climate change contrarians use a segment of the recent record of global temperature that masks the strongly increasing trend that is obvious when looking at a longer time period (Figure 10.1). This selective use of evidence is dishonest because contrarians don't acknowledge that their starting point was unusually warm because of the 1998 El Niño.

Here's another illustration of dishonesty by climate change contrarians. Justin Gillis is a reporter who writes about climate change and other environmental issues for the *New York Times*. He wrote a story in the *Times* on June 10, 2013 called "What to Make of a Warming Plateau." Gillis described the situation as follows: "The rise in the surface temperature of earth has been markedly slower [recently] . . . even as greenhouse gases have accumulated in the atmosphere at a record pace. The slowdown is a bit of a mystery to climate scientists." On the same day that Gillis's story appeared, Noel Sheppard blogged about the story for *NewsBusters*, highlighting the above sentences at the beginning of Gillis's story. Several other bloggers did the same. None of these climate change contrarians, however, mentioned the nuance in the rest of Gillis's story, where he pointed out that "scientists . . . reject . . . selective use of numbers, . . . conclud[ing] that it continues to warm through time." Gillis also reported that there is good evidence that the additional heat accumulating on Earth due to anthropogenic global climate change is warming water deep in the oceans, so is not reflected in current measurements of surface temperature. Gillis ended his story with a reminder that previous apparent pauses in global warming have been followed by rapid increases in temperature and that we can expect stronger and more frequent bouts of extreme weather as a consequence of climate change. Contrarian bloggers either ignored these parts of Gillis's news story or dismissed them without explanation.

Replicability

Science involves a continuous cycle of making and reporting new discoveries and testing the credibility of these discoveries. As you saw in Chapter 9, attempts to replicate new results are a key process in testing credibility, which failed in the case of chronic fatigue syndrome because several labs could not reproduce the original finding by Judy Mikovits and her colleagues of a virus in tissue samples of chronic fatigue patients. In the logic section of this chapter, I introduced the faulty claim that because the Medieval Warm Period resulted from natural variation in climate, contemporary warming must also result solely from natural variation in climate. Climatologists as well as climate change contrarians have been interested in past climatic conditions such as the Medieval Warm Period, but climatologists haven't been satisfied with historical reports. Instead, they've tried to estimate actual temperatures using proxy data as described on page 236.

Michael Mann and two colleagues published the first estimates of global average temperatures based on proxy data for the past 1,000 years in 1999 (Plate 18). Plate 18 soon became known as the hockey stick graph because it is shaped like a hockey stick lying on its side with the blade pointing up at the right. The hockey stick was controversial among scientists for three main reasons: (i) it involved combining different kinds of proxy data—tree rings, corals, lake sediments, and ice cores—to draw inferences; (ii) these data were sparse for early time periods in the last millennium; and (iii) the authors did some fairly complex statistical analyses to get a meaningful overall picture. The hockey stick was especially controversial for climate change contrarians because it undermined the justification for their argument that the Medieval Warm Period was just as warm as today. Contrarians challenged the hockey stick on various grounds, ranging from alternative statistical analyses (later shown to be incorrect) to accusations of fraud on the part of Mann and his coauthors (also thoroughly refuted).

The story of the hockey stick is a long, complex, and interesting one. For several years, climate change contrarians focused on it more than was justified by its importance, since there are many other manifestations of human influences on contemporary climate besides global average temperature. For illustrating how science really works, however, the most important aspect of the hockey stick story has been attempts by other groups of scientists to replicate the results of Mann and his colleagues. As with chronic fatigue syndrome, scientists are especially motivated to try to replicate controversial and potentially significant discoveries. For the hockey stick, this has meant gathering more proxy data to extend the analysis further back in time and provide denser coverage of Earth's surface and using refined statistical techniques to test the reliability of estimated patterns of global average temperature. All of the attempted replications since 1998 have shown higher temperatures today than during the Medieval Warm Period. The most recent and most extensive attempt at replicating and extending the results of Mann and his colleagues was published in April 2013 by a group of 78 researchers from 24 countries who used 511 data sets involving sediments, ice cores, tree rings, corals, stalagmites, pollen, historical documents, and direct measurements. This group concluded that "there were no globally synchronous multi-decadal warm or cold intervals that define a worldwide Medieval Warm Period or Little Ice Age," but there has been a long-term cooling trend for the past 2,000 years, only reversed in recent decades. From 1971 to 2000, "average reconstructed temperature was higher than any other time in nearly 1,400 years."

Sufficiency

James Lett's final principle of critical thinking is sufficiency, which he describes as "the evidence offered in support of any claim must be adequate to establish the truth of that claim." Lett suggests that believers in paranormal phenomena ignore this principle when they claim that their belief must be true unless every single manifestation of that belief has been disproven. For example, suppose that I believe in unidentified flying objects (UFOs). If you are a good skeptic, then you give several examples of supposed UFOs that were shown to have natural, earthly causes. I reply that not all UFO sightings have been explained by natural, earthly causes, so at least some UFOs piloted by aliens must exist. In other words, the evidence that I'm using to support my belief in UFOs is that not all UFO sightings have been disproven. This isn't very convincing evidence for my belief, especially since my belief is quite dramatic. As Lett says, "Extraordinary claims demand extraordinary evidence."

Arguments of contrarians violate the principle of sufficiency by promoting one-sided skepticism about the scientific consensus on climate change. As illustrated by Wylie's poem on page 230, absolute certainty isn't possible in science. Nevertheless, there may be sufficient evidence supporting an hypothesis that we can be justified in believing the hypothesis, just as virtually all climatologists are confident that humans contribute to global climate change by burning fossil fuels. Denial of this consensus is an extraordinary claim, demanding extraordinary evidence.

One of the favorite tactics of climate change contrarians is to highlight uncertainties in the evidence for climate change. They selectively cite individual pieces of information to claim that anthropogenic climate change is false, while ignoring the large body of evidence to the contrary. For example, contrarians in recent years have been fixated on the supposed plateau in average global surface temperature since 1998, while ignoring the evidence that deep ocean waters have absorbed most of the excess heat gained by the Earth during this time period because of the greenhouse effect.

Naomi Oreskes and Erik Conway described these tactics in their 2010 book *Merchants of Doubt*. Highlighting the uncertainties that occur in any scientific field is part of a larger strategy to undermine the science, in part by sowing confusion in the general public. So far this strategy has been successful in the case of climate science: despite the overwhelming consensus among climate scientists about the basic facts of anthropogenic climate change, surveys consistently show that many US citizens either don't understand or doubt these facts.

Oreskes and Conway also describe the long history of this strategy of encouraging doubt about well-established scientific ideas, starting with the campaign by the tobacco industry to cast doubt on the hypothesis that cigarette smoking causes lung cancer. There have been similar campaigns against the ideas that emissions from power plants caused acid rain, that compounds used in refrigerators and aerosol cans caused depletion of the protective layer of ozone in the atmosphere, and that DDT caused a variety of environmental and health problems. Industry groups funded these campaigns to avoid regulations that might depress their profits. The most interesting things discovered by Oreskes and Conway in their extensive historical research were that some individuals with impressive scientific credentials and backgrounds were involved in developing and implementing each of these campaigns. For example, Frederick Seitz, William Nierenberg, and Robert Jastrow founded the George C. Marshall Institute in 1984. All were distinguished physicists, as well as staunch anti-communists and supporters of free enterprise. Seitz worked for the R. J. Reynolds Tobacco Company from 1979 to 1985, despite having no background in medical science or epidemiology. Nierenberg chaired a panel created by President Reagan to evaluate the risks of acid rain despite no experience in ecology. Jastrow, Seitz, and Nierenberg published a report in 1989 claiming that increased global temperatures in the twentieth century were due to increased output of solar radiation, with increased greenhouse gases playing no role. This report was published by the Marshall Institute without peer review; had it gone through the standard review process, a reviewer might have noticed that Jastrow and colleagues based their claim on a misrepresentation of the results of climatologist James Hansen and colleagues published in *Science* in 1981. Hansen's group compared empirical data on global warming through 1980 to predictions of climate models. They considered three factors—volcanic activity, variation in solar radiation, and increases in CO_2 in the atmosphere. They found a good fit to empirical data for a model with all three factors plus an ocean in which there was deep exchange of heat with the atmosphere (Figure 10.5). In their report, Jastrow's group ignored this conclusion and showed only the results of a model with a shallow ocean and increase in atmospheric CO_2, but no variation in volcanic activity or solar radiation. From cherry-picking the results in this way (top left panel of Figure 10.5), Jastrow's group concluded that there was little evidence for an influence of greenhouse gases on global climate. In fact, Hansen's team had shown that predictions of a model

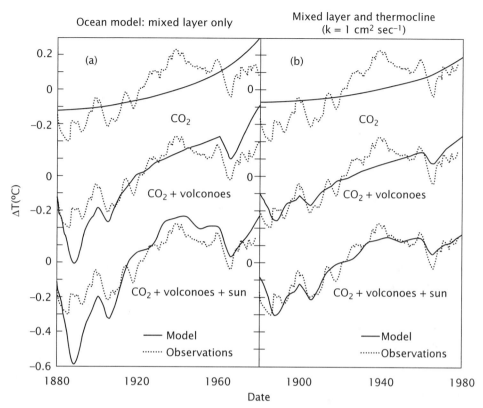

FIGURE 10.5 James Hansen and colleagues used this figure to illustrate an early comparison of a global climate model to actual global temperature from 1880 to 1980. They tested three versions of the model. One included only increased CO_2 in the atmosphere due to use of fossil fuels, one included increased CO_2 plus variation in volcanic activity between 1880 and 1980, and one included these two factors plus variation in output of solar energy between 1880 and 1980. Each of these versions was paired with either limited, shallow exchange of heat between the atmosphere and the oceans (a) or deep exchange (b). Which version of the model shows the closest match to the observed temperature record?

incorporating greenhouse gases in combination with volcanic activity and variation in solar output closely matched global average temperature between 1880 and 1980 (bottom right panel of Figure 10.5). This conclusion has been confirmed by more recent analyses (Plate 17; see Question 6).

The "merchants of doubt" who have promoted uncertainty about climate science in the general public have included business leaders wanting to maintain profits, whether from selling cigarettes or fossil fuels, as well as scientists, sometimes but not always motivated by partisan agendas. As described by Michael Mann, these are the professionals, but the community of climate change contrarians also includes a host of amateurs, from talk radio hosts to newspaper columnists to bloggers. In particular, the tremendous growth of the Internet has given sustenance to the contrarian movement, just as for the anti-vaccination movement described in Chapter 6. I'm asking you to be a true skeptic who critically considers all available evidence about climate change and other important scientific issues. As Michael Mann explained, "True skepticism . . . demands that one subject all sides of a scientific contention or dispute to equal scrutiny and weigh the totality of evidence without prejudice" (Mann 2012:26). Oreskes and Conway are optimistic about this eventually happening for climate change, although worried that too much damage will have been done before it happens:

> Of the many cases of doubt-mongering that we have studied, most ended for the better. At a certain point, the companies manufacturing chlorofluorocarbons (CFCs), admitted their link to ozone depletion and did the right thing by committing to phasing them out. The public is now firmly convinced of the link between cigarettes and cancer. Inductive reasoning implies that the same should happen with climate change: the consensus scientific view will eventually win public opinion. But in the meantime irreversible damage is being done—to the planet, and to the credibility of science.
>
> (ORESKES AND CONWAY 2010:687)

Three General Principles of Climate Change

The fifth assessment report of the Intergovernmental Panel on Climate Change was published in 2013–2014. This is a consensus document involving multiple stages of writing and review by large teams of scientific experts and, ultimately, representatives of countries that are members of the United Nations. This report, just like the first four reports, includes sections on the physical science of climate change; impacts, adaptation, and vulnerability; and mitigation of climate change. It also includes a synthesis report with a summary for policymakers. Parts of the IPCC Fifth Assessment Report describe and discuss projections of climate changes over the next century based on various scenarios about human population growth, economic activity, and actions to mitigate climate change by reducing use of fossil fuels. Because of the consensus nature of the process, including buy-in by representatives of a broad diversity of world governments, projections are likely to be conservative. For example, the Fourth Assessment Report published in 2007 projected minimal increases in sea level because it did not include contributions from melting of the ice caps on Greenland or Antarctica. By 2013, sea level had already risen 80% more than projected by the Fourth Assessment Report.

To keep informed about the evolving science of climate change, you'll need to deploy the tools for critical thinking that I've outlined in this chapter. It will also be important to understand three general principles that are fundamental to how climate change will play out over the rest of your life. These are the principles of inertia, feedback, and tipping points; here is a brief explanation of how these principles apply to climate change.

Inertia

A dictionary definition of inertia is the resistance of an object to change. How does inertia affect the climate system? Four key greenhouse gases are CO_2, methane, nitrous oxide, and water vapor. The two most abundant of these in the atmosphere are water vapor, with an average concentration of about 10,000 parts per million, and CO_2, with a concentration of about 400 parts per million. As you've learned, humans influence the concentration of CO_2, as well as methane and nitrous oxide, by burning fossil fuels and cutting forests. We don't, however, directly influence the concentration of water vapor. The basic reason for this is that water cycles rapidly between the atmosphere, oceans, and terrestrial environments (including lakes and rivers) by evaporation and precipitation. The average residence time for a molecule of water in the atmosphere is only nine days. Imagine a water molecule that evaporates from the middle of the Atlantic Ocean. Once in the atmosphere, it will be carried by winds for several days but eventually fall back to Earth in rain or snow. For some molecules at some times, this will happen in fewer than nine days, in other cases it will take more than nine days, but the average residence time for a water molecule in the atmosphere is nine days.

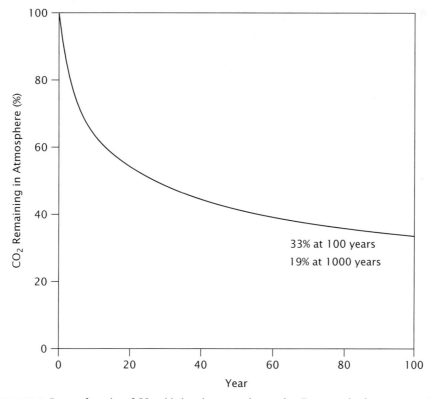

FIGURE 10.6 Decay of a pulse of CO_2 added to the atmosphere today. For example, the average resident of the United States emitted 17 tons of CO_2 in 2011, more than residents of any other country except Australia, Saudi Arabia, and the United Arab Emirates. Of these 17 tons, 5.6 tons (33%) will still be present in 2111 and 3.2 tons (19%) will still be present in 3111.

By contrast, some of the CO_2 added to the atmosphere by burning fossil fuels stays in the atmosphere for a very long time. There is exchange of CO_2 with ocean water that begins immediately, but about one-third of a pulse of CO_2 added to the atmosphere today will still be present in 100 years and 19% will still be present in 1,000 years (Figure 10.6). This represents strong inertia in atmospheric concentration of an important greenhouse gas. In simple terms, our actions today will have consequences for centuries.

Feedback

The second general principle that's important for understanding climate change is feedback. A dictionary definition of feedback is that the results of some process influence the process itself. There are two basic forms of feedback, negative and positive. Negative feedback promotes stability, while positive feedback promotes runaway effects, or instability. Our bodies have several examples of both negative and positive feedback. Levels of glucose in the blood are normally regulated within a narrow range by negative feedback. If blood glucose drops, a hormone called glucagon is released by the pancreas. Glucagon travels to the liver and causes liver cells to convert glycogen to glucose, raising the level of glucose in the blood. If blood glucose increases above the equilibrium level, after eating for example, insulin is released by the pancreas. Insulin stimulates liver, muscle, and fat cells to take up glucose from the circulating blood, restoring equilibrium. By contrast, cancer involves positive feedback when genetic changes in cells cause them to produce excessive amounts of growth factors, leading to runaway cell division.

One basic form of feedback in the climate system involves water vapor. As we burn fossil fuels, we increase atmospheric concentrations of CO_2 and other greenhouse gases, leading to warmer temperatures. This in turn increases the rate of evaporation of water into the atmosphere, increasing the greenhouse effect. As I wrote above, human activity doesn't *directly* influence the concentration of water vapor in the atmosphere, but it does have this indirect effect through positive feedback.

The positive feedback of water vapor on global warming is complicated by clouds. More water in the atmosphere contributes to formation of clouds and the effect of these clouds depends on where they form. High-level clouds enhance the greenhouse effect by trapping long wave infrared radiation emitted from Earth's surface, while the tops of low-level clouds reflect some incoming sunlight back to space, counteracting some of the greenhouse effect. One of the biggest uncertainties of global climate models has been whether increasing greenhouse gases will cause more low-level or high-level clouds. If most clouds are at low levels, their feedback effect will be negative; if most clouds are at high levels, their feedback effect will be positive. Recent satellite measurements for 2000–2010 imply that the net feedback from clouds is positive but relatively small.

Other forms of feedback in global climate change are strongly positive. I introduced one in Appendix 4 when I described what happens to solar energy hitting the Arctic Ocean. During much of the year, the Arctic Ocean is covered by ice, with a layer of snow on top. This white surface reflects most of the incoming solar energy. Historically, this was true year-round, even in summer when it's light most of the day in the Arctic. One signal of recent global climate change, however, has been melting of more and more of the ice pack covering the Arctic Ocean in summer (Figure 10.2). This exposes dark water, which absorbs most of the incoming solar energy, enhancing warming through positive feedback.

Another example of strong positive feedback in the climate system also involves the Arctic. Most of the Arctic across Canada, Alaska, and Siberia has permanently frozen soil, or *permafrost*, below depths of 2 feet to 13 feet depending on location. The soil above permafrost freezes in winter but melts in summer, allowing growth of giant pumpkins in Alaska during the short time of near-constant sunlight (J. D. Megchelsen of Nikiski, Alaska, holds the current record for largest pumpkin at 1,287 pounds). In some parts of Siberia, permafrost extends to depths of nearly 5,000 feet. As trees and other plants die in the Arctic, they get incorporated in the permafrost where they remain for centuries. With global warming, however, the upper layers of permafrost are beginning to thaw in many parts of the Arctic. As this happens, plant material begins to decompose, releasing CO_2 and methane. The runaway global warming that would result from large-scale thawing of permafrost in the Arctic is potentially devastating. In her 2006 book *Field Notes from a Catastrophe*, Elizabeth Kolbert describes a field trip that she took with Vladimir Romanovsky, a permafrost expert at the University of Alaska. Romanovsky estimated that most of the permafrost in Alaska has been in place for 120,000 years. According to Romanovsky, "It's really a very interesting time" watching the permafrost melt after such a long period of stability (Kolbert 2006:17).

Finally, one biological process involving CO_2 might slow the rate of global warming through negative feedback. This is CO_2 "fertilization," referencing the fact that green plants remove CO_2 from the atmosphere in photosynthesis. If increased CO_2 stimulates plant growth, then increased uptake of CO_2 in photosynthesis may result, slowing the rate of further increases of CO_2 in the atmosphere. This potential negative feedback process might put a brake on the rate of global warming. Therefore scientists at various research stations have devoted lots of attention to comparing rates of uptake of CO_2 by plants in control plots to rates of uptake in experimental plots surrounded by pipes that blow air enriched with CO_2 over the plants.

While there is some evidence of CO_2 fertilization from these experiments, this negative feedback in the climate system won't be a panacea that stabilizes atmospheric CO_2 in the face of continued burning of fossil fuels. Increased growth of trees in response to increased CO_2 in the atmosphere will eventually be limited by other nutrients, like nitrogen; by lack of water, as climate change leads to more severe droughts in some regions; or by insect pests that benefit from warmer temperatures. Climate change is also likely to increase the scope and intensity of forest fires, which add large amounts of CO_2 to the atmosphere as trees burn and as dead wood decomposes after fires. In short, there are some definite positive feedbacks (melting of ice in the Arctic Ocean and permafrost in

Arctic soils) that will accelerate global climate change and a potential negative feedback (increased plant growth resulting from CO_2 fertilization) that may slow the rate of climate change somewhat but will be constrained by other consequences of climate change that influence the growth of plants.

Tipping Points

The third general principle is the possibility that climate change will cause the Earth to cross tipping points—transitions to new and dramatically different conditions that are likely to remain for hundreds of future generations. Inertia and positive feedbacks in the climate system, as described above, can contribute to tipping points. As Jim Hansen explains: "Our home planet is dangerously near a tipping point at which human-made greenhouse gases reach a level where major climate changes can proceed mostly under their own momentum" (Hansen 2008:7–8).

Melting of the Greenland ice sheet is one example of a potential tipping point. This would raise sea level around the world by about 7 meters (23 feet), last seen during the Eemian interglacial period that ended about 125,000 years ago. Partial melting of the Greenland or West Antarctic ice sheet caused sea level during the Eemian to be 6 to 9 meters higher than today, yet temperature was only a few tenths of a degree Celsius warmer than today. Average global temperature has increased 0.8°C (1.4°F) since 1880; according to a recent model, the Greenland ice sheet will reach a tipping point if temperature increases by another 0.8°C, well within the range of possibilities if greenhouse gases produced by burning fossil fuels continue to accumulate in the atmosphere. This estimate accounts for positive feedbacks that accelerate melting of ice on Greenland. In fact, researchers have used precise satellite measurements to show a sharp and steady decline in the mass of the Greenland ice sheet between 2002 and 2012 (Figure 10.7). Surface ice typically melts around the periphery of Greenland in summer and is rebuilt with winter snows. Between July 8 and July 12, 2012, however, surface melting increased from its usual peripheral locations to 97% of the total area of Greenland, including sites in the interior at about 10,000 feet elevation.

These comparisons with Eemian conditions, satellite measurements of changes in the Greenland ice sheet over the past 10 years, and modeling studies imply that we may be close to committing ourselves to complete melting of the ice sheet, although it may take a few hundred years for this to occur. This change would not technically be irreversible, if it cooled again for a long enough time, but sea levels would be much higher than today for hundreds to thousands of generations. Rising sea levels, in turn would cause major disruptions of human societies because about 10% of the world population currently lives within 30 feet of sea level and two-thirds of cities with more than 5 million people are within 30 feet of sea level. There would certainly be mass migrations away from current coastlines, leading to huge numbers of refugees. These migrations, in turn, might trigger major food shortages, spread of infectious diseases, and widespread warfare.

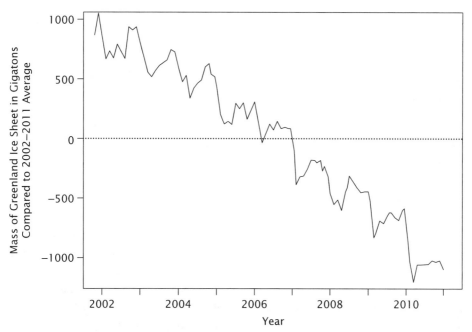

FIGURE 10.7 Change in estimated mass of the Greenland ice sheet in billions of metric tons from 2002 to 2011. The mass is plotted relative to the average over this 10-year interval, so the total loss of ice from 2002 to 2011 was about 2,000 gigatons. Estimates based on satellite altimetry (precise measurements of elevation at many points on the surface of the ice sheet) and satellite gravimetry (precise measurements of gravitational fields at many points).

Dieback of the Amazon rainforest is another example of a tipping point that might result from global climate change. Tropical rainforests occur in Southeast Asia, sub-Saharan Africa, Central and South America, and many islands in the tropical Pacific Ocean, but about half of all remaining tropical rainforest worldwide is in the Amazon. These areas have more species of plants and animals than any other habitats on Earth. For example, 35 acre plots of tropical rainforest in the Amazon have 700 to 1,000 different species of native trees, compared to 700 species of native trees in *all* of the United States and Canada.

Cutting forests in the Amazon and elsewhere for agriculture, mining, and other development contributes to global climate change by converting trees from a sink for CO_2 (live trees use CO_2 in photosynthesis) to a source (when dead trees decay, CO_2 is released). This makes tropical rainforest an endangered habitat, but this threat may be exacerbated by climate change because higher temperatures and more persistent El Niño conditions likely to result from climate change will cause less rainfall in the Amazon, with death of trees that depend on ample moisture. An element of positive feedback may influence this process as well, because a large part of precipitation in the Amazon is recycled—it evaporates from the ground and vegetation in one part of the forest and then falls as rain in another part. The ultimate result could be loss of much of the tree cover in the Amazonian rainforest, with devastating consequences not only for people living in this environment but also for global biodiversity. Many species of plants and animals dependent on trees in the Amazon would go extinct; since biodiversity in the Amazon is higher than in any other region of the world, this would mean extinction of a large percentage of species on Earth. Saddest of all, this change would be irreversible—once a species is extinct, it's gone forever.

Conclusion—A Problem at the Intersection of Science, Politics, Economics, and Ethics

This is a book about how science works, especially about tools for critical thinking used in science. You will have opportunities to use these tools yourself in decisions that you make in the coming years. Some of these will be personal decisions, affecting just you and your family. Other decisions will concern matters of public policy. I used the science of global climate change for most of the examples in this chapter because climate change has broad and deep impacts on policies throughout the world. Because of these impacts, climate change involves more than science. In discussing critical thinking I've alluded to several ways in which the science of climate change intersects politics, so I won't elaborate on this connection here. Look at the news practically any day and you will see examples of the intersection between climate and politics; I encourage you to consider these examples with an appropriate level of skepticism.

The intersection between climate science and economics is too large and complex to develop in more than a cursory way here. Some argue that taking action now to mitigate climate change—by phasing out use of fossil fuels, for example—would be too costly. Others argue that such actions to mitigate climate change would spur economic growth and employment by encouraging development of innovative new technologies. Much of this basic disagreement depends on forecasts of future economic growth, which in turn depends on cost–benefit analysis. Economic forecasts are beset with major sources of uncertainty—not only difficulty visualizing future innovations but also estimating the future economic value of today's investments. If anything, future economic conditions are less predictable than future climatic conditions.

You should consider politics and economics in addition to science in thinking about future climate change, but the most important consideration is ethics because ethics should be the foundation of the other three. Science works as well as it does because (most) scientists have a basic commitment to honesty that is grounded in ethics. Different economic beliefs are based on alternative values; for example, the value placed on individuals versus communities. Even politics is rooted in ethics; consider the US Declaration of Independence or Bill of Rights. What are the ethical implications of scientific knowledge about global climate change?

The Perfect Storm was a large, devastating storm that developed in the North Atlantic in late October 1991. It began as a nor'easter but eventually absorbed Hurricane Grace from the south and became a powerful cyclone. The storm weakened as it traveled southwest, then looped back to the northeast, and became a strong hurricane. The Perfect Storm caused $200 million in damages in seven states and accounted for at least 13 deaths, 6 on a fishing boat called the *Andrea Gail*. In 1997, Sebastian Junger wrote a book about the storm, which was made into a movie starring George Clooney and featuring the capsizing of the *Andrea Gail*. I tell this story not because the Perfect Storm was a harbinger of effects of climate change on extreme weather but because the philosopher Stephen Gardiner used the Perfect Storm as a metaphor for "the ethical tragedy of climate change" in his 2011 book *A Perfect Moral Storm* (the quoted phrase is his subtitle).

Gardiner used this metaphor because the ethical tragedy of climate change entails "the unusual intersection of . . . [three] serious, and mutually reinforcing, problems, which creates an unusual and perhaps unprecedented challenge" (Gardiner 2011:7). The first two problems are that climate change is both a global issue and an intergenerational issue. Being a philosopher, Gardiner describes the third problem as theoretical, by which he really means that we don't have a well-developed understanding of ethical principles necessary to solve the first two problems.

The global nature of climate change makes it a challenging problem for several reasons. First, the causes are widely dispersed rather than localized—they include you and me driving gasoline-powered vehicles to and from work and play, coal-fired power plants in China, deforestation in Brazil, and much more. Second, the poor in third-world countries who contribute least to climate change are likely to be the most victimized by climate change. Third, we live in a world of many different nations, with different forms of government, different economic systems, and different cultures, making it very difficult to mount a coordinated, mutually acceptable, and effective global response.

The intergenerational nature of climate change is also challenging for several reasons. Just as there is an asymmetry between rich and poor people in different parts of the world today, there is an asymmetry between current and future generations. Our comfortable lifestyles as residents of the United States and other developed countries currently depend on relatively profligate use of fossil fuels. Because of inertia, positive feedbacks, and tipping points in the climate system, much of the cost resulting from these benefits for us will be borne by future generations. Another intergenerational challenge is uncertainty about how great these costs will be. Finally, we can easily imagine our own children, grandchildren, or even great-grandchildren (some of us may already have these descendants), but global changes in climate already in the pipeline because of inertia and potential tipping points will affect many generations beyond these. As we continue to add greenhouse gases to the atmosphere, these irrevocable changes will become even more extreme.

Gardiner describes a side effect of the intergenerational problem that exacerbates the perfect moral storm of climate change. He calls this side effect *moral corruption*. Our personal connection to our own children and grandchildren will probably inspire us to protect this handful of individuals from harm. But it's difficult to imagine extending this protectiveness to contemporaries of our children and grandchildren, much less to members of succeeding generations. Humans can be altruistic, but for most people altruism isn't an abstract commitment that extends to all other people who live at any time in the future. Moral corruption, then, is a tendency to justify our current lifestyles, including lack of action on climate change, because we can only relate in a vague, abstract way to the people most victimized by climate changes in future generations. As Gardiner says, the current generation "may be attracted to weak or deceptive arguments that appear to license buck-passing [to future generations], and so give them less scrutiny than it ought" (Gardiner 2011:45). According to Gardiner, moral corruption fosters distraction, complacency, selective attention, unreasonable doubt, delusion, pandering, and hypocrisy. Do these sound like some of the breakdowns in critical thinking that I described at the beginning of this chapter?

Here is the fundamental challenge: "The dominant discourses about the nature of the climate threat are scientific and economic. But the deepest challenge is ethical. What matters most is what we do to protect those vulnerable to our actions and unable to hold us accountable, especially the global poor, future generations, and nonhuman nature" (Gardiner 2011:xii). Writing this book has been part of my contribution to meeting this challenge. I hope that you combine what you've learned here about science and critical thinking with your moral sensibility to embrace this most important challenge of our time.

Questions to Ponder

1. Which examples in this book were the most interesting to you? Which were the least interesting? Why? Which concepts were the most challenging? Why?

2. In 2012, James Lawrence Powell updated Oreskes' 2004 study by examining almost 14,000 peer-reviewed scientific articles published between 1991 and 2012. He found that only 24 "clearly reject global warming or endorse a cause other than CO_2 emissions for observed warming." Powell found 10,855 additional peer-reviewed articles on climate change published in 2013, only two of which rejected human activity as a cause of global warming. Powell's analysis was not published in a peer-reviewed scientific journal as of November 2014, but he gives a detailed description of his methodology and provides a spreadsheet with information on a large subset of the articles published in 2013 on his website (http://www.jamespowell.org/index.html). This means that you can duplicate his analysis and reach your own conclusions. John Cook and several colleagues also updated Oreskes' study, publishing their results in *Environmental Research Letters* in 2013. Cook's group considered 11,944 peer-reviewed papers about climate change published between 1991 and 2011. They found that about two-thirds of the papers took no position on human contributions to global climate change, while 97% of the papers that did take a position agreed with the scientific consensus that we contribute substantially to climate change.

Compare the results of these two studies by Powell and Cook's team. What are the key similarities and differences? What do you think accounts for the differences between the two studies? Which study is more credible and why?

3. How do you interpret these two parts of Wylie's sonnet: (i) " 'If . . . , then . . . ,' this only I assert"; and (ii) "for it is vain to ask If what I postulate be justified, or what I prove possess the stamp of fact"? How do these lines relate to the scientific process?

4. Does the failure of global climate models to predict accurately the decline in summer extent of ice in the Arctic Ocean as described here refute the hypothesis of anthropogenic global climate change? Why or why not?

5. Why does the hypothesis of anthropogenic global warming predict warming of the troposphere but cooling of the stratosphere, while the hypothesis of warming by increased solar radiation predict warming of both the troposphere and stratosphere? Think about how the greenhouse effect works in answering this question (see Appendix 4 for an overview of the greenhouse effect).

6. What happens to the apparent plateau in average global temperature in recent years if we begin at a different year than the 1998 starting point preferred by contrarians? Suppose we start a year or two earlier or later than 1998? Describe the pattern of average global temperature since each of these alternative starting points.

7. Compare and contrast the following three cases of lapses in critical thinking: Brian Martin's promotion of the tainted polio vaccine hypothesis for the origin of AIDS discussed in Chapter 9, the treatment by climate change contrarians of Gillis's news story about the apparent plateau in global warming in the first decade of the twenty-first century discussed on pages 238–239, and Jastrow, Seitz, and Nierenberg's misrepresentation of the comparison by Hansen and colleagues of various versions of a global climate model with observed temperature records illustrated in Figure 10.5.

8. I described two potential crossings of tipping points that might result from anthropogenic climate change, both with disastrous consequences. Melting of the Greenland ice sheet and loss of the Amazon rainforest aren't the only possibilities of tipping points due to climate change but are two of the most dramatic possibilities. Can you imagine any tipping points that might have positive effects, in contrast to these two examples? Consider potential tipping points involving social, economic, or political factors.

9. Global climate change is not just a scientific issue but also involves politics, economics, and ethics. I argued that the ethical dimension is most important. How do the politics and economics of climate change interact with the ethics of climate change? Are there political and economic factors that reinforce or counteract Gardiner's "perfect moral storm"?

Resources for Further Exploration

1. Climate Wizard is a website developed by the Nature Conservancy, the University of Washington, and the University of Southern Mississippi to display maps of past and projected future changes in average temperature and precipitation for the United States and the world as a whole. It is available at http://www.climatewizard.org.

2. Gapminder is a truly amazing website for graphical exploration of changes in energy use, CO_2 emissions, and many socioeconomic and educational variables for different parts of the world since the beginning of the Industrial Revolution. It is available at http://www.gapminder.org/ (go to the Gapminder World tab).

3. Climate Interactive is a set of simulation models developed by a group at the Massachusetts Institute of Technology to help people learn about global climate change. These models and related resources are available at http://climateinteractive. org/. Try the Climate Momentum Simulation for an overview of possible changes in atmospheric CO_2, temperature, and sea level during the rest of this century, first with no significant action to avoid climate change and then with more and more serious control measures. Then try C-Learn Simulator for more detailed analysis of how new policies may mitigate climate change.

4. Several websites provide valuable information about the scientific and policy aspects of global climate change. Here are five of the most useful.

    ClimateProgress offers daily postings on all aspects of climate change but especially issues relating to policy: http://thinkprogress.org/climate/issue.

    Skeptical Science has extensive current information about all scientific aspects of climate change: http://www.skepticalscience.com/. The authors thoroughly debunk the most common myths about climate change.

    Climate Science Watch mostly reports on policy issues, including politics and economics: http://www.climatesciencewatch.org/.

    The National Center for Science Education has basic educational information and news about climate change as well as evolution: http://ncse.com/. The mission of the NCSE is to combat efforts across the United States to undermine teaching about evolution and climate change in public schools.

    RealClimate is a site developed and maintained by several climatologists at the forefront of research on climate change. It has the most technical and detailed information about climate change of the sites listed here: http://www.realclimate.org/.

Units of Measurement Used in This Book

All scientists use the metric system of measurement, as do nonscientists in most of the world. In keeping with this standard practice, I use metric measurements in this book, but provide English equivalents for readers in the United States and other countries that still use feet, miles, pounds, and so on. Figures 3.1 to 3.3 show both metric and English units for sizes of people and other animals to familiarize you with metric units if you use the English system in everyday life (Table A1.1).

TABLE A1.1 English equivalents of metric measurements used in this book.

Category	Metric Unit	English Unit	Notes
Distance	centimeter (cm)	1 cm = 0.39 inches, 1 inch = 2.54 cm	
	meter (m)	1 m = 39.37 inches = 3.28 feet = 1.09 yards	One meter is a little longer than one yard.
	kilometer (km)	1 km = 0.62 miles, 1 mile = 1.61 km	
	nanometer (nm)	1 nm = 10^{-9} meter = 0.000000001 m	See Appendix 3 for use of nanometers.
Area	hectare (ha)	1 ha = 2.47 acres, 1 acre = 0.4 ha	There are 640 acres in one square mile and 100 hectares in one square kilometer. This illustrates the rationality of the metric system—you can convert from ha to km^2 in your head, but not from acres to $miles^2$.

(Continued)

TABLE A1.1 Continued

Category	Metric Unit	English Unit	Notes
Mass[a]	gram (g)	1 g = 0.035 ounces, 1 ounce = 28.3 g	A US dime weighs about 2.3 grams.
	kilogram (kg)	1 kg = 2.2 pounds, 1 pound = 0.45 kg	
	metric ton = 1,000 kg	1 metric ton = 1.10 US tons	
	gigaton = 1,000 metric tons = 1,000,000 kg		
Volume	milliliter (ml)	1 ml = 0.034 fluid ounces, 1 fluid ounce = 29.6 ml	
	liter (l)	1 l = 0.26 gallons, 1 gallon = 3.8 l	A liter is a little smaller than a quart.
Temperature	degrees Celsius (°C)	°C = (5/9) × (°F−32), °F = 1.8 × °C + 32	The boiling point of water is 100°C, which equals 212°F; the freezing point of water is 0°C, which equals 32°F. To translate changes in temperature from Celsius to Fahrenheit, multiply by 1.8.

[a]Mass is a measure of the amount of material in an object, while weight is a measure of the force of gravity acting on the object. This means that the same object would have the same mass on Earth and on the moon, but lower weight on the moon because the force of gravity is less on the moon. In ordinary conversation, we often use "mass' and "weight" interchangeably, and I will follow this tradition here.

How Does Evolution Work?

Evolution is the central unifying principle of biology. As expressed by the geneticist Theodosius Dobzhansky in 1973, "Nothing in biology makes sense except in the light of evolution." If anything, this sentiment is even more true today than it was 40 years ago. For example, there was little appreciation in those days for the implications of evolution for medical practice, but this is a very active area of research and teaching today.

Evolutionary biologists study two major topics, the history of life on Earth and causes and consequences of genetic changes in populations of organisms, from microbes to plants, animals, and humans. We touched on both of these topics in Chapter 1 and will see more of them in later chapters, but here I give a more complete introduction to natural selection as the primary mechanism of evolutionary change. Natural selection causes most of the important genetic changes in populations and is involved in the longer-term changes that lead to new species and, ultimately, the whole history of life. Here, we'll focus on the role of natural selection in changes within populations.

In Chapter 1 we discussed how monarch butterflies navigate during migration and why long-distance migration may have evolved in monarchs. A group of researchers at the University of Massachusetts Medical School recently determined the complete genome of monarchs, identifying genes that may be involved in vision, circadian rhythms, and perhaps even sun-compass navigation. However, the evolution of migration happened millions of years ago, so this work on the genetics of monarchs can't give a detailed picture of the genetic changes in monarchs that led to migration. Instead, a different example with a relative of monarch butterflies called the peppered moth can illustrate this process of genetic change in a population that is one of the major features of evolution.

Peppered moths fly at night and rest on branches and trunks of trees during the day. They don't migrate like monarchs, but they vary in their color patterns, specifically in how well they match the bark of trees where they rest. Coloration of peppered moths is primarily determined by a single gene, like eye color in humans. There are different forms of these genes in both cases, so people have brown or blue eyes or less common variations, and moths are usually dark or light, but sometimes intermediate in color. Each form of a gene for color occasionally mutates to another form, and the ultimate source of genetic variation in populations is this process of rare, random mutation.

What kinds of genetic changes have occurred in populations of peppered moths? The story of these changes takes place mostly in England, although peppered moths are widespread and researchers have seen similar changes in North America. England is our setting for this story, however, because the English have been collecting moths for a long time so evolutionary changes in coloration of moths are very well documented. All moths collected before 1848 were the light form, called *typica*. This doesn't necessarily mean that dark forms, *carbonaria*, never occurred, but it does imply that mutations of the color gene from light to dark were very rare. In hindsight, it also suggests that dark forms might have been at a significant disadvantage in the woodland homes of peppered moths before 1848. The trees in these woods were covered with whitish lichens; note how difficult it is to see a *typica* moth against this background compared to a dark *carbonaria* moth (Figure A2.1a). Imagine that a hungry bird flies by. Which moth is most likely to become a meal for the bird?

The first *carbonaria* moth was collected near Manchester, England in 1848. By 1864, most moths near Manchester were dark, and by 1898, 98% were dark. In just 50 years, the population of peppered moths near Manchester had changed from 100% light to almost 100% dark. Similar changes occurred downwind of other industrial cities in England, which is why this process is called *industrial melanism*. One consequence of the Industrial Revolution in the 1800s was an increase in pollution by soot and smoke. These pollutants settled on trees in woodlands occupied by peppered moths, darkening the surfaces where they rest. Figure A2.1b shows dark and light moths on a soot-covered tree trunk. Which would our foraging bird be most likely to take?

The thought experiments that I suggested you do in the last two paragraphs lead to an hypothesis about the cause of genetic change in the population of peppered moths near Manchester between 1848 and 1898. Remember that our basic information is about frequencies of light and dark moths at different times. These colors are part of the *phenotypes* of the moths, where the phenotype includes all aspects of morphology, physiology, and behavior. However, we are really interested in changes in gene frequencies, so we need to consider the relationship between phenotype and *genotype*, or complete genetic makeup of an individual. In this case, there is a straightforward relationship between phenotypes and genotypes, based on breeding experiments in the lab, so we know that there was a big change in gene frequencies in the last half of the nineteenth century. One hypothesis for this change is that, after industrialization, dark moths in one generation passed more genes to the next generation than light moths, and this continued for generation after generation until almost all moths were dark by 1898. A shorthand way of describing this difference is to say that dark moths had greater *genetic fitness* than light

FIGURE A2.1 Light (*typica*) and dark (*carbonaria*) peppered moths resting on a lichen-covered tree trunk (top panel) and on a soot-covered tree trunk (bottom panel). Look carefully to see the *typica* moth on the lichen-covered trunk.

moths in polluted environments. More specifically, if dark moths were better protected from predation by birds when resting on the matching background of soot-covered branches, then they would be more likely to survive and reproduce than light moths. This is an example of *natural selection*, defined as differential reproductive success of genetically different individuals in a population. Some genes may influence reproductive success directly, for example, by increasing the number of eggs laid. Other genes may influence reproductive success indirectly, by increasing an individual's chances of survival, hence the number of times it has an opportunity to reproduce. Our hypothesis in this case is that genes that determine coloration of peppered moths influence survival, so have an indirect effect on reproductive success.

The hypothesis that natural selection by bird predation caused evolutionary change in coloration of peppered moths in environments altered by industrialization is consistent with geographic variation in the frequencies of light and dark moths. Dark moths were most common downwind of industrial cities in England, but light moths were more common in less polluted areas. However, it wasn't until the early 1950s that H. B. D. Kettlewell tested the predation hypothesis experimentally.

Kettlewell did his first experiment in 1953 in polluted woods near Birmingham, England. He marked a large number of light and dark moths with small spots of paint on their abdomens and released them to rest on tree trunks during the day. For the next several nights, he recaptured some of them as they flew around his study site. He recaptured about 13% of the light moths and 28% of the dark moths, and he interpreted this difference to mean that dark moths survived better than light ones in this polluted environment where trees were covered with soot.

Kettlewell repeated his experiment in 1955 near Dorset, an unpolluted site where trees were covered with whitish lichens. Here he recaptured 14% of the light moths and 5% of the dark moths that he had released. The results of these two experiments are consistent with the hypothesis that dark moths survive better in polluted woods while light moths survive better in unpolluted woods, perhaps because of differences in effectiveness of their camouflage in these two environments, although these experiments don't provide evidence that predation by birds was responsible for differential survival. Therefore Kettlewell and Niko Tinbergen did another experiment at both sites in which they placed light and dark moths on tree trunks during the day and watched from a blind while native birds captured some of the moths. At Birmingham, redstarts were the only birds that preyed on moths and they ate three times as many light ones as dark ones. At Dorset, five species of birds preyed on moths, and they ate six times as many dark ones as light ones.

Another chapter in this story happened after Kettlewell did his experiments. With decreased use of coal and implementation of pollution controls, woods in England returned to a more natural state and dark forms of the peppered moth decreased in frequency at sites that had been heavily polluted. In Manchester, for example, nearly 100% of moths were dark in 1960 but only about 10% were dark in 2000.

In summary, alternative forms of a single gene determine coloration of peppered moths. Before industrialization, dark moths were rare or nonexistent. Within 50 years of the first record of a dark moth near Manchester, virtually the entire population was dark. This remained true from 1900 until 1960, when effects of pollution on forests were reversed. In the next 40 years, light moths increased in frequency, almost completely replacing dark moths. It's possible that pollution had a direct effect on reproduction or survival of moths but more likely, based on the experiments of Kettlewell and others, that the agent of natural selection was differential predation by birds on moths that were most conspicuous when resting on trees. These would have been light moths on soot-darkened bark between about 1850 and 1960 and dark moths on lichen-covered bark after 1960. With mutation producing genetic variation in peppered moths, differential reproductive success of these genetically different individuals caused evolution of populations of moths—in one direction while forests were heavily polluted and the opposite direction when pollution was reduced. In this case, differential reproductive success was probably due to conspicuous moths being eaten by birds before they had much if any chance to pass their genes to the next generation.

The peppered moth story is a good example of evolution in action because we have direct evidence for all the elements of this fundamental process of biology—genetic variation, natural selection, and change in gene frequencies over time. Although there are many similar examples in nature and the lab, evidence may be incomplete or indirect in other cases. For instance, biologists may not be able to document changes in frequencies of specific genes in a population over time but may see differences between populations living in different environments that can be sensibly interpreted as adaptations to the different environments. A human gene that enables digestion of milk provides an interesting example.

Milk was the defining innovation in the evolution of mammals—the name of our class of vertebrates comes from the Latin root *mamma*, meaning breast, site of the mammary glands that produce milk used by all mammal mothers to nourish their young. Milk contains a rich supply of proteins, fats, and carbohydrates including lactose, or milk sugar. Lactose must be broken down in the digestive tract to be absorbed, and nursing babies use an enzyme called lactase to do this. In most mammals, however, lactase is inactivated after weaning, meaning adults can't digest milk. This is true of most humans too, which isn't to say that adults can't consume any dairy products, but that they might have symptoms of digestive distress if they consumed as much as children.

This story may be a little puzzling to you if your ancestors came from northern Europe, because you may drink milk regularly without any problems. It may make more sense if you are Chinese or Japanese and don't drink much milk. Although most humans worldwide have limited ability to absorb lactose, some can do so just fine. This difference in lactose tolerance is due to a mutation in a gene that regulates the production of the enzyme lactase. The primitive form of the gene in humans was like that in other mammals—it caused lactase production to diminish after weaning. The mutated form of the gene allowed lactase production to continue into adulthood.

A mutation for lactose tolerance was first reported in humans in northern Europe, where cattle were domesticated by members of the Funnel-Beaker Culture about 9,000 years ago. Three different African populations also domesticated cattle, and these people now drink milk and tolerate lactose. In each of these four cases, the mutation for lactose tolerance is different, implying independent evolution of the ability to use milk as adults. Gabrielle Bloom and Paul Sherman analyzed the prevalence of lactose malabsorption, or *inability* to digest lactose in adulthood, in 100 populations around the world (Figure A2.2). They found that lactose malabsorption was more common in areas with extreme climates or where cattle were subject to serious infectious diseases, that is, areas "where it is impossible or dangerous to maintain dairy herds" (Bloom and Sherman 2005:301.e1). Where cattle could be maintained, on the other hand, adults able to digest lactose likely had greater genetic fitness than intolerant adults because the lactose-tolerant ones could use milk as well as meat as a source of calories. In this human example, evolution of lactose tolerance occurred in conjunction with cultural evolution as well as features of the physical environment that influenced the ability to domesticate cattle.

These two examples and many others illustrate the creative power of natural selection. If random mutation introduces useful genetic variation into populations, then

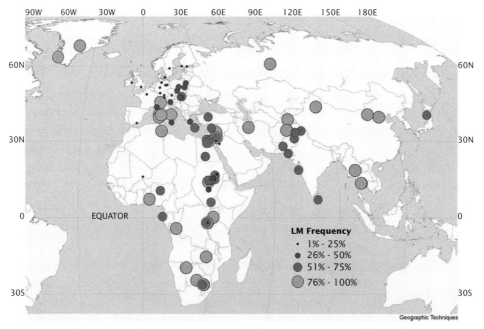

FIGURE A2.2 Frequencies of lactose malabsorption in 100 human populations. The small dots are populations where more than 75% of adults can absorb lactose because they have a mutation that enables continued production of the enzyme lactase.

natural selection can cause evolutionary change in those populations that improves the ability of individuals to survive and reproduce in existing or new environments; that is, individuals become better adapted to their environments. It may seem that this process is critically dependent on chance, since mutation provides the raw material, but keep in mind that with large populations or long periods of time, as in the examples discussed here, there is much opportunity for useful mutations to occur. Once they do, natural selection may cause rapid changes in gene frequency if fitnesses of different genotypes are sufficiently different.

Sensory Worlds of Humans and Other Animals

We humans see objects that emit, transmit, or reflect light in what we call the visible portion of the electromagnetic spectrum, ranging from 380 nanometers to 760 nanometers in wavelength (760 nanometers is about one-hundredth the width of a human hair). Visible light includes the colors of the rainbow—violet, blue, green, yellow, orange, and red—but we can distinguish thousands to millions of specific colors within this range, from vermilion to bisque to puce and beyond. As primates, we can recognize a greater range of colors than most other mammals, yet other animals have even better color vision. More generally, different animals live in different sensory worlds even though they may occupy the same physical space. If you have normal hearing, you can detect sound waves with frequencies between about 20 Hertz and 20,000 Hertz. Your dog, however, can detect sounds up to 40,000 Hertz, while some bats and marine mammals can detect even higher frequency sounds of 100,000 to 200,000 Hertz. Bats, dolphins, and whales use these ultrasonic sounds in echolocation—emitting high frequency sounds that bounce off objects in the environment and then locating these objects by hearing the echoes. Other animals communicate with infrasonic sounds that have frequencies less than 20 Hertz. Likewise, humans can detect a wide variety of different smells, but our canine companions and other animals have much greater olfactory sensitivity.

I introduced you to the electromagnetic spectrum as a basis for learning about a potential relationship between use of cell phones and brain cancer, so let's use part of that spectrum to see how scientists learn about the different sensory worlds of animals. What do various animals see as they go about their daily activities? How different is their visual world from ours, and how can we study their visual world using our eyes?

Honeybees are tremendously important to humans because they provide us food and pollinate many of our crops. For these reasons, scientists have studied many aspects of their biology, especially their sensory systems and foraging behavior, which are keys to pollination and honey production. If we are interested in an animal's sensory world, one of the first questions we might ask is whether or not they see in color. In 1913, a German ophthalmologist named von Hess published a long report on experiments that he did with fish, insects (including bees), and other invertebrates that led him to conclude that these animals are incapable of seeing color. But another German scientist, Karl von Frisch, was skeptical of this conclusion, at least for bees, because it didn't make sense to him that the flowers visited by bees would be brightly colored if the bees couldn't see those colors. Bees and flowers have a mutually beneficial relationship—flowers produce nectar that bees carry back to their hive to feed the brood; when bees gather nectar, grains of pollen stick to their bodies; these pollen grains are deposited on the stigmas of other flowers as the bees continue to forage; and pollination ensues. Von Frisch thought that the various colors of different flowers must play a role in this mutualistic relationship, perhaps enabling bees to recognize certain kinds of flowers, thereby decreasing the chance that a foraging bee delivers pollen from a bright blue flower of species A to a pale yellow flower of species B where pollination would be impossible.

Bees make great subjects for learning experiments for at least two reasons: they respond to simple rewards—a little honey or sugar water—and they don't get bored—there's always a need for more food back at the hive. Von Frisch decided to test the color vision of honeybees by capitalizing on their learning ability. He first set up a table with a small supply of honey to attract bees from a nearby hive. He then placed the bait on a piece of blue cardboard and waited until several bees from the hive had made multiple visits to collect the honey. Then he removed the bait, placed a new blue card and a red card on the table, and watched where the bees landed when there was no honey to attract them by smell. All bees landed on the blue card.

Von Frisch quickly realized that there was a problem with this experiment. The blue and red cards differed not only in color but in brightness, and the bees may have picked the blue card because of its greater brightness rather than its different color. So he did a more complex experiment. He started with a blue card in the middle of an array of gray cards ranging in brightness from white to black. He placed a small glass container on each card, with sugar water in the container on the blue card and the others empty. The bees quickly found the sugar water and continued going to the blue card containing the bait even when von Frisch rearranged the positions of the various cards to be sure learning by the bees was based on some feature of the blue card rather than its position in the middle of the array.

The critical test in this experiment was to remove the bait and use a clean blue card. When von Frisch did this, the bees continued to land on the blue card, rather than landing half the time on the gray card that matched the blue one in brightness. Von Frisch even considered the possibility that the blue card had some unique odor detectable to bees but not to us, so he covered the whole array with glass to block this hypothetical odor and the bees now landed on the glass above the blue card. This convinced von Frisch that honeybees truly were responding to color in this experiment.

Von Frisch tested other colors in the same way and discovered that bees could learn to forage for sugar water on orange, yellow, green, and violet cards but not red ones. Apparently the color vision of bees differs from us at the red end of the spectrum; unlike us, bees see red as black. By modifying his experiment slightly, von Frisch discovered another difference between bee vision and human vision. Once he had trained bees to seek food on a blue card, he tested them with cards of all colors rather than gray ones of different shades. He found that bees trained on blue sometimes landed on violet cards in this experiment. When he trained them to seek food on a yellow card, they landed on orange and green as well as yellow in the testing phase of the experiment. Honeybees seemed unable to discriminate colors as finely as humans in the orange-yellow-green and the blue-violet parts of the spectrum. However, about 40 years after von Frisch published these results in 1915, another researcher devised a different apparatus in which he could train and test bees individually using specific, single wavelengths of light or mixtures of two wavelengths. Under these conditions, bees *could* distinguish orange, yellow, and green, suggesting that their color vision is not that different from ours after all.

Yet another experiment, however, revealed one big difference between color vision of bees and humans. In 1927, a researcher named Kühn modified von Frisch's experiment with colored cards by using a light source that could emit a specific wavelength to illuminate the place on the feeding table where the bait was located. In addition to using wavelengths in the visible spectrum, Kühn did trials with ultraviolet light. He found that bees could learn to discriminate ultraviolet light from all shades of gray. In summary, the main difference between our color vision and that of honeybees is in the range of colors that humans and bees detect—red to violet for us, orange to ultraviolet for bees.

Von Frisch began these experiments because he thought the colors of flowers must mean something to the bees that pollinate them. Flowers probably look roughly the same to honeybees as they did to von Frisch, except that bees see red flowers as black, and they see ultraviolet as a color as well as orange through violet. This leads to two further questions—what animals pollinate red flowers, and how do bees use their ability to see ultraviolet? The first question was easy to answer based on observations of natural history made long before von Frisch's time: most red flowers are pollinated by hummingbirds and butterflies. The second question was trickier. Flowers might facilitate pollination by reflecting ultraviolet light in patterns attractive to bees, but, if so, we can't see these patterns. Alternatively, ultraviolet vision in bees might be used in some other kind of behavior, such as their social interactions or communication. We can't test either of these ideas directly, because we can't see ultraviolet light.

Von Frisch was inspired by the discovery that bees could see in the ultraviolet range to check for ultraviolet patterns on flowers pollinated by bees (he also did extensive work on social behavior and communication in bees, discovering the famous dance language that scout bees use to communicate locations of new food sources to foragers at the hive for which he shared the Nobel Prize for Physiology or Medicine in 1973). By examining ultraviolet photos of flowers, he found that many look quite different than under white

FIGURE A3.1 Black-eyed Susan photographed by A. Davidhazy with visible light (left) and ultraviolet light (right).

light (Figure A3.1). Not only do bee-pollinated flowers often have markings that show up in ultraviolet light, these markings usually form nectar guides that "point the way" to the center of the flower where nectar and pollen are found. A remarkable example of a mutualistic relationship between a plant and its pollinators is seen in poppies. Poppy flowers appear red to us; if they looked the same to bees, they would appear black instead of red and so would not stand out against the background because bee vision does not extend into the red part of the spectrum. But the poppy is one of the few red flowers that is pollinated by bees, and the center of the flower containing nectar and pollen reflects ultraviolet light that is visible to bees.

Thinking like a biologist means learning how biologists answer questions about ourselves and about the natural world in general. It means thinking critically about different kinds of evidence that we use to test hypotheses about how nature works—simple observations, comparisons and correlations, and experiments. I believe that learning how to think like a biologist can help you make better decisions in your own life because you will learn to think more carefully and critically than you might otherwise. But this discussion of vision in honeybees illustrates another benefit of the biological way of thinking—appreciating that different animals live in different sensory worlds than we do. Being sensitive to this possibility opens up a range of fascinating new discoveries about the lives of animals and their interactions with the natural environment.

Global Climate Change: How Can Amateurs Comprehend Complex Models?

Scientists work with many models that are much more complex than the examples discussed in Chapter 6 yet have vital implications for public policy. Models of global climate change are the most important examples of our generation because they deal with issues of worldwide scope that demand responses from individuals and governments. How should an ordinary citizen with little training in science and none in mathematical modeling of complex systems respond to these models? For that matter, how should a biologist like me with some experience in modeling but only basic training in the chemistry, physics, and advanced calculus used to build complex models of climate respond to these models? My purpose in this appendix is to offer some tools to help you evaluate these models without fully understanding the technical details.

I need to provide some key background information before describing the basic nature of models of global climate change. Contrarians about climate change are those who deny a role of humans in climate change or admit a role of humans but deny the seriousness of the problem. Contrarians argue that natural variation in solar radiation received by Earth drives changes in climate; since humans can't influence solar radiation, we must not be able to influence climate. There is an element of truth in the first part of this claim—the sun is indeed the ultimate source of energy for the Earth. However, this argument assumes that the Earth is a passive participant in the process. Your own experience belies this assumption. Think about how you would feel on a warm sunny day if you were wearing dark or light clothes. A black shirt would absorb more solar radiation than a white shirt, so you might feel uncomfortably hot in a black shirt but OK in a white shirt. Now expand your perspective to a large area of the Earth's surface, like the Arctic

Ocean. The same principle applies. If the Arctic Ocean is covered by white ice and snow, as most of it is during most of the year, much of the solar energy that hits the surface is reflected back into space. But when the ice melts, as more and more does each summer (Figure 10.2 on page 232), the exposed ocean surface is much darker, absorbing most of the incoming solar radiation. In short, while incoming solar radiation varies over centuries and millennia, the amount of energy absorbed by the Earth also varies as the color of the surface varies across the centuries and around the globe (dark ocean waters and dark evergreen forests to light snow-covered mountains and light desert soils).

The Earth is actively involved in its own energy budget in another way that is even more important than the proportions of its surface that are light and dark. As incoming solar radiation warms the surface of Earth, some of this energy is converted to thermal infrared radiation, which travels back toward outer space. Part of this infrared radiation escapes the Earth but some is trapped by gases in the atmosphere and therefore retained in Earth's environment. These gases are called greenhouse gases and include water vapor, carbon dioxide (CO_2), methane, and other gases (but not nitrogen and oxygen, which are the main constituents of the atmosphere). Based on distance from the Sun to the Earth and reflectivity of the Earth's surface, scientists estimate that the average temperature of a hypothetical Earth without greenhouse gases would be about $0°F$, compared to the actual average of about $57°F$. If it existed at all, life on an Earth with an atmosphere but no greenhouse gases would be very different from life as we know it, and probably wouldn't include humans.

Many natural sources produce greenhouse gases. Volcanic eruptions inject CO_2 and methane into the atmosphere. Respiration by animals puts CO_2 into the atmosphere (plants respire too, but they use more CO_2 in photosynthesis than they give off in respiration). Cows and other ruminants give off methane from both ends of their digestive tracts. Two main human activities, in addition to keeping cows for meat and milk, add greenhouse gases to the atmosphere. We have been cutting and burning forests and other natural vegetation for a long time but have increased these activities in recent decades, especially in the tropics. Some intact forests remove CO_2 from the atmosphere, but burning forests releases CO_2, as does decomposition of dead plant material left lying on the ground. Burning fossil fuels—coal, oil, and natural gas—dates to the beginning of the Industrial Revolution in about 1780, but has also greatly accelerated in recent decades. Because CO_2 is a small proportion of the atmosphere, the amount of CO_2 in the atmosphere is quite sensitive to burning of fossil fuels and deforestation. Researchers have made continuous direct measurements of atmospheric CO_2 at a facility on the top of Mauna Loa in Hawaii since 1958; levels at that time were 315 parts per million compared to almost 400 parts per million in 2012. Based on less direct but still credible estimates for earlier times, atmospheric CO_2 was 280 parts per million in 1750, for an increase of 42% in the past 260 years.

You need to know one more feature of CO_2 before I give you a brief introduction to models of global climate change. Contrarians about human impacts on climate change can't deny the changes described above nor the role of fossil fuel use in these changes. But they can't imagine how such tiny amounts of CO_2 in the atmosphere can influence climate. They point out, correctly, that water vapor is 10 times as abundant as CO_2 in the atmosphere and contributes about three times as much to the greenhouse effect as CO_2. What contrarians ignore is that each molecule of CO_2 added to the atmosphere by burning fossil fuels or any other process has a much longer lifespan in the atmosphere than each molecule of water vapor added by evaporation from the ocean or land surface or the leaves of plants—30 to 95 years on average for CO_2, a few days for water vapor. What's more, almost 20% of CO_2 added to the atmosphere will remain for more than 1,000 years (Figure 10.6 on page 244). Human activity has little direct influence on the amount of water vapor in the atmosphere but direct and long-lasting influence on the amount of CO_2.

This discussion of greenhouse gases sets the stage for my rudimentary description of models of climate change because it shows that what happens in the atmosphere is important for heating and cooling of the Earth. The first global climate models (GCMs; also called general circulation models) were built in the 1970s and included only physical processes of the atmosphere such as solar radiation and wind patterns. By the 1990s, GCMs included processes happening in sea ice and the upper layer of the oceans as well as the atmosphere and more realistic representation of clouds than the earliest models. Researchers made increasingly realistic and more complex models as time went on, so that contemporary models include physical, chemical, and biological processes in the atmosphere, oceans, and terrestrial environments.

GCMs share one feature with our predator–prey model in dividing the environment into cells and keeping track of what happens in each individual cell (Plate 11). In the case of GCMs, however, the environment is not a single meadow or lake but the atmosphere as a whole plus Earth's land surface plus the top layers of the ocean. Imagine standing someplace and looking straight up into the sky. If you were at one corner of a square grid cell for the atmospheric part of a GCM, the opposite corner would be about 200 kilometers (120 miles) away. Unlike the NetLogo model, the GCM is three dimensional, so there would be up to 40 cubes stacked on top of each other extending from where you were standing to the top of the atmosphere. And this is just the atmospheric part of the model; there are cells for the land and oceans as well.

GCMs typically keep track of several variables in each of these cells at half hour intervals for up to 100 years. These variables include things like solar radiation, temperature, water content, CO_2 content, cloud cover, and wind speed. The variables are interconnected through equations based on detailed understanding of the physics and chemistry of the atmosphere and oceans. Each cell influences neighboring cells, for example, through transmission of incoming solar radiation from upper to lower levels of the atmosphere and through wind patterns that move clouds horizontally. Thus GCMs include hundreds to thousands of equations and require months to run on the fastest supercomputers. Several research groups in the United States and elsewhere use GCMs to do simulation experiments, like our experiment with predators and prey using NetLogo but at a much larger scale. These researchers also continuously refine their models, including more details of the climate system at smaller and smaller resolutions to make predictions of localized climate patterns in addition to global and continental-scale patterns.

What is the nature of these simulation experiments done with GCMs? The Intergovernmental Panel on Climate Change (IPCC) is an international group of scientists established by the United Nations in 1988 to synthesize climate research for policymakers at periodic intervals. The IPCC produced its Fourth Assessment Report in 2007 and completed its Fifth Report in 2014. IPCC Reports include extensive discussion of evidence for climate change that has already occurred as well as projections of potential climate change in the next 100 years under various scenarios. These projections come from experiments with GCMs. For example, one scenario assumes that energy use continues to depend heavily on fossil fuels, a second assumes that we transition fully to energy sources that don't produce greenhouse gases, and a third assumes that that we use a balanced mixture of these two sources of energy. All three of these scenarios assume that the human population peaks at 9 billion in 2050 and then gradually declines and that there is rapid technological development throughout this century. Several research groups around the world used their GCM models to do experiments representing these three scenarios. These experiments led to projections of increases in average global temperatures by 2100 of 4.3^0F to 11.5^0F for continued reliance on fossil fuels, 3^0F to 7.9^0F for a mixed strategy, and 2.5^0F to 6.8^0F for complete transition to nonpolluting energy sources. By contrast, if atmospheric concentrations of greenhouse gases could be maintained at their 2000 level, projected temperature increase by 2100 would be 0.5^0F to 1.6^0F. This is a purely hypothetical scenario, however, since CO_2 in the atmosphere has already increased by almost 7% from 2000 to 2012.

Many people are uneasy about these kinds of projections, and you should be too. The projections come from extremely complex models, and the greater the complexity of a model, the more room for error. Even if the assumptions underlying each scenario are true, projections of temperature in 2100 are uncertain, reflected in the ranges reported in the last paragraph. In fact, the IPCC reports these ranges because different research groups in different countries using different models get different results, even with the same assumptions. More fundamentally, things may happen in the next several decades that are totally unanticipated and not incorporated in the models. In introducing models of disease in Chapter 6, I mentioned the possibility of spillover of a disease from another animal to humans. Suppose a worldwide pandemic of a new disease kills a substantial part of the human population? If so, the population might be much less throughout this century than the scenarios specify, leading to lower energy use and confounding projections of the models. Or suppose researchers genetically engineer a plant to use atmospheric CO_2 much more efficiently than current plants, and this species can be planted widely to withdraw CO_2 from the atmosphere. This would make projections of climate models meaningless. Incidentally, and not to put too fine a point on it, results of climate models are called projections rather than predictions because of these unanticipated possibilities. A projection is really a "what if" exercise, asking the question, what do we expect to happen if our assumptions about the future hold true? By contrast, a prediction is something we'd be willing to bet money will actually happen.

Healthy skepticism about GCMs is appropriate for the reasons outlined above, but this shouldn't lead to outright rejection of the models, as it does for many contrarians, and it certainly shouldn't lead to rejection of the whole idea that humans contribute to global climate, which is based on a wide range of empirical evidence in addition to climate modeling. We'll discuss some of this empirical evidence in Chapter 10, but I want to conclude this brief overview of the complex process of modeling global climate change by describing an interesting test of these models against real temperature records.

In addition to forecasting climate for the next hundred years, modelers can see how well their models match climate conditions that have already occurred. Rather than starting a climate model with conditions in 2000 and projecting forward to 2100, they can start with conditions in 1860 and project to 2000 and then compare their projection with what actually happened. But they can do something even more interesting and useful by thinking like an experimenter. Specifically, climate modelers have compared three versions of their models run from 1860 to 2000—a version containing only natural forcings such as variations in solar radiation and volcanic eruptions, a version containing only anthropogenic (human-caused) forcings such as greenhouse gas emissions and deforestation, and a version with both natural and anthropogenic forcings. These factors are called forcings because they force changes in the energy balance of Earth, leading to changes in temperature. Plate 17 shows the results. Models with only natural forcings (Plate 17a) overestimated actual temperature from 1860 to 1880 and underestimated temperature from 1960 to 2000. Models with only anthropogenic forcings (Plate 17b) underestimated temperature from about 1920 to 1970. Models with both natural and anthropogenic forcings (Plate 17c) gave the closest fit to actual temperatures from 1860 to 2000. This doesn't prove that climate projections into the future will be correct, but it does suggest that contemporary climate models include enough detail and realism to paint a reasonably accurate picture of changes in average global temperature over a period of 140 years. More important, Plate 17 implies that long-term changes in climate are influenced by natural and anthropogenic factors acting together.

The Mystery of Missing Heritability

Advances in technology contribute to progress in all sciences, but in recent years nowhere more than in genetics. The Human Genome Project initiated in 1990 and completed in 2003 was a milestone in this process. The Human Genome Project cost about $1 billion in public and private funds and involved hundreds of researchers in the United States and other countries who identified the 3 billion base pairs in the 20,000 genes of our genetic code. The hardware and software developed and refined in this and subsequent projects has been used to determine complete genetic sequences of many additional species, revolutionizing our ability to study evolutionary relationships between organisms. Researchers can now determine sequences of multiple individuals of a species quickly and cheaply using tiny amounts of DNA. For example, between 2008 and 2012 a group of researchers in the United States, the United Kingdom, and China sequenced the genomes of 1,000 people of diverse ethnic groups to study genetic variation in humans. In 2012, researchers in the United Kingdom began the 100,000 Genomes Project to sequence the genomes of patients of the National Health Service with various diseases and members of the patients' families (http://www.genomicsengland.co.uk/the-100000-genomes-project/).

These advances in technology have produced several new analytical tools dependent on collecting large amounts of data quickly and cheaply and, just as important, managing these data in a form that facilitates large-scale statistical analyses. For example, researchers have developed genome wide association studies (GWASes) to study traits influenced by multiple unknown genes. The goal of GWASes is to identify many of these unknown genes and thereby begin to understand the biochemical and physiological mechanisms involved in the development and maintenance of traits such as height, weight, IQ, aspects of personality, and especially susceptibility to a host of diseases.

Genome wide association studies depend on a particular type of genetic variation called a single-nucleotide polymorphism (SNP, pronounced SNiP). In describing Watson and Crick's scale model of DNA in Chapter 6, I explained that the backbone of a molecule of DNA is a sequence of nucleotides (Figure 6.2). Only four types of nucleotides occur in DNA—adenine (A), guanine (G), thymine (T), and cytosine (C), but the DNA in a chromosome is a very long molecule, and the nucleotides can occur in any sequence, so a huge amount of variation is possible. SNPs, however, are variations in single nucleotides that occur at exactly the same position on a particular chromosome in different individuals of a species. For example, if one person has a single-stranded stretch of DNA on chromosome 13 like this... ATTCGCGAA... and another person has... ATTTGCGAA ... at exactly the same position on his chromosome 13, then the different nucleotides in position 4 of these sequences (C in the first person and T in the second person) are a SNP. SNPs usually exist in two alternate forms, or alleles, in a population. Since each individual has two copies of each chromosome, he can be homozygous for a SNP if he has the same allele on both chromosomes or heterozygous if he has alternative alleles on the two chromosomes.

Although many nucleotide locations are identical in all individuals of a species, researchers have used "next generation sequencing" to identify thousands of SNPs in humans and other species. This sets the stage for a genome wide association study in which researchers look for correlations between particular SNPs and various traits in a population. Do people with colon cancer have different SNPs than people without colon cancer? What SNPs are more common in taller people than in short people? Are certain SNPs associated with schizophrenia? If a particular SNP is associated with a trait to an extent that can't be explained by chance, this suggests the presence of a gene located near the SNP that influences the trait. This means that the gene is not only on the same chromosome as the SNP but also physically close enough to the SNP that the gene and the SNP are inherited together rather than being separated in the dance of the chromosomes that occurs generation after generation in meiosis (see Chapter 4).

It's important to emphasize that genome wide association studies don't always identify specific genes associated with a particular trait, but only suggest where in the genome to look for such genes. This is because the associations highlighted in a GWAS are between SNPs and traits, and SNPs aren't genes but just single nucleotides. In Chapter 7 I discussed one example of a specific gene that *was* identified by a GWAS: the *FTO* gene associated with risk of obesity. Researchers knew about this gene in humans because it is homologous to one in mice. When the GWAS identified an association between risk of obesity and a mutant allele of *FTO* in humans, the function of the normal allele wasn't known. Further studies of animal models, both mice and rats, as well as broader biochemical comparisons across organisms from fish to algae and bacteria suggested that *FTO* might be involved in control of feeding and energy use.

This background sets the stage for describing "the mystery of missing heritability." I used height as a main example for developing the concept of heritability of human traits; we found that heritability of height is about 80% in most populations that have been studied. Human geneticists have understood for at least 100 years that variation in height among people is strongly related to genetic variation and that multiple genes must be involved. In 2007, Michael Weedon and his colleagues were the first researchers to use a GWAS to look for genes associated with variation in height. They studied 365,000 SNPs in an initial sample of almost 5,000 individuals and then replicated their study with about 19,000 more individuals. They found an allele of one gene, called *HGMA2*, that had a clear effect on height. People with one copy of this allele were 0.4 centimeters taller on average than those with no copies of the allele; people with two copies of the "tall" allele were 0.8 centimeters taller than those with no copies. This translates to a difference of about 1/3 inch for two copies of the allele, or less than 1% of normal variation in height.

In their large-scale GWAS, Weedon's group found 43 other genes associated with height variation in humans, and other researchers increased the total number to 50 genes by 2009. However, the collective contribution of variation in these genes to variation in normal height was less than 10%. This is much less than the estimated 80% contribution of genetic variation to height variation derived from twin studies. What accounts for the remaining 70% of the heritability? This is the mystery of missing heritability—how to reconcile relatively large estimates of heritability from studies of twins and other relatives with results of genome wide association studies that account for a small proportion of total variation in traits despite using huge numbers of SNPs and very large samples of people.

There are several possible solutions to this mystery. There may be many more than 50 common genes with small individual effects on complex traits like height, or there may be some relatively rare genes that have large effects but haven't been discovered yet. Discovering either of these possibilities will require even larger numbers of SNPs and people willing to provide DNA and be measured for height or another trait of interest.

Another possible solution involves more sophisticated statistical analysis of GWAS data. One problem with using huge sample sizes for these or other statistical analyses is that researchers must account for the possibility that an association between a SNP and height, for example, is just due to chance. Weedon's group used 365,000 SNPs. Each of these has two alternative alleles—call them *A* and *a*. So each of the 24,000 people they studied could be *AA* or *Aa* or *aa* for the first allele, *AA* or *Aa* or *aa* for the second allele, and so on. For each SNP, individuals of the three genotypes will differ in height. For some of the SNPs, the differences between *AA*, *Aa*, and *aa* may be substantial simply by chance, just as if you flipped 10 coins 365,000 times you might get 10 heads or 10 tails a few times even if the coins were fair. So initial GWAS analyses used a high threshold to accept an association between a SNP and variation in a phenotypic trait as real; for example, researchers might only accept it if there was less than one chance in a billion that it was simply random. In 2010, Peter Visscher and several colleagues described a new statistical approach to these data that was more sensitive for detecting genes with small effects without producing false positive results. They analyzed almost 300,000 SNPs in about 4,000 individuals and were able to account for 45% of variation in height of these individuals by differences in their SNP alleles. Based on this and other studies, researchers have identified at least 180 genes contributing to variation in height of humans. Many of these are involved in networks of biochemical processes that influence growth of the skeleton, although the specific functions of some genes associated with height variation remain unknown. However, much of the genetic variation in height based on heritability estimates still can't be attributed to specific genes.

This discussion so far has been based on the premise that heritability estimated by classical methods like twin studies is accurate, and therefore missing heritability represents a limitation of genome wide association studies. Based on this premise, if GWAS researchers can't account for all of the heritability estimated from classical methods, they either need to use more SNPs, larger samples of people, or better statistical methods. An alternative possibility, however, is that classical methods often overestimate heritability, and this accounts for the gap between estimates from twin studies and estimates from genome analyses using molecular techniques. This is a real and perhaps likely possibility because the classical methods described in Chapter 7 make three key assumptions that may be unjustified. First, they assume that the effects of different alleles of a gene that influences a trait are independent of each other. For some genes, however, one allele may be dominant to other alleles of the same gene. If there are multiple alleles of a gene in a population, *homozygous* individuals will have the same allele on both members of the pair of chromosomes that contain that gene, while *heterozygous* individuals will have different alleles on these chromosomes. If one allele is dominant to the other, then the heterozygote will have the same phenotype as the homozygote with two copies of the dominant allele; the only different phenotype will belong to the homozygote with two copies of the recessive allele. I illustrated dominance with two examples—color of peas in Mendel's breeding experiments and PKU in humans.

The second way in which effects of genes may not be independent applies to different genes at different loci, perhaps even on different, unpaired chromosomes. This is called *epistasis* and relates to the fact that many genes code for enzymes that function in sequences of chemical reactions. If a mutated allele of a gene prevents an early step in such a biochemical pathway, then alleles of other genes that influence later steps in the pathway don't matter, because the pathway is blocked before those alleles have a chance to act. Coat color in cats illustrates epistasis. A gene called agouti influences whether hairs are banded: *AA* and *Aa* cats have a pattern of stripes or swirls caused by light and dark bands on individual hairs, while *aa* cats are solid black. Another gene called orange causes reddish/orange coloration instead of black in cats that have *OO* or *Oo* genotypes. But the orange allele, *O*, is epistatic to the black allele, *a*, so *OOaa* and *Ooaa* cats have banded orange hairs even though their alleles at the agouti locus would make them uniformly colored otherwise. In short, because of epistatis, there are no solid orange cats.

For each of the thousands of genetic loci in a person's genome, she inherits one allele from her mother and one from her father. However, the dominance relationships between alleles at one locus and the epistatic relationships between alleles at different loci are established anew in each generation. Therefore, strong dominance and epistatic relationships between alleles reduce heritability because, unlike the alleles themselves, dominance and epistasis are genetic effects that aren't passed directly from parents to offspring. Analyses of twin data can account for shared environmental variance, as we did in Chapter 7, or variance due to dominance but not both, although analyses that include other kinds of relatives like parents and offspring can be used to distinguish these effects. Accounting for epistatic effects is even more complex, and methods to do this are just beginning to be developed.

The third key assumption of classical methods for estimating heritability from data on twins and other related individuals is that genes and environments have separate and independent effects on phenotypes. You saw several examples where this assumption is false, most notably the study by Turkheimer and his colleagues showing that IQ is much more heritable in families of high socioeconomic status than in families of low socio-economic status. These kinds of interactions between genetic and environmental effects are likely to be common if not universal in studies of human traits because of the tremendous genetic *and* environmental heterogeneity of human populations. This contrasts with studies of domestic animals, where researchers can raise a set of genetically variable individuals in a uniform environment and make practical use of heritability estimates for selective breeding to enhance economically desirable traits (see also Box A5.1).

BOX A5.1 Unraveling Complexities of Causation with New Research on Twins

In Chapter 7 and this appendix, I wrote as if there is a clear separation between genetic and environmental factors that influence traits of organisms, although I made a point to illustrate interactions between these two kinds of factors in plants, birds, and humans. Each individual organism has a unique environment—our diets, families, and friends, for example. The genetic material in each of our cells also has an environment, and recent technological advances have enabled detailed studies of these environments, opening up new opportunities to understand complexities of causation. Environments of genes on chromosomes include several elements that influence whether these genes are active or inactive in a particular cell. If a gene is active, its genetic code is used in building a specific protein in that cell; if inactive, that protein is not built. One such element that can regulate expression of genes involves addition of a small group of atoms called a methyl group to cytosine, which is one of the four nucleotides in DNA that comprise the genetic code (see Figure 6.2). This process is called *DNA methylation*; by controlling whether certain genes in a particular type of tissue are expressed or not, DNA methylation can have profound effects on a person's phenotype.

(Continued)

BOX A5.1 Continued

The study of DNA methylation and other processes regulating expression of genes is *epigenetics*, and it is one of the most active areas of research in genetics today. Researchers are beginning to do epigenome wide association studies to understand cellular and molecular processes that contribute to diseases by inactivating particular genes. In 2010, the Department of Twin Research and Epidemiology at King's College London and the Beijing Genomics Institute began a large-scale epigenome wide association study called EpiTwin on 5,000 pairs of twins, 3,000 of which are monozygotic. By searching for genes that differ in methylation in monozygotic twins that differ in their status for various diseases, the researchers hope to discover genes that make important contributions to these diseases. This, in turn, may lead to better understanding of the causes of the diseases and eventually, perhaps, to more effective treatments. There is already some evidence for specific differences in DNA methylation in twin pairs with and without schizophrenia, bipolar disorder, and type 1 diabetes, although sample sizes are too small to be definitive. With the large number of twin pairs in the EpiTwin project, the London and Beijing researchers should be able to clarify the picture. Despite having the same DNA, monozygotic twin pairs differ more often than not in traits ranging from autism to breast cancer and colon cancer. Differences in their epigenomes may be one reason for these differences in their diseases.

Eric Turkheimer wrote a very interesting essay entitled "Three Laws of Behavior Genetics and What They Mean." His first law is that "all human behavioral traits are heritable." Turkheimer argues, however, that despite the evidence for heritability of traits like IQ and personality, the development of all human behavior involves a complex web of causation. One reason for the broad evidence for heritability of traits ranging from height and IQ to popularity and political activism is that we can easily estimate heritability by comparing monozygotic twins who share all their genes to dizygotic twins who share 50% of their genes. There are no parallel groups of people who share 100% or 50% of their environmental circumstances that we can use to directly estimate environmental contributions to variation in human traits. The most interesting and important challenge in studying behavior and other human traits is learning how they develop over time as a consequence of interactive effects of genes and environments. Genomic studies using contemporary technology will contribute to this process but not provide immediate or easy answers.

Mapping Arguments to Aid Critical Thinking about the Weight of Evidence

Tim van Gelder is an Australian philosopher who received the 2001 Eureka Prize for Critical Thinking from the Australian Museum. While teaching at the University of Melbourne in 2004, van Gelder founded Austhink Software to develop computer programs to help students learn tools for critical thinking. In 2005, he published a paper outlining six key lessons about critical thinking. Van Gelder's first lesson is that "critical thinking is hard." Critical thinking is hard because it doesn't come naturally and it takes plenty of practice with lots of different applications, informed by an understanding of basic principles. One particular challenge to critical thinking is our inherent propensity to interpret information so as to maintain our current beliefs. Confirmation bias is an important element of such belief preservation that we discussed in Chapter 2.

Van Gelder argues that we can improve our critical thinking skills by making diagrams of arguments about issues that engage us. These diagrams provide a useful summary of the evidence on both sides of an argument, but, more important, they help reveal the logic of the pro and con positions and any hidden assumptions that underlie these positions. Through Austhink Software, van Gelder released a program called Rationale™ for making diagrams of arguments called *argument maps*. This program includes extensive learning aids such as practice exercises and tutorials; see the Resources for Further Exploration for more information.

Argument mapping can be used for all kinds of purposes, including weighing the pros and cons of personal decisions and evaluating the merits of alternative arguments about issues of public policy. In a review of *Rationale* published in *The Atlantic* in 2007, James Fallows wrote:

Van Gelder's initial sales target was schools and universities, but he increasingly sells to consulting firms, government agencies, and other groups wrestling with decisions, as well as to individuals. The strongest interest has come from U.S. intelligence agencies, which are using the software to train analysts to think critically about intelligence claims.

For scientific questions, argument mapping is especially useful when we don't have conclusive evidence favoring a certain hypothesis but have to rely on the weight of lots of different lines of evidence to decide between alternative hypotheses. This was the case for the examples discussed in Chapter 8, including the question of why the population of sea otters in the Aleutian Islands declined precipitously in the 1990s. Plates 19 to 23 map some of the arguments about answers to this question.

Plates 19 to 23 include all of the hypotheses proposed by Estes and his colleagues in their 1998 discussion of the dramatic decline in number of sea otters in the Aleutian Islands in the 1990s and by Kuker and Barrett-Leonard in their 2010 critique of the killer whale predation hypothesis favored by Estes's group. The plates also include much, but not all, of the evidence and arguments in these papers as well as in an unpublished rebuttal of Kuker and Barrett-Leonard by the Estes group.

The building blocks of an argument map are sentences called claims, contentions, or premises depending on where they are used. Each claim appears in its own box in Plates 19 to 23. The claims are connected in a tree-like structure, with the most general claim at the top and increasingly more specific claims lower in the tree. In reading an argument map, we start at the box at the very top of the tree, which contains the main claim that is the subject of the map. In this case, I use a question instead of a statement—What caused the population of sea otters in the Aleutian Islands to crash in the 1990s?

The next level below the question in the tree contains three larger boxes, each with two claims. The main claims in each of these boxes are general hypotheses to account for the decline of sea otters in the Aleutians—emigration, reduced birth rate, or increased death rate. The secondary claims in each of these boxes are called *co-premises*; that is, they work together with the main premise (another name for a claim) to make a coherent argument. Co-premises often represent assumptions required for the argument to be valid. For example, one hypothesis for the drop in the sea otter population is that many otters moved elsewhere in the 1990s, but this would only cause a decline if there wasn't a compensating increase in birth rate of the remaining otters.

Working down in the tree, there are levels in parts of the tree that contain sub-hypotheses, for example, specific factors that might cause mortality to increase in Plates 19 to 23, and even sub-sub-hypotheses in Plates 22 and 23. Boxes in levels below these show evidence and arguments used for evaluating the alternative hypotheses, sub-hypotheses, and sub-sub-hypotheses. Some of these boxes are shaded green if they support the claim above to which they are connected; others are pink if they oppose the claim above or orange if they rebut an opposing argument.

One purpose of an argument map for a scientific question is to lay out a comprehensive set of alternative hypotheses to answer a question together with supporting and conflicting evidence and arguments for each hypothesis. However, you can also use an argument map to evaluate an overall argument. After reading the map from the top down to understand its structure and components, you need to read it from the bottom up in your evaluation. How credible is each piece of evidence in the bottom layer of the tree? Does it include supporting data, or is it simply a description of an observation or a statement of an opinion? Is the source of the evidence reliable, for example, someone with appropriate expertise or experience? Does the evidence logically support or refute the claim to which it is connected in the tree? Are there hidden assumptions that weaken the value of a particular piece of evidence? Think about this argument map in reaching your own conclusion about how strongly the weight of evidence supports or refutes the hypothesis that predation by killer whales accounted for the population crash of sea otters in the Aleutian Islands in the 1990s.

Resources for Further Exploration

1. I used Rationale™ software to make an argument map of alternative explanations for the crash of the sea otter population in the Aleutian Islands (Plates 19–23). You can learn more about this program and use a free trial version or purchase a copy for a relatively modest price at http://rationale.austhink.com/. The software includes extensive examples, tutorials, and exercises for learning the details of argument mapping. I introduced the process in this appendix, but there's a lot more depth that you can explore with Rationale™.

What Are the Benefits of Organic Farming? The Weight of Evidence

Many questions in science aren't resolved as cleanly as the question of whether chronic fatigue syndrome is related to the XMRV virus (Chapter 9). This was a question of great interest to patients and physicians, so the initial report of XMRV in chronic fatigue syndrome patients by Mikovits and her colleagues immediately stimulated several attempts to replicate the result, culminating in the test organized by Lipkin that caused even Mikovits to abandon the connection between XMRV and chronic fatigue syndrome. In other cases, attempts at replication give inconsistent results. Over time, a set of studies of the same question by different research teams accumulates. These studies may answer the question in different ways—yes, no, or maybe. Some studies are more convincing than others because they are designed better (e.g., an experiment may be more convincing than a strictly observational study) or because they have larger sample sizes. It's often impossible, however, to pick out one study that trumps all the others and answers the question with certainty. Therefore researchers need to consider the weight of evidence from all relevant studies to reach a conclusion, and they often use a statistical technique called meta-analysis to do this. As described by Kenneth Chang, meta-analyses "seek out robust nuggets in studies of disparate designs and quality that offer confounding and often conflicting findings, especially in nutrition and medicine" (Chang 2012). In one classic example, a large group of researchers used a meta-analysis of 287 experimental studies involving more than 200,000 patients to show that low doses of aspirin protect against heart attacks and strokes.

I can't describe meta-analysis in detail in this book (see Jenkins 2004 if you want to learn more). However, the news media often report results of meta-analytic studies of a wide variety of topics, so I will use a recent and controversial example to help you interpret such stories. On September 3, 2012, the *New York Times* reported that "Stanford scientists cast doubt on advantages of organic meat and produce." This headline mirrored those in other national news outlets and stimulated a flurry of commentary by advocates of organic farming.

The Stanford scientists published their analysis in *Annals of Internal Medicine* with the title "Are Organic Foods Safer or Healthier than Conventional Alternatives?" Their basic methods were similar to those of all meta-analyses: searching bibliographic databases for published studies addressing their question, selecting studies that met pre-specified criteria for quality, extracting a quantitative measure of the answer to their question from each of these studies, and finally combining the answers from the individual studies into an overall measure. The quantitative measures used in meta-analysis are called *effect sizes*; in this case, the effect size was the difference in nutrient or contaminant level between the same crop grown conventionally and organically. The overall estimate of the size of an effect isn't a simple average of the effect sizes from individual studies but an average that gives more weight to effect sizes that are measured more precisely, either because of research designs with better control of other variables that might confound the results or because of larger sample sizes.

The Stanford researchers, Crystal Smith-Spangler and her colleagues, used only peer-reviewed studies reported in English. They found 17 comparisons of people eating organic and conventionally grown food for short periods of time and 223 comparisons of nutrients or contaminants in these types of food. I'll focus on the contaminant comparisons to illustrate how meta-analysis works since these were more straightforward than the nutrient comparisons. About two-thirds of the food studies compared produce (fruits, vegetables, or grains), and one-third compared animal products (meat, poultry, milk, or eggs). Half of the produce studies were experiments in which the same type of food was grown on the same farm using separate plots with organic and conventional methods, but only 11% of the studies of animal products were experimental. The remaining studies were comparative but did not qualify as experiments for various reasons; for example, farmers did not assign fields randomly to the organic or conventional treatment. The main differences in the two types of treatments were the use of inorganic fertilizer and pesticides on conventional fields and organic fertilizer with no pesticides on organic fields.

Since organic farmers don't use pesticides, crops grown organically should have fewer pesticide residues than crops grown using conventional practices. On the other hand, crops grown organically might have greater contamination by bacteria such as the common intestinal bacterium *Escherichia coli* because organic farmers use organic fertilizer, that is, manure. What did the Stanford researchers show with their meta-analysis? They found nine studies that compared pesticide residues in fruits, vegetables, or grains grown organically or conventionally. Sample sizes varied from 10 to more than 4000 in these studies. One group of researchers found no pesticide residues in any of the organically or conventionally grown samples, but they only had 10 of each. In the remaining eight cases, more conventionally grown than organically grown samples had residues of one or more pesticides. On average, conventionally grown fruits, vegetables, and grains were 32% more likely than organically grown produce to have pesticide residues. The 95% confidence limit on this estimate was 25% to 39%; that is, there was only a 5% chance that the real difference was less than 25% or greater than 39%.

The results were quite different for *E. coli*. Smith-Spangler and her coworkers found five comparisons of contamination by *E. coli* of organically and conventionally grown produce. In four of the comparisons, organically grown food had somewhat more *E. coli*; in the fifth, the opposite occurred. Overall, they found no significant difference in contamination by *E. coli* of organically and conventionally grown produce. They also found no differences in contamination by two other kinds of bacteria in chickens raised organically and conventionally.

One of the most interesting results of the Stanford study involved antibiotic-resistant bacteria. In conventional farming, poultry and livestock are typically fed low doses of antibiotics to stimulate growth. This, together with excessive use of antibiotics in human medicine, can cause bacteria that live in cows, pigs, chickens, and people to evolve resistance to antibiotics, limiting use of these antibiotics for treating disease. If a person is infected by a bacterium that is resistant to several antibiotics, this makes it especially difficult to treat the infection. Smith-Spangler and colleagues found five comparisons of bacteria resistant to three or more antibiotics in organically and conventionally raised meat, four comparisons for chicken, and one for pork. In all cases, more conventionally grown samples had bacteria resistant to three or more antibiotics.

The Stanford researchers found clear evidence that organic food was less likely to contain pesticide residues and antibiotic-resistant bacteria than conventionally grown food although they downplayed the significance of these results because they found little direct evidence of health benefits of organic diets. Some commentators criticized their cautious interpretation of these results because the Stanford team only considered *presence* of pesticide residues in food, not amounts or toxicities, and they ignored the possibility that conventionally grown foods might be more likely to contain residues of multiple pesticides with synergistic effects. The Stanford researchers also ignored a large amount of data collected by the Pesticide Data Program of the US Department of Agriculture (USDA), presumably because these data weren't reported in the peer-reviewed literature. However, the USDA uses a rigorous protocol to collect and analyze food samples and posts annual summaries of the results on the Web. The USDA also posts all of the data collected each year, so any interested person can download these data and do their own analysis if they aren't satisfied with the analysis by USDA scientists. One limitation of these data is that they rely on marketing claims reported by farmers. How reliable are claims by farmers that their food was grown organically or free of pesticides? However, these uncertainties probably also bedevil some of the non-experimental but peer-reviewed studies used by the Stanford team in their meta-analysis, so this isn't necessarily a good reason to exclude the USDA data from the meta-analysis, as done by the Stanford team.

Mark Bittman is a food columnist for the *New York Times*. He quotes Crystal Smith-Spangler, lead author of the Stanford study, as saying, "Some believe that organic food is always healthier and more nutritious. We were a little surprised that we didn't find that." Bittman's response is "How can something that reduces your exposure to pesticides and antibiotic-resistant bacteria not be 'more nutritious' than food that doesn't?" (Bittman 2012). According to Bittman, the main point of eating organic food is to reduce exposure to pesticides; the Stanford researchers found clear evidence supporting this, so he's puzzled about why they downplayed this evidence and why his own newspaper, for example, headlined the story with "Stanford Scientists Cast Doubt on Advantages of Organic Meat and Produce."

The Stanford researchers also compared nutrients such as vitamin C, calcium, iron, and protein in organic and conventionally grown food, finding few differences; this was the basis for the statement by Smith-Spangler to which Bittman objected. Kristen Brandt and three other British researchers had published a meta-analysis of nutritional content of fruits and vegetables in 2011, a year before publication of the Stanford study. The Stanford and British researchers reviewed many of the same papers and considered several of the same nutrients but reported some different results and came to markedly different conclusions. For example, the British team reported more vitamin C in organically grown produce while the Stanford team did not, and the British team emphasized the nutritional benefits of eating organically grown food while the Stanford team did not. Another difference is that the British study received much less attention in the news media than the Stanford study. Although Charles Benbrook of the Center for Sustaining Agriculture and Natural Resources at Washington State University argued that the British study was more rigorous than the Stanford study, I found strengths and weaknesses in both. For example, the British researchers accounted more sensibly for effects of confounding factors like weather on nutrient levels in produce grown organically and conventionally, while the Stanford researchers used more appropriate statistical methods for weighting individual studies in estimating average effect sizes. Despite the relatively large number of studies that have asked whether organic foods are more nutritious than those grown using conventional methods, we don't yet have a clear answer to this question. There is good evidence, however, that organic foods have lower pesticide levels and harbor fewer antibiotic-resistant bacteria.

Finally, how should you evaluate an issue like this without having enough statistical background to understand the details of meta-analyses that are the basis for news reports? If you learned about the Stanford study from reading Tom Philpott's column in *Mother Jones* on September 5, 2012, or Mark Bittman's column in the *New York Times* on October 2, 2012, you should look for assessments from other journalists not committed to organic farming. You might look up the original scientific reports by Kirsten Brandt and colleagues in 2011 and Crystal Smith-Spangler and colleagues in 2012. Even if you don't understand all of the technical details of their papers, you can get a sense of what issues were most important to the two teams. In 2013, *Annals of Internal Medicine* published a series of letters critiquing the Stanford study, with a rebuttal by part of Smith-Spangler's team; this exchange of views is not too technical and is fascinating reading (all of these sources are listed in the bibliography). The main thing is to think critically about whatever you read on this or another topic of interest. If you have a prior preference for a particular perspective—for example, that organic foods are safer and healthier—try especially hard to think critically about arguments favoring this perspective. This might cause you to change your preference, but it might also strengthen it by helping you understand it better (see Question 1).

Questions to Ponder

1. (a) Smith-Spangler and her colleagues at Stanford used published, peer-reviewed studies in their meta-analysis of possible health benefits of organic food, while Brandt and her English colleagues included "conference proceedings and other non-reviewed publications." How might this difference contribute to the different conclusions of the two research teams? Could it contribute to the greater emphasis by the British on nutritional differences between organic and conventionally grown food? Explain your answer.

   (b) People who do meta-analyses worry about the "file-drawer problem," that is, the possibility that studies relevant to the question they are trying to answer exist, but haven't been published. How might this bias the results of a meta-analysis? Think about a concrete example to develop your answer, perhaps the example of nutritional differences between organic and conventionally grown food. Is there any feature of the results of such studies that would cause the authors not to publish these studies? If so, what would be the consequence of a meta-analysis that didn't use the unpublished studies?

2. On December 9, 2013, Kenneth Chang of the *New York Times* reported that a new study "is the most clear-cut instance of an organic food's offering a nutritional advantage over its conventional counterpart." The study compared the fatty acid composition of milk from cows raised using conventional and organic methods, and the authors of the study showed that milk from organically raised cows had more omega-3 fatty acids, which are considered beneficial for cardiovascular health. The main reason for the difference was that cows raised using organic methods spend more time feeding on pasture grasses while cows raised using conventional methods eat more corn, which has more omega-6 and less omega-3 fatty acids. Nutritionists agree that omega-3 fatty acids are beneficial, but disagree about the detrimental effects of omega-6 fatty acids. Read Chang's report and assess the evidence for the nutritional benefits of drinking milk from cows raised organically. What further studies are needed to fully evaluate the relative nutritional value of milk from organic and conventional sources?

3. Stories about science in books like this are often incomplete because new research adds to the stories or changes their conclusions. For example, a group of 18 researchers published a new meta-analysis comparing organic and conventionally grown food in September 2014, several months after I wrote this appendix and while the book was in production. Scan this paper by Barański and colleagues to see how the conclusions differed from those of Smith-Spangler's group and Brandt's group. Can you get a sense for why the conclusions of these three studies differed? Is this new study more credible than the earlier ones?

Global Climate Change: Evaluating Expert Opinion

From reading this book, you know that I recommend thinking about the details of what scientists do to understand and evaluate their conclusions. This is more difficult in some cases than others, especially when complex models play an important role, as they do for global climate change. In these cases, nonprofessionals may have to rely on opinions of experts without a full understanding of the scientific basis for these opinions. For scientific questions with important policy implications, we may ask whether researchers who study these questions have reached a consensus about the answers. If so, we can focus on debating the best policies to address major risks identified by scientists rather than debating the science itself.

Contributions of humans to global climate change and projected consequences of these contributions are complex scientific questions with profound policy implications. You and I are not climatologists; therefore we must rely in part on their expert opinions about these questions. We need some tools to evaluate expert opinion about issues involving complex science, especially to decide whether there is a strong consensus among experts about such issues. These tools complement other tools that we've discussed in this book for evaluating evidence from experiments, comparisons and correlations, and so on. Tools for evaluating expert opinion can be applied in many areas of science and may be relevant to your personal decisions about matters such as health or nutrition.

One challenge is figuring out who counts as an expert. You actually have some experience with this already. Say you twisted your ankle. You'd probably ask a doctor if you broke it. If instead your car was making a funny noise, you'd ask an auto mechanic for a diagnosis. It's not so easy to find out who counts as an expert on an issue like climate change, partly because there may be people who claim to be experts to advance a particular political agenda. These self-proclaimed experts may even be scientists but may not be knowledgeable or objective about climate change.

Two potential sources of information about scientific consensus on particular issues are reports by professional organizations and by think tanks, advocacy groups, and lobbyists. An advocacy group may be an untrustworthy source for evaluating expert opinion because it is an organization devoted to promoting a particular opinion, possibly (though not always) at the expense of scientific objectivity. For example, the Vaccine Risk Awareness Network is an organization "designed to provide you with more information about risks and potential side-effects of vaccines, to get support if you or someone you know may have suffered adverse reactions from vaccines, and to foster a multi-disciplinary approach to child and family health [which] continues the work of The Committee Against Compulsory Vaccination." The mission of this organization is clear; it would be naïve to expect them to provide a dispassionate account of the evidence on benefits as well as risks of vaccination. Huge numbers of websites are maintained by various advocacy groups, so searching the Web for objective information about controversial issues is challenging. In some cases, an advocacy group or a think tank may use a name that obscures its political agenda. Your only recourse is to be ever vigilant, especially when relying on the Web for information.

Professional organizations also study various controversial issues and may provide more objective analyses of evidence because their work is not driven by a specific political agenda, but rather by the desire to maintain their credibility as professionals. In the United Sates, the most prestigious organization of scientists, encompassing all the sciences, is the National Academy of Sciences (NAS). Current members elect new members based on distinguished records of research; as of 2009 the NAS had about 2,000 members, 10% of whom had won Nobel Prizes. The NAS typically produces reports in response to requests by federal agencies or members of Congress. For example, in 2001 President George W. Bush requested a report on "the areas in the science of climate change where there are the greatest certainties and uncertainties," including an evaluation of the Third Assessment Report of the Intergovernmental Panel on Climate Change. The committee of scientists appointed by the NAS to prepare this report concluded that the Intergovernmental Panel on Climate Change was largely on the mark in their assessment of the causes and potential consequences of global climate change.

Another source of information about the degree of consensus about climate change comes from opinion surveys of experts. Although surveys of experts may include some of the same questions as surveys of the general public, their purpose is much different. Our goal in this appendix is to learn some tools for evaluating expertise, so we'll ignore public opinion here because it is influenced by many factors besides knowledge of the scientific evidence. The opinions of experts aren't immune to these other factors, but one would expect that they are more influenced by the scientific evidence itself than the opinions of nonprofessionals.

Two recent surveys with contrasting results illustrate what you and I must deal with in deciding whether there is a convincing consensus of experts about the essential features of climate change. Stacy Rosenberg and three colleagues published the results of one survey in the journal *Climatic Change* in 2010; Lianne Lefsrud and Renate Meyer published the results of the other survey in *Organization Studies* in 2012. Rosenberg's group surveyed 468 scientists, while Lefsrud and Meyer surveyed 1,077. The researchers asked different questions and analyzed their results differently, making direct comparison difficult. Despite these differences, there were some dramatic contrasts in the general patterns of results.

Rosenberg's group asked three key questions: whether scientists are certain that global warming is happening, whether scientists are certain that humans contribute to global warming, and whether there is enough certainty about global climate change that we need to take action now to deal with climate change (they did their survey in 2005, so "now" means a decade ago). Among respondents to this survey, 94% agreed that scientists are certain that global warming is happening, 88% agreed that humans contribute to global warming, and 91% agreed that immediate action is necessary.

Lefsrud and Meyer classified their respondents in five groups based on their answers to several multiple-choice questions as well as written comments. They described these groups as "Comply with Kyoto," "Nature Is Overwhelming," "Economic Responsibility," "Fatalists," and "Regulation Activists." Kyoto refers to the Kyoto Protocol to the United Nations Framework Convention on Climate Change, which is an international treaty designed to reduce emissions of greenhouse gases. The Kyoto Protocol has been ratified by all members of the United Nations except Andorra, Canada, South Sudan, and the United States (Canada initially ratified the treaty but withdrew in 2011). In Lefsrud and Meyer's analysis, the "Comply with Kyoto" group corresponds most closely with the strong majority of respondents to Rosenberg's survey who believed there is sufficient evidence of human impacts on climate to take action to mitigate these impacts. Thirty-six percent of Lefsrud and Meyer's respondents were in this group. The "Nature Is Overwhelming" group in their study largely dismissed human effects on climate; this group comprised 24% of their respondents. The "Economic Responsibility" group comprised 10% of respondents who believed that the causes of climate change were uncertain and did not justify policies for mitigation. "Fatalists" were 17% of respondents who believed that both natural and human factors contribute to climate change but that policy responses were unjustified. Finally, 5% of respondents were regulation activists who believed that the science is uncertain but favored environmental regulations, although were skeptical of the efficacy of the Kyoto Protocol. The remaining 9% of respondents could not be clearly categorized. This analysis was highlighted by an op-ed published in *Forbes Magazine* under the headline "Peer-Reviewed Survey Finds Majority of Scientists Skeptical of Global Warming Crisis."

Although questions asked in the two surveys were different, three were similar enough for answers to be compared. While 88% of scientists surveyed by Rosenberg's group agreed that scientists are certain that humans contribute to global climate change, only 24% of scientists surveyed by Lefsrud and Meyer agreed that the scientific debate on causes of recent climate change is resolved. While 91% of those surveyed by Rosenberg and colleagues thought immediate action on climate change is necessary, 54% of those surveyed by Lefsrud and Meyer thought that climate change is a "significant risk to public safety and interest." Finally, 77% of respondents to the survey by Rosenberg's group believed that projections of climate models are "moderately accurate," but only 19% of respondents to the survey by Lefsrud and Meyer agreed.

Scientists surveyed by Lefsrud and Meyer were clearly much more skeptical than those surveyed by Rosenberg's team about scientific consensus on the causes and consequences of climate change. What accounts for these differences? Should you and I, as nonexperts about climatology, give more credence to one of the two studies, or is there some way to reconcile the results?

I left out one key detail in this story that may help you answer these questions—the pool of individuals surveyed by each research group. Rosenberg and colleagues sent their survey to all authors of research articles published between 1995 and 2004 in 13 scientific journals that deal with climatology. These journals publish peer-reviewed articles, that is, articles reviewed by scientists who are experts in the particular topics of the articles before they are accepted for publication (see Chapter 9). If reviewers identify deficiencies, editors either reject the articles or require revision to correct the deficiencies before publication. Most of the respondents to this survey were trained in atmospheric science (43%), ecology (11%), oceanography (9%), or physics (8%).

By contrast, Lefsrud and Meyer surveyed members of the Association of Professional Engineers and Geoscientists of Alberta (Canada) at the request of the leadership of this organization. Most of the respondents were professional engineers (70%) or geologists (14%); the rest were engineers or geologists in training. Many respondents were employed by the petroleum industry, which is the largest private-sector source of investment in Canada. Only 24% worked professionally on climate change.

Is there a consensus among relevant experts about the existence, causes, and potential consequences of global climate change? These surveys of two groups of scientists provide one kind of evidence for answering this question; the literature reviews by Oreskes, Powell, and Cook and colleagues described in Chapter 10 provide another kind. Which of these kinds of evidence is more credible? Whatever decision you reach when you are faced with the challenge of evaluating expertise about scientific questions, I recommend that you engage the evidence with an active and curious mind, just as you do for the other kinds of evidence discussed in this book.

Bibliography

PREFACE

Lett, J. 1990. "A Field Guide to Critical Thinking." *Skeptical Inquirer* 14(4): 153–160.

Van Gelder, T. 2005. "Teaching Critical Thinking: Some Lessons from Cognitive Science." *College Teaching* 53:41–46.

CHAPTER 1. DISCOVERY AND CAUSATION

Introduction

Dobson, A. P., M. Borner, A. R. E. Sinclair, P. J. Hudson, T. M. Anderson, G. Bigurube, T. B. B. Davenport, et al. 2010. "Road Will Ruin Serengeti." *Nature* 467:272–273.

Egevang, C., I. J. Stenhouse, R. A. Phillips, A. Petersen, J. W. Fox, and J. R. D. Silk. 2010. "Tracking of Arctic Terns *Sterna paradisaea* Reveals Longest Animal Migration." *Proceedings of the National Academy of Sciences USA* 107:2078–2081.

Norment, C. J. 1994. "Breeding Site Fidelity in Harris' Sparrows, *Zonotrichia querula*, in the Northwest Territories." *Canadian Field-Naturalist* 108:234–236.

Norment, C. 2007. *Return to Warden's Grove: Science, Desire, and the Lives of Sparrows.* Iowa City: University of Iowa Press.

Discovering the Winter Home of Monarch Butterflies

Taylor, C. 2013. "Monarch Watch." http://www.monarchwatch.org/ (accessed November 8, 2013).

McNeil, D. G., Jr. 2006, October 3. "Fly Away Home." *New York Times.*

Urquhart, F. A. 1976. "Discovering the Monarch's Mexican Haven." *National Geographic* 150(2):160–173.

Urquhart, F. A. 1987. *The Monarch Butterfly: International Traveler.* Chicago: Nelson-Hall.

What Did Dinosaurs Look Like?

Li, Q., K. Q. Gao, J. Vinther, M. D. Shawkey, J. A. Clarke, L. D'Alba, Q. Meng, D. E. G. Briggs, and R. O. Prum. 2010. "Plumage Color Patterns of an Extinct Dinosaur." *Science* 327:1369–1372.

Mayr, E. 1961. "Cause and Effect in Biology." *Science* 134:1501–1506.

Tinbergen, N. 1963. "On Aims and Methods of Ethology." *Zeitschrift für Tierpsychologie* 20:410–433.

Zimmer, C. 2010, February 4. "Evidence Builds on Color of Dinosaurs." *New York Times.*

Proximate Causes of Migration by Monarch Butterflies

Goehring, L., and K. S. Oberhauser. 2002. Effects of Photoperiod, Temperature, and Host Plant Age on Induction of Reproductive Diapause and Development Time in *Danaus plexippus.*" *Ecological Entomology* 27:674–685.

Herman, W. S., and M. Tatar. 2001. "Juvenile Hormone Regulation of Longevity in the Migratory Monarch Butterfly." *Proceedings of the Royal Society B* 268:2509–2514.

Proximate Causes and Navigation
Mouritsen, H., and B. J. Frost. 2002. "Virtual Migration in Tethered Flying Monarch Butterflies Reveals Their Orientation Mechanisms." *Proceedings of the National Academy of Sciences USA* 99:10162–10166.
Perez, S. M., O. R. Taylor, and R. Jander. 1997. "A Sun Compass in Monarch Butterflies." *Nature* 387:29.

How Do Monarchs Tell Time?
Pennisi, E. 2003. "Monarchs Check Clock to Chart Migration Route." *Science* 300:1216–1217.
Potvin, C. L., and J. Bovet. 1975. "Annual Cycle of Patterns of Activity Rhythms in Beaver Colonies (*Castor canadensis*)." *Journal of Comparative Physiology* 98:243–256.
Reppert, S. M., R. J. Gegear, and C. Merlin. 2010. "Navigational Mechanisms of Migrating Monarch Butterflies." *Trends in Neurosciences* 33:399–406.
Stensmyr, M. C., and B. S. Hansson. 2011. "A Genome Befitting a Monarch." *Cell* 147:970–972.

Ultimate Causes of Migration by Monarch Butterflies
Altizer, S., R. Bartel, and B. A. Han. 2011. "Animal Migration and Infectious Disease Risk." *Science* 331:296–302.
Altizer, S. M., K. S. Oberhauser, and L. P. Brower. 2000. "Associations between Host Migration and the Prevalence of a Protozoan Parasite in Natural Populations of Adult Monarch Butterflies." *Ecological Entomology* 25:125–139.
Bradley, C. A., and S. Altizer. 2005. "Parasites Hinder Monarch Butterfly Fight: Implications for Disease Spread in Migratory Hosts." *Ecology Letters* 8:290–300.
Ffrench-Constant, R. H. 2014. "Genomics: Of monarchs and migration." *Nature* 514:314–315.

Broadening Our Perspective on Proximate and Ultimate Causation
Austad, S. N. 1997. *Why We Age: What Science Is Discovering about the Body's Journey through Life.* New York: Wiley.
Locke, R. 2006. "The Oldest Bat." *BATS Magazine* 24(2). http://www.batcon.org/index.php/media-and-info/bats-archives.html?task=viewArticle&magArticleID=164 (accessed November 8, 2013).

Questions to Ponder
Estes, R. D. 1976. "The Significance of Breeding Synchrony in the Wildebeest." *East African Wildlife Journal* 14:135–152.
Sinclair, A. R. E., S. A. R. Mduma, and P. Arcese. 2000. "What Determines Phenology and Synchrony of Ungulate Breeding in the Serengeti?" *Ecology* 81:2100–2111.
Thompson, K. V. 1995. "Flehmen and Birth Synchrony among Female Sable Antelope, *Hippotragus niger*." *Animal Behaviour* 50:475–484.

CHAPTER 2. OBSERVATIONS AS EVIDENCE
Introduction
National Academy of Sciences. 2008. *Science, Evolution, and Creationism.* Washington, DC: National Academies Press.
University of California Museum of Paleontology. 2013. "Understanding Science." http://undsci.berkeley.edu/ (accessed November 11, 2013).

Two Case Studies of Observations as Evidence
Brown, D., and E. Pianin. 2005, 29 April. "Extinct? After 60 Years, Woodpecker Begs to Differ." *Washington Post.*
Collinson, J. M. 2007. "Video Analysis of the Escape Flight of Pileated Woodpecker *Dryocopus pileatus*: Does the Ivory-Billed Woodpecker *Campephilus principalis* Persist in Continental North America?" *BMC Biology* 5:8.

Dalton, R. 2005. "A Wing and a Prayer." *Nature* 437:188–190.

Dalton, R. 2010. "Still Looking for That Woodpecker." *Nature* 463:718–719.

Fitzpatrick, J. W., M. Lammertink, M. D. Luneau, Jr., T. W. Gallagher, B. R. Harrison, G. M. Sparling, K. V. Rosenberg, et al. 2005. "Ivory-Billed Woodpecker (*Campephilus principalis*) Persists in Continental North America." *Science* 308:1460–1462.

Fitzpatrick, J. W., M. Lammertink, M. D. Luneau, Jr., T. W. Gallagher, and K. V. Rosenberg. 2006. "Response to Comment on 'Ivory-Billed Woodpecker (*Campephilus principalis*) Persists in Continental North America.'" *Science* 311:1555b.

Jackson, J. A. 2002. "Ivory-Billed Woodpecker (*Campephilus principalis*)." The Birds of North America Online.

Jackson, J. A. 2006. "Ivory-Billed Woodpecker (*Campephilus principalis*): Hope, and the Interfaces of Science, Conservation, and Politics." *The Auk* 123:1–15.

Moriarty, K. M., W. J. Zielinski, A. G. Gonzales, T. E. Dawson, K. M. Boatner, C. A. Wilson, F. V. Schlexer, K. L. Pilgrim, J. P. Copeland, and M. K. Schwartz. 2009. "Wolverine Confirmation in California after Nearly a Century: Native or Long-Distance Immigrant?" *Northwest Science* 83:154–162.

Sibley, D. A., L. R. Bevier, M. A. Patten, and C. S. Elphick. 2006. "Comment on 'Ivory-Billed Woodpecker (*Campephilus principalis*) Persists in Continental North America.'" *Science* 311:1555a.

Observations in Court

Davis, D., and E. F. Loftus. 2012. "The Dangers of Eyewitnesses for the Innocent: Learning from the Past and Projecting into the Age of Social Media." *New England Law Review* 46:769–809.

Gawande, A. 2001, January 8. "Under Suspicion: The Fugitive Science of Criminal Justice." *The New Yorker*, 50–53.

Gross, S. R., and M. Shaffer. 2012. "Exonerations in the United States, 1989–2012: Report by the National Registry of Exonerations." National Registry of Exonerations.

Nickerson, R. S. 1998. "Confirmation Bias: A Ubiquitous Phenomenon in Many Guises." *Review of General Psychology* 2:175–220.

Steblay, N. K., J. E. Dysart, and G. L. Wells. 2011. "Seventy-Two Tests of the Sequential Lineup Superiority Effect: A Meta-Analysis and Policy Discussion." *Psychology, Public Policy, and Law* 17:99–139.

Wise, R. A., M. A. Safer, and C. M. Maro. 2011. "What U.S. Law Enforcement Officers Know and Believe about Eyewitness Factors, Eyewitness Interviews and Identification Procedures." *Applied Cognitive Psychology* 25:488–500.

Confirmation Bias in Normal Science

Dubos, R. J. 1998. *Pasteur and Modern Science*. Washington, DC: ASM Press.

Farley, J. 1977. *The Spontaneous Generation Controversy from Descartes to Oparin*. Baltimore: Johns Hopkins University Press.

Polya, G. 1954. *Mathematics and Plausible Reasoning*, Volume I. Princeton, NJ: Princeton University Press.

Observations as Evidence in Medicine

Allchin, D. 2003. "Scientific Myth-Conceptions." *Science Education* 87:329–351.

Aschwanden, C. 2010, April 20. "Convincing the Public to Accept New Medical Guidelines." *Pacific Standard*. http://www.psmag.com/health/convincing-the-public-to-accept-new-medical-guidelines-11422/ (accessed November 13, 2013).

McAnulty, S. R., J. T. Owens, L. S. McAnulty, D. C. Nieman, J. D. Morrow, C. L. Dumke, and G. L. Milne. 2007. "Ibuprofen Use during Extreme Exercise." *Medicine & Science in Sports & Exercise* 39:1075–1079.

Newman, T. B. 2003. "The Power of Stories over Statistics." *British Medical Journal* 327:1424–1427.

Nieman, D. C., D. A. Henson, C. L. Dumke, K. Oley, S. R. McAnulty, J. M. Davis, E. A. Murphy, et al. 2006. "Ibuprofen Use, Endotoxemia, Inflammation, and Plasma Cytokines during Ultramarathon Competition." *Brain, Behavior, and Immunity* 20:578–584.

Small, D. H., and R. Cappai. 2006. "Alois Alzheimer and Alzheimer's Disease: A Centennial Perspective." *Journal of Neurochemistry* 99:708–710.

Wikipedia contributors. 2013, November 9. "Alexander Fleming." Wikimedia Foundation, Inc.

CHAPTER 3. FROM OBSERVATIONS TO DATA
Introduction
Guevara-Aguirre, J., P. Balasubramanian, M. Guevara-Aguirre, M. Wei, F. Madia, C.-W. Cheng, D. Hwang, et al. 2011. "Growth Hormone Receptor Deficiency Is Associated with a Major Reduction in Pro-Aging Signaling, Cancer, and Diabetes in Humans." *Science Translational Medicine* 3:70ra13.

Guinness World Records Ltd. 2012. Guinness World Records. http://www.guinnessworldrecords.com/ (accessed August 21, 2012).

Laron, Z., and J. Kopchick, eds. 2011. *Laron Syndrome—From Man to Mouse. Lessons from Clinical and Experimental Experience.* New York: Springer.

Tenenbaum, D. J. 2011, February 17. "Dwarf Gene Cuts Both Ways." The Why? Files: Science Behind the News. http://whyfiles.org/2011/genetic-solution to cancer diabetes/ (accessed February 17, 2011)

Wade, N. 2011, February 16. "Ecuadorean Villagers May Hold Secret to Longevity." *New York Times.*

Describing and Analyzing Human Heights and Weights
Reed, M. P. 2013. "Anthropometric Data." http://mreed.umtri.umich.edu/mreed/downloads.html (accessed November 18, 2013).

Sizes of Animals
Burness, G. P., J. Diamond, and T. Flannery. 2001. "Dinosaurs, Dragons, and Dwarfs: The Evolution of Maximal Body Size." *Proceedings of the National Academy of Sciences USA* 98:14518–14523.

Goldbogen, J. A. 2010. "The Ultimate Mouthful: Lunge-Feeding in Rorqual Whales." *American Scientist* 98:124–131.

Jones, K. E., J. Bielby, M. Cardillo, S. A. Fritz, J. O'Dell, C. D. L. Orme, K. Safi, et al. 2009. "PanTHERIA: A Species-Level Database of Life History, Ecology, and Geography of Extant and Recently Extinct Mammals." *Ecology* 90:2648.

Rittmeyer, E. N., A. Allison, M. C. Gründler, D. K. Thompson, and C. C. Austin. 2012. "Ecological Guild Evolution and the Discovery of the World's Smallest Vertebrate." *PLoS One* 7:e29797.

Ruxton, G. D. 2011. "Why Are Whales Big?" *Nature* 469:481.

Schmidt-Nielsen, K. 1984. *Scaling, Why Is Animal Size So Important?* Cambridge: Cambridge University Press.

Secor, S. M., and J. Diamond. 1997. "Determinants of the Postfeeding Metabolic Response of Burmese Pythons, *Python molurus.*" *Physiological Zoology* 70:202–212.

Sieg, A. E., M. P. O'Connor, J. N. McNair, B. W. Grant, S. J. Agosta, and A. E. Dunham. 2009. "Mammalian Metabolic Allometry: Do Intraspecific Variation, Phylogeny, and Regression Models Matter?" *American Naturalist* 174:720–733.

CHAPTER 4. EXPERIMENTS: THE GOLD STANDARD FOR RESEARCH
Two Experimental Studies of Medicinal Use of Marijuana
Abrams, D. I., C. A. Jay, S. B. Shade, H. Vizoso, H. Reda, S. Press, M. E. Kelly, M. C. Rowbotham, and K. L. Petersen. 2007. "Cannabis in Painful HIV-Associated Sensory Neuropathy: A Randomized Placebo-Controlled Trial." *Neurology* 68:515–521.

Dawkins, R. 2011. "The Double-Blind Control Experiment." *Edge World Question Center: What Scientific Concept Would Improve Everybody's Cognitive Toolkit?* http://edge.org/q2011/q11_17.html (accessed November 5, 2014).

Elphick, M. R., and M. Egertová. 2001. "The Neurobiology and Evolution of Cannabinoid Signalling." *Philosophical Transactions of the Royal Society B* 356:381–408.

Grant, I., J. H. Atkinson, A. Mattison, and T. J. Coates. 2010. *Center for Medicinal Cannabis Research: Report to the Legislature and Governor of the State of California Presenting Findings pursuant to SB847 Which Created the CMCR and Provided State Funding.* San Diego: University of California San Diego.

Hecht, P. 2012, July 12. "California Pot Research Backs Therapeutic Claims." *Sacramento Bee.*

Jenkins, M. 2011, August 8. "Ganjanomics: Bringing Humboldt's Shadow Economy into the Light." *High Country News.*

Joy, J. E., S. J. Watson, Jr., and J. A. Benson, Jr. (eds.). 1999. *Marijuana and Medicine: Assessing the Science Base*. Washington, DC: National Academies Press.

Kupperschmidt, K. 2013, November 21. "Painkillers May Curb Memory Loss from Medical Marijuana." *Science*. http://news.sciencemag.org/brain-behavior/2013/11/painkillers-may-curb-memory-loss-medical-marijuana.

National Institute of Drug Abuse. 2012, July. "Is Marijuana Medicine?" http://www.drugabuse.gov/publications/drugfacts/marijuana-medicine.

Russo, E. B. 2011. "Taming THC: Potential Cannabis Synergy and Phytocannabinoid-Terpenoid Entourage Effects." *British Journal of Pharmacology* 163:1344–1364.

Simon, S. D. 2006. *Statistical Evidence in Medical Trials: What Do the Data Really Tell Us?* Oxford: Oxford University Press.

Tewksbury, J. J., and G. P. Nabhan. 2001. "Seed Dispersal—Directed Deterrence by Capsaicin in Chillies." *Nature* 412:403–404.

Di Tomaso, E., M. Beltramo, and D. Piomeli. 1996. "Brain Cannabinoids in Chocolate." *Nature* 382:677–678.

Wallace, M., G. Schulteis, J. H. Atkinson, T. Wolfson, D. Lazzaretto, H. Bentley, B. Gouaux, and I. Abramson. 2007. "Dose-Dependent Effects of Smoked Cannabis on Capsaicin-Induced Pain and Hyperalgesia in Healthy Volunteers." *Anesthesiology* 107:785–796.

How Is Sex Determined? Observational and Experimental Evidence

Austad, S. N., and M. E. Sunquist. 1986. "Sex-Ratio Manipulation in the Common Opossum." *Nature* 324:58–60.

Cameron, E. Z. 2004. "Facultative Adjustment of Mammalian Sex Ratios in Support of the Trivers-Willard Hypothesis: Evidence for a Mechanism." *Proceedings of the Royal Society B* 271:1723–1728.

Cameron, E. Z., P. R. Lemons, P. W. Bateman, and N. C. Bennett. 2008. "Experimental Alteration of Litter Sex Ratios in a Mammal." *Proceedings of the Royal Society B* 275:323–327.

Cameron, E. Z., and W. L. Linklater. 2007. "Extreme Sex Ratio Variation in Relation to Change in Condition around Conception." *Biology Letters* 3:395–397.

Cameron, E. Z., W. L. Linklater, K. J. Stafford, and C. J. Veltman. 1999. "Birth Sex Ratios Relate to Mare Condition at Conception in Kaimanawa Horses." *Behavioral Ecology* 10:472–475.

Hardy, I. C. W. 2010. "Sex Allocation, Sex Ratios, and Reproduction." *Encyclopedia of Animal Behavior* 3:146–151.

Henneke, D. R., G. D. Potter, J. L. Kreider, and B. F. Yeates. 1983. "Relationship between Condition Score, Physical Measurements and Body Fat Percentage in Mares." *Equine Veterinary Journal* 15:371–372.

Le Boeuf, B. J. 1974. "Male-Male Competition and Reproductive Success in Elephant Seals." *American Zoologist* 14:163–176.

Sunquist, M. E., and J. F. Eisenberg. 1993. "Reproductive Strategies of Female *Didelphis*." *Bulletin of the Florida Museum of Natural History, Biological Sciences* 36:109–140.

Trivers, R. L., and D. E. Willard. 1973. "Natural Selection of Parental Ability to Vary the Sex Ratio of Offspring." *Science* 179:90–92.

Wilson, K., and I. C. W. Hardy. 2002. "Statistical Analysis of Sex Ratios: An Introduction," in *Sex Ratios: Concepts and Research Methods*, ed. I. C. W. Hardy. Cambridge: Cambridge University Press, 48–92.

Questions to Ponder and Resources for Further Exploration

Cameron, E. Z., and F. Dalerum. 2009. "A Trivers-Willard Effect in Contemporary Humans: Male-Biased Sex Ratios among Billionaires." *PLoS One* 4:e4195.

Gelman, A., and D. Weakliem. 2009. "Of Beauty, Sex and Power." *American Scientist* 97:310–316.

Kean, S. 2012. "Reinventing the Pill: Male Birth Control." *Science* 338:318–320.

Miller, A. S., and S. Kanazawa. 2008. *Why Beautiful People Have More Daughters: From Dating, Shopping, and Praying to Going to War and Becoming a Billionaire*. New York: Perigee.

Wilcox, A. J., and D. D. Baird. 2011. "Natural versus Unnatural Sex Ratios—A Quandary of Modern Times." *American Journal of Epidemiology* 174:1332–1334.

CHAPTER 5. CORRELATIONS, COMPARISONS, AND CAUSATION

Storks and Babies

Höfer, T., H. Przyrembel, and S. Verleger. 2004. "New Evidence for the Theory of the Stork." *Paediatric and Perinatal Epidemiology* 18:88–92.

Matthews, R. 2000. "Storks Deliver Babies ($p = 0.008$)." *Teaching Statistics* 22:36–38.

Sies, H. 1988. "A New Parameter for Sex Education." *Nature* 332:495.

Cell Phones and Brain Cancer: A Correlational Study

Boniol, M., J. F. Doré, and P. Boyle. 2011. "Re. Lehrer S, Green S, Stock RG (2011) Association between Number of Cell Phone Contracts and Brain Tumor Incidence in Nineteen U.S. States. J Neurooncol 101:505–507." *Journal of Neurooncology* 103:433–434.

Lehrer, S. 2011. "Response to Boniol et al." *Journal of Neurooncology* 105:435.

Lehrer, S., S. Green, and R. G. Stock. 2011. "Association between Number of Cell Phone Contracts and Brain Tumor Incidence in Nineteen U.S. States." *Journal of Neurooncology* 101:505–507.

Mukherjee, S. 2011, April 13. "Do Cellphones Cause Brain Cancer?" *New York Times.*

Cell Phones and Brain Cancer: A Comparative Study

Ahlbom, A., and M. Feychting. 2011. "Mobile Telephones and Brain Tumours." *BMJ* 343:d6605–d6605.

Elinder, M., and O. Erixson. 2012. "Gender, Social Norms, and Survival in Maritime Disasters." *Proceedings of the National Academy of Sciences USA* 109:13220–13224.

Parker-Pope, T. 2011, June 6. "Piercing the Fog around Cellphones and Cancer." *New York Times.*

Repacholi, M. H., A. Lerchl, M. Röösli, Z. Sienkiewicz, A. Auvinen, J. Breckenkamp, G. d'Inzeo, et al. 2012. "Systematic Review of Wireless Phone Use and Brain Cancer and Other Head Tumors." *Bioelectromagnetics* 33:187–206.

Rothman, K. J. 2009. "Health Effects of Mobile Telephones." *Epidemiology* 20:653–655.

Saracci, R., and J. Samet. 2010. "Commentary: Call Me on My Mobile Phone . . . Or Better Not?—A Look at the INTERPHONE Study Results." *International Journal of Epidemiology* 39:695–698.

Swerdlow, A. J., M. Feychting, A. C. Green, L. Kheifets, D. A. Savitz, and International Commission for Non-Ionizing Radiation Protection Standing Committee on Epidemiology. 2011. "Mobile Phones, Brain Tumors, and the Interphone Study: Where Are We Now?" *Environmental Health Perspectives* 119:1534–1538.

The INTERPHONE Study Group. 2010. "Brain Tumour Risk in Relation to Mobile Telephone Use: Results of the INTERPHONE International Case-Control Study." *International Journal of Epidemiology* 39:675–694.

Trotter, L. 2012, April 2. "Are Cell Phones a Possible Carcinogen? An Update on the IARC Report." http://www.sciencebasedmedicine.org/index.php/are-cell-phones-a-possible-carcinogen-an-update-on-the-iarc-report/ (accessed December 3, 2013).

Lead and Crime Rates

Cecil, K. M., C. J. Brubaker, C. M. Adler, K. N. Dietrich, M. Altaye, J. C. Egelhoff, S. Wessel, et al. 2008. "Decreased Brain Volume in Adults with Childhood Lead Exposure." *PLoS Medicine* 5:e112.

Drum, K. 2013, February. "America's Real Criminal Element: Lead." *Mother Jones.*

Gilbert, S. G., and B. Wiess. 2006. "A Rationale for Lowering the Blood Lead Action Level from 10 to 2 μg/dL." *Neurotoxicology* 27:693–701.

Mielke, H. W., and S. Zahran. 2012. "The Urban Rise and Fall of Air Lead (Pb) and the Latent Surge and Retreat of Societal Violence." *Environment International* 43:48–55.

Nevin, R. 2000. "How Lead Exposure Relates to Temporal Changes in IQ, Violent Crime, and Unwed Pregnancy." *Environmental Research* 83:1–22.

Nevin, R. 2007. "Understanding International Crime Trends: The Legacy of Preschool Lead Exposure." *Environmental Research* 104:315–336.

Reyes, J. W. 2007, May. "Environmental Policy as Social Policy? The Impact of Childhood Lead Exposure on Crime." Working Paper, National Bureau of Economic Research. http://www.nber.org/papers/w13097 (accessed December 3, 2013).

Wright, J. P., K. N. Dietrich, M. D. Ris, R. W. Hornung, S. D. Wessel, B. P. Lanphear, M. Ho, and M. N. Rae. 2008. "Association of Prenatal and Childhood Blood Lead Concentrations with Criminal Arrests in Early Adulthood." *PLoS Medicine* 5:e101.

CHAPTER 6. THE DIVERSE USES OF MODELS IN BIOLOGY
Introduction
Stevenson, R. L. 1930. *Treasure Island*. New York: George H. Doran Company. (first published in 1883)

Two Concrete Scale Models
Giere, R. N., J. Bickle, and R. F. Mauldin. 2006. *Understanding Scientific Reasoning*, 5th ed. Belmont, CA: Thomson Higher Education.

Watson, J. D. 1980. *The Double Helix: A Personal Account of the Discovery of the Structure of DNA*, edited by G. S. Stent. New York: Norton.

Watson, J. D., and F. H. C. Crick. 1953. "A Structure for Deoxyribose Nucleic Acid." *Nature* 171:737–738.

Weisberg, M. 2013. *Simulation and Similarity: Using Models to Understand the World*. New York: Oxford University Press.

A Mathematical Model of the Spread of Disease
Allen, A. 2013. "The Pertussis Paradox." *Science* 341:454–455.

Gross, L. 2009. "A Broken Trust: Lessons from the Vaccine–Autism Wars." *PLoS Biology* 7:e1000114.

Jansen, V. A. A., N. Stollenwerk, H. J. Jensen, M. E. Ramsay, W. J. Edmunds, and C. J. Rhodes. 2003. "Measles Outbreaks in a Population with Declining Vaccine Uptake." *Science* 301:804.

May, R. M. 1983. "Parasitic Infections as Regulators of Animal Populations." *American Scientist* 71:36–45.

Morens, D. M., G. K. Folker, and A. S. Fauci. 2004. "The Challenge of Emerging and Re-Emerging Infectious Diseases." *Nature* 430:242–249.

Quammen, D. 2012. *Spillover: Animal Infections and the Next Human Pandemic*. New York: Norton.

Rho, H. 2013, February 21. "What's the Matter with Vermont?" *Slate* http://www.slate.com/articles/health_and_science/medical_examiner/2013/02/pertussis_epidemic_how_vermont_s_anti_vaxxer_activists_stopped_a_vaccine.html (accessed December 6, 2013).

A Different Kind of Mathematical Model of Predators and Prey
Caltagirone, L. E., and R. L. Doutt. 1989. "The History of the Vedalia Beetle Importation to California and Its Impact on the Development of Biological Control." *Annual Review of Entomology* 34:1–16.

Carson, R. 1962. *Silent Spring*. Boston: Houghton Mifflin.

DeBach, P. 1991. *Biological Control by Natural Enemies*, 2nd ed. Cambridge: Cambridge University Press.

Gotelli, N. J. 2001. *A Primer of Ecology*, 3rd ed. Sunderland, MA: Sinauer Associates.

Weisberg, M., and K. Reisman. 2008. "The Robust Volterra Principle." *Philosophy of Science* 75:106–131.

CHAPTER 7. GENES, ENVIRONMENTS, AND THE COMPLEXITY OF CAUSATION
The Problem of Nature and Nurture and the Complexity of Causation
Pinker, S. 2002. *The Blank Slate: The Modern Denial of Human Nature*. New York: Viking.

Pinker, S. 2004. "Why Nature & Nurture Won't Go Away." *Daedalus* 133(4):5–17.

Turkheimer, E. 2013. "The Nature-Nurture Question," in *Noba Textbook Series: Psychology*, eds. R. Diener-Biswas and E. Diener. http://nobaproject.com/chapters/the-nature-nurture-question (accessed December 20, 2013).

Traits with a "Simple" Genetic Basis
Kuehn, B. M. 2013. "After 50 Years, Newborn Screening Continues to Yield Public Health Gains." *JAMA* 309:1215–1217.

Learning from Twins
Loughry, W. J., and C. M. McDonough. 2002. "Phenotypic Variability within and between Litters of Nine-Banded Armadillos." *Southeastern Naturalist* 1:287–298.

Monozygotic Twins Reared Apart
Miller, P. 2012. "A Thing or Two about Twins." *National Geographic* 221(1):38–65.
Segal, N. L. 2013. "Twins: The Finest Natural Experiment." *Personality and Individual Differences* 49:317–323.

Correlations of Height between Twins; A Model for Using Twins to Study Nature and Nurture
Bouchard, T. J., D. T. Lykken, M. McGue, N. L. Segal, and A. Tellegen. 1990. "Sources of Human Psychological Differences: The Minnesota Study of Twins Reared Apart." *Science* 250:223–228.
Eaves, L. J., A. C. Heath, N. G. Martin, M. C. Neale, J. M. Meyer, J. L. Silberg, L. A. Corey, K. Truett, and E. Walters. 1999. "Biological and Cultural Inheritance of Stature and Attitudes," in *Personality and Psychopathology*, ed. C. R. Cloninger. Washington, DC: American Psychiatric Press, 269–308.
Evans, D. M., and N. G. Martin. 2000. "The Validity of Twin Studies." *GenoScreen* 1:77–79.
Komlos, J., and B. E. Lauderdale. 2007. "The Mysterious Trend in American Heights in the 20th Century." *Annals of Human Biology* 34:206–215.
Silventoinen, K., S. Sammalisto, M. Perola, D. I. Boomsma, B. K. Cornes, C. Davis, L. Dunkel, et al. 2003. "Heritability of Adult Body Height: A Comparative Study of Twin Cohorts in Eight Countries." *Twin Research* 6:399–408.

More Examples of Heritability
Fowler, J. H., C. T. Dawes, and N. A. Christakis. 2009. "Model of Genetic Variation in Human Social Networks." *Proceedings of the National Academy of Sciences USA* 106:1720–1724.
Jackson, M. O. 2009. "Genetic Influences on Social Network Characteristics." *Proceedings of the National Academy of Sciences USA* 106:1687–1688.
Klass, P. 2012, December 10. "Understanding How Children Develop Empathy." *New York Times*.

How Heritable Is IQ?
Devlin, B., M. Daniels, and K. Roeder. 1997. "The Heritability of IQ." *Nature* 388:468–471.
Herrnstein, R. J., and C. A. Murray. 1994. *The Bell Curve: Intelligence and Class Structure in American Life.* New York: Free Press.
Kirp, D. L. 2006, July 23. "After the Bell Curve." *New York Times*.
McGue, M. 1987. "The Democracy of the Genes." *Nature* 388:417–418.
Zimmer, C. 2008. "The Search for Intelligence." *Scientific American* 299(4):68–75.

Some Limitations of Heritability
Fowler, J. H., and D. Schreiber. 2008. "Biology, Politics, and the Emerging Science of Human Nature." *Science* 322:912–914.
Keller, E. F. 2010. *The Mirage of a Space between Nature and Nurture.* Durham, NC: Duke University Press.

Nature, Nurture, and the Complexity of Causation
Clausen, J., D. D. Keck, and W. M. Hiesey. 1948. *Experimental Studies on the Nature of Species. III. Environmental Responses of Climatic Races of* Achillea. Carnegie Institution of Washington Publication No. 581, Washington, DC.
Frayling, T. M., M. J. Timpson, M. N. Weedon, E. Zeggini, R. M. Freathy, C. M. Lindgren, J. R. B. Perry, et al. 2007. "A Common Variant in the *FTO* Gene Is Associated with Body Mass Index and Predisposes to Childhood and Adult Obesity." *Science* 316:889–894.
Kilpeläinen, T., L. Qi, S. Brage, S. J. Sharp, E. Sonestedt, E. Demeranth, T. Ahmad, et al. 2011. "Physical Activity Attenuates the Influence of *FTO* Variants on Obesity Risk: A Meta-Analysis of 218,166 Adults and 19,268 Children." *PLoS Medicine* 8:e1001116.
Rampersaud, E., B. D. Mitchell, T. I. Pollin, M. Fu, H. Shen, J. R. O'Connell, J. L. Ducharme, et al. 2008. "Physical Activity and the Association of Common *FTO* Variants with Body Mass Index and Obesity." *Archives of Internal Medicine* 168:1791–1797.

Schuett, W., S. R. X. Dall, A. J. Wilson, and N. J. Royle. 2013. "Environmental Transmission of a Personality Trait: Foster Parent Exploration Behaviour Predicts Offspring Exploration Behaviour in Zebra Finches." *Biology Letters* 9:20130120.

Turkheimer, E., A. Haley, M. Waldron, B. D'Onofrio, and I. I. Gottesman. 2003. "Socioeconomic Status Modifies Heritability of IQ in Young Children." *Psychological Science* 14:623–628.

Yang, J. 2012. "*FTO* Genotype Is Associated with Phenotypic Variability of Body Mass Index." *Nature* 490:267–273.

CHAPTER 8. FROM CAUSES TO CONSEQUENCES: CONSIDERING THE WEIGHT OF EVIDENCE
Introduction, Hurricane Katrina, Challenges in Dissecting Webs of Causation

Freedman, A. 2013, July 8. "Hurricanes Likely to Get Stronger & More Frequent: Study." http://www.climatecentral.org/news/study-projects-more-frequent-and-stronger-hurricanes-worldwide-16204 (accessed January 3, 2014).

Krieger, N. 1994. "Epidemiology and the Web of Causation: Has Anyone Seen the Spider?" *Social Science and Medicine* 39:887–903.

Nuccitelli, D. 2013, April 29. "New Research Shows Humans Causing More Strong Hurricanes." http://www.skepticalscience.com/grinsted-hurricane-stronger.html (accessed January 3, 2014).

Piltz, R. 2013, June 9. "Proposed Ending of Federal Gray Wolf Protection: Another Case of Setting Science Aside?" http://www.climatesciencewatch.org/2013/06/09/proposed-ending-federal-gray-wolf-protection/ (accessed January 3, 2014).

Keystone Species in Food Webs—Analyzing Complex Causation in Ecology

Foster, J. 2013, August 29. "How Much Would Sea Otters Fetch on the Carbon Market?" http://thinkprogress.org/climate/2013/08/29/2553471/sea-otters-carbon-market/ (accessed January 3, 2014).

Paine, R. T. 1969. "A Note on Trophic Complexity and Community Stability." *American Naturalist* 103:91–93.

Paine, R. T. 1974. "Intertidal Community Structure: Experimental Studies on the Relationship between a Dominant Competitor and Its Principal Predator." *Oecologia* 15:93–120.

Power, M. E., D. Tilman, J. A. Estes, B. A. Menge, W. J. Bond, L. S. Mills, G. Daily, J. C. Castilla, J. Lubchenco, and R. T. Paine. 1996. "Challenges in the Quest for Keystones." *Bioscience* 46:609–620.

Is the Sea Otter a Keystone Species? What Does the Weight of Evidence Suggest?

Duggins, D. O. 1980. "Kelp Beds and Sea Otters: An Experimental Approach." *Ecology* 61:447–453.

Estes, J. A., and D. O. Duggins. 1995. "Sea Otters and Kelp Forests in Alaska: Generality and Variation in a Community Ecological Paradigm." *Ecological Monographs* 65:75–100.

Estes, J. A., and C. Harrold. 1988. "Sea Otters, Sea Urchins, and Kelp Beds: Some Questions of Scale," in *The Community Ecology of Sea Otters*, eds. G. R. VanBlaricom and J. A. Estes. Berlin: Springer-Verlag, 116–150.

Estes, J. A., and J. F. Palmisano. 1974. "Sea Otters: Their Role in Structuring Nearshore Communities." *Science* 185:1058–1060.

Foster, M. S., and D. R. Schiel. 1988. "Kelp Communities and Sea Otters: Keystone Species or Just Another Brick in the Wall?," in *The Community Ecology of Sea Otters*, eds. G. R. VanBlaricom and J. A. Estes. Berlin: Springer-Verlag, 92–115.

Perkins, S. 2013, 12 July. "What Role Do Beavers Play in Climate Change?" *Science*. http://news.sciencemag.org/2013/07/what-role-do-beavers-play-climate-change.

Riedman, M. L., and J. A. Estes. 1990. *The Sea Otter* (Enhydra lutris): *Behavior, Ecology, and Natural History*. Washington, DC: Fish and Wildlife Service, US Department of the Interior.

Stolzenburg, W. 2008. *Where the Wild Things Were: Life, Death, and Ecological Wreckage in a Land of Vanishing Predators*. New York: Bloomsbury.

Wilmers, C. C., J. A. Estes, M. Edwards, K. L. Laidre, and B. Konar. 2012. "Do Trophic Cascades Affect the Storage and Flux of Atmospheric Carbon? An Analysis of Sea Otters and Kelp Forests." *Frontiers in Ecology and the Environment* 10:409–415.

The Plot Thickens for Sea Otters

Doroff, A. M., J. A. Estes, M. T. Tinker, D. M. Burn, and T. J. Evans. 2003. "Sea Otter Population Declines in the Aleutian Archipelago." *Journal of Mammalogy* 84:55–64.

Estes, J. A., M. T. Tinker, T. M. Williams, D. Doak, J. L. Bodkin, K. L. Laidre, W. Jarman, D. Monson, S. L. Reese, and B. B. Hatfield. 2011. "Where Have All the Otters Gone? Reply to Kuker and Barrett-Lennard (2010)." Unpublished manuscript.

Estes, J. A., M. T. Tinker, T. M. Williams, and D. F. Doak. 1998. "Killer Whale Predation on Sea Otters Linking Oceanic and Nearshore Ecosystems." *Science* 282:473–476.

Kuker, K., and L. Barrett-Lennard. 2010. "A Re-Evaluation of the Role of Killer Whales *Orcinus orca* in a Population Decline of Sea Otters *Enhydra lutris* in the Aleutian Islands and a Review of Alternative Hypotheses." *Mammal Review* 40:103–124.

Schrope, M. 2007. "Killer in the Kelp." *Nature* 445:703–705.

Williams, T. M., J. A. Estes, D. F. Doak, and A. M. Springer. 2004. "Killer Appetites: Assessing the Role of Predators in Ecological Communities." *Ecology* 85:3373–3384.

The Stage Expands for Killer Whales

Dalton, R. 2005. "Is This Any Way to Save a Species?" *Nature* 436:14–16.

DeMaster, D. P., A. W. Trites, P. Clapham, S. Mizroch, P. Wade, R. J. Small, and J. Ver Hoef. 2006. "The Sequential Megafaunal Collapse Hypothesis: Testing with Existing Data." *Progress in Oceanography* 68:329–342.

Paine, R. T., D. W. Bromley, M. A. Castellini, L. B. Crowder, J. A. Estes, J. M. Grebmeier, F. M. D. Gulland, et al. 2002. *The Decline of the Steller Sea Lion in Alaskan Waters: Untangling Food Webs and Fishing Nets.* Washington, DC: National Academies Press.

Springer, A. M., J. A. Estes, G. B. van Vliet, T. M. Williams, D. F. Doak, E. M. Danner, K. A. Forney, and B. Pfister. 2003. "Sequential Megafaunal Collapse in the North Pacific Ocean: An Ongoing Legacy of Industrial Whaling?" *Proceedings of the National Academy of Sciences USA* 100:12223–12228.

Trites, A. W., V. B. Deecke, E. J. Gregr, J. K. B. Ford, and P. F. Olesiuk. 2007. "Killer Whales, Whaling, and Sequential Megafaunal Collapse in the North Pacific: A Comparative Analysis of the Dynamics of Marine Mammals in Alaska and British Columbia Following Commercial Whaling." *Marine Mammal Science* 23:751–765.

Wade, P. R., V. N. Burkanov, M. E. Dahlheim, N. E. Friday, L. W. Fritz, T. R. Loughlin, S. A. Mizroch, et al. 2007. "Killer Whales and Marine Mammal Trends in the North Pacific—A Re-Examination of Evidence for Sequential Megafauna Collapse and the Prey-Switching Hypothesis." *Marine Mammal Science* 23:766–802.

Wade, P. R., J. M. Ver Hoef, and D. P. DeMaster. 2009. "Mammal-Eating Killer Whales and Their Prey-Trend Data for Pinnipeds and Sea Otters in the North Pacific Ocean Do Not Support the Sequential Megafaunal Collapse Hypothesis." *Marine Mammal Science* 25:737–747.

Wolves and Elk in Yellowstone—Another Trophic Cascade?

Beschta, R. L., and W. J. Ripple. 2013. "Are Wolves Saving Yellowstone's Aspen? A Landscape-Level Test of a Behaviorally Mediated Trophic Cascade: Comment." *Ecology* 94:1420–1425.

Kauffman, M. J., J. F. Brodie, and E. S. Jules. 2010. "Are Wolves Saving Yellowstone's Aspen? A Landscape-Level Test of a Behaviorally Mediated Trophic Cascade." *Ecology* 91:2742–2755.

Kauffman, M. J., J. F. Brodie, and E. S. Jules. 2013. "Are Wolves Saving Yellowstone's Aspen? A Landscape-Level Test of a Behaviorally Mediated Trophic Cascade: Reply." *Ecology* 94:1425–1441.

Leopold, A. 1968. *A Sand County Almanac, and Sketches Here and There.* New York: Oxford University Press.

Ripple, W. J., and R. L. Beschta. 2005. "Linking Wolves and Plants: Aldo Leopold on Trophic Cascades." *Bioscience* 55:613–621.

Ripple, W. J., and R. L. Beschta. 2012. "Trophic Cascades in Yellowstone: The First 15 Years after Wolf Reintroduction." *Biological Conservation* 145:205–213.

CHAPTER 9. SCIENCE AS A SOCIAL PROCESS

Introduction; One Researcher, Two Examples of the Everyday Practice of Science

Grinnell, F. 2009. *Everyday Practice of Science: Where Intuition and Passion Meet Objectivity and Logic.* New York: Oxford University Press.

Jenkins, S. H. 1980. "A Size-Distance Relation in Food Selection by Beavers." *Ecology* 61:740–746.

Krebs, J. R., J. T. Erichsen, M. I. Webber, and E. L. Charnov. 1977. "Optimal Prey Selection in the Great Tit (*Parus major*)." *Animal Behaviour* 25:30–38.

Longland, W. S., S. H. Jenkins, S. B. Vander Wall, J. A. Veech, and S. Pyare. 2001. "Seedling Recruitment in *Oryzopsis hymenoides*: Are Desert Granivores Mutualists or Predators?" *Ecology* 82:3131–3148.

Schoener, T. W. 1979. "Generality of the Size-Distance Relation in Models of Optimal Feeding." *American Naturalist* 114:902–914.

Schoener, T. W. 1987. "A Brief History of Optimal Foraging Ecology," in *Foraging Behavior*, eds. A. C. Kamil, J. R. Krebs, and J. R. Pulliam. New York: Plenum Press, 5–67.

Science Is a Social Process

Committee on the Conduct of Science, National Academy of Sciences. 1989. "On Being a Scientist." *Proceedings of the National Academy of Sciences USA* 86:9053–9074.

Kata, A. 2010. "A Postmodern Pandora's Box: Anti-Vaccination Misinformation on the Internet." *Vaccine* 28:1709–1716.

National Research Council. 2009. *On Being a Scientist: A Guide to Responsible Conduct in Research*, 3rd ed. Washington, DC: National Academies Press.

Oreskes, N. 1996. "Objectivity or Heroism? On the Invisibility of Women in Science." *Osiris* 11:87–113.

Peer Review

Mayr, E. 1982. *The Growth of Biological Thought: Diversity, Evolution, and Inheritance*. Cambridge, MA: Belknap Press.

Peer Review and Controversial Ideas

Springer, A. M., J. A. Estes, G. B. van Vliet, T. M. Williams, D. F. Doak, E. M. Danner, K. A. Forney, and B. Pfister. 2003. "Sequential Megafaunal Collapse in the North Pacific Ocean: An Ongoing Legacy of Industrial Whaling?" *Proceedings of the National Academy of Sciences USA* 100:12223–12228.

Stolzenburg, W. 2008. *Where the Wild Things Were: Life, Death, and Ecological Wreckage in a Land of Vanishing Predators*. New York: Bloomsbury.

Peer Review and the Origin of AIDS

Hahn, B. H., G. M. Shaw, K. M. De Cock, and P. M. Sharp. 2000. "AIDS as a Zoonosis: Scientific and Public Health Implications." *Science* 287:607–614.

Hooper, E. 1999. *The River: A Journey to the Source of HIV and AIDS*. Boston: Little, Brown and Company.

Martin, B. 1993. "Peer Review and the Origin of AIDS: A Case Study in Rejected Ideas." *Bioscience* 43:624–627.

Martin, B. 2010. "How to Attack a Scientific Theory and Get Away with It (Usually): The Attempt to Destroy an Origin-of-AIDS Hypothesis." *Science as Culture* 19:215–239.

Moore, J. 2004. "The Puzzling Origins of AIDS." *American Scientist* 92:540–547.

Quammen, D. 2012. *Spillover: Animal Infections and the Next Human Pandemic*. New York: W. W. Norton & Company.

Zimmer, C. 2009, July 22. "AIDS and the Virtues of Slow-Cooked Science." http://blogs.discovermagazine.com/loom/2009/07/22/aids-and-the-virtues-of-slow-cooked-science/#.Us3ND_sliRM (accessed January 8, 2014).

Correction of Faulty Results after Publication

Alter, H. J., J. A. Mikovits, W. M. Switzer, F. W. Ruscetti, S.-C. Lo, N. Klimas, A. L. Komaroff, et al. 2012. "A Multicenter Blinded Analysis Indicates No Association between Chronic Fatigue Syndrome/Myalgic Encephalomyelitis and Either Xenotropic Murine Leukemia Virus-Related Virus or Polytropic Murine Leukemia Virus." *mBio* 3:e00266–12.

Callaway, E. 2011. "Fighting for a Cause." *Nature* 471:282–285.

Callaway, E. 2012, 18 September. "The Scientist Who Put the Nail in XMRV's Coffin." *Nature*. http://www.nature.com/news/the-scientist-who-put-the-nail-in-xmrv-s-coffin-1.11444.

Cohen, J., and M. Enserink. 2011. "False Positive." *Science* 333:1694–1701.

Enserink, M. 2012, September 18. Final study confirms: virus not implicated in chronic fatigue syndrome. *Science*. http://news.sciencemag.org/2012/09/final-study-confirms-virus-not-implicated-chronic-fatigue-syndrome?ref=hp.

Johnson, H. 2013, March 7. "Chasing the Shadow Virus: Chronic Fatigue Syndrome and XMRV." *Discover Magazine*. http://discovermagazine.com/2013/march/17-shadow-virus#.Us3Pj_sliRO.

Lombardi, V. C., F. W. Ruscetti, J. Das Gupta, M. A. Pfost, K. S. Hagen, D. L. Peterson, S. K. Ruscetti, et al. 2009. "Detection of an Infectious Retrovirus, XMRV, in Blood Cells of Patients with Chronic Fatigue Syndrome." *Science* 326:585–589.

Rehmeyer, J. 2011, December 2. "Stolen Notebooks and a Biochemist in Chains." *Slate*. http://www.slate.com/articles/health_and_science/medical_examiner/2011/12/judy_mikovits_in_prison_what_does_it_mean_for_research_on_chronic_fatigue_syndrome_.single.html#pagebreak_anchor_2 (accessed January 8, 2014).

Zimmer, C. 2011, May 31. "The Chronic Fatigue Virus: De-Discovered?" http://phenomena.nationalgeographic.com/2011/05/31/the-chronic-fatigue-virus-de-discovered/ (accessed January 8, 2014).

Zimmer, C. 2012, September 18. "The Slow, Slow Road to De-Discovery." http://phenomena.nationalgeographic.com/2012/09/18/the-slow-slow-road-to-de-discovery/ (accessed January 8, 2014).

Conclusions

Chamberlin, T. C. 1965. "The Method of Multiple Working Hypotheses. *Science* 148:754–759. (reprinted from Science [1890] old series 15:92–96)

CHAPTER 10. CRITICAL THINKING ABOUT CLIMATE CHANGE
FiLCHeRS—An Introduction to Six Tools for Critical Thinking

Cook, J., D. Nuccitelli, S. A. Green, M. Richardson, B. Winkler, R. Painting, R. Way, et al. 2013. "Quantifying the Consensus on Anthropogenic Global Warming in the Scientific Literature." *Environmental Research Letters* 8:024024.

Lett, J. 1990. "A Field Guide to Critical Thinking." *Skeptical Inquirer* 14(4):153–160.

Mann, M. E. 2012. *The Hockey Stick and the Climate Wars: Dispatches from the Front Lines*. New York: Columbia University Press.

Oreskes, N. 2004. "The Scientific Consensus on Climate Change." *Science* 306:1686.

Falsifiability

Cook, J. 2010, March 29. "The Human Fingerprint in Global Warming." Skeptical Science. http://www.skepticalscience.com/human-fingerprint-in-global-warming.html (accessed January 14, 2014).

Hardin, G. 1976. "Vulnerability—The Strength of Science." *American Biology Teacher* 38:465, 483.

Milton, J., and R. Wiseman. 1999. "Does Psi Exist? Lack of Replication of an Anomalous Process of Information Transfer." *Psychological Bulletin* 125:387–391.

Schmidt, G. 2013, February 7. 2012. "Updates to Model-Data Comparisons." RealClimate. http://www.realclimate.org/index.php/archives/2013/02/2012-updates-to-model-observation-comparions/ (accessed January 14, 2014).

Wayne, G. P. 2013, August 1. "Empirical Evidence That Humans Are Causing Global Warming." Skeptical Science. http://www.skepticalscience.com/empirical-evidence-for-global-warming.htm (accessed January 14, 2014).

Wylie, C. R., Jr. 1948. "Paradox." *Scientific Monthly* 67:63.

Logic

Mann, M. E., R. S. Bradley, and M. K. Hughes. 1999. "Northern Hemisphere Temperatures during the Past Millennium: Inferences, Uncertainties, and Limitations." *Geophysical Research Letters* 26:759–762.

Mooney, C. 2013, May 9. "The Most Controversial Chart in History, Explained." Climate Desk. http://climatedesk.org/2013/05/the-most-controversial-chart-in-history-explained/ (accessed January 14, 2014).

Comprehensiveness and Honesty

Gillis, J. 2013, June 10. "What to Make of a Warming Plateau." *New York Times*. http://www.nytimes. com/2013/06/11/science/earth/what-to-make-of-a-climate-change-plateau.html.

Sheppard, N. 2013, June 10. "*New York Times* Shocker: Global Warming Plateaued Last 15 Years Despite Rapid CO2 Rise." NewsBusters. http://newsbusters.org/blogs/noel-sheppard/2013/06/10/new-york-times-shocker-global-warming-plateaued-last-15-years-despite (accessed January 14, 2014).

Replicability

PAGES 2k Consortium. 2013. "Continental-Scale Temperature Variability during the Past Two Millennia." *Nature Geoscience* 6:339–346.

Rudolph, J. C. 2010, September 23. "The 'Hockey Stick' Lives." *New York Times*. http://green.blogs.nytimes. com/2010/09/23/the-hockey-stick-lives/.

Sufficiency

Hansen, J., D. Johnson, A. Lacis, S. Lebedeff, P. Lee, D. Rind, and G. Russell. 1981. "Climate Impact of Increasing Atmospheric Carbon Dioxide." *Science* 213:957–966.

Oreskes, N., and E. M. Conway. 2010. *Merchants of Doubt: How a Handful of Scientists Obscured the Truth on Issues from Tobacco Smoke to Global Warming*. New York: Bloomsbury Press.

Principles of Climate Change: Inertia

Archer, D., M. Eby, V. Brovkin, A. Ridgwell, L. Cao, U. Mikolajewicz, K. Caldeira, et al. 2009. "Atmospheric Lifetime of Fossil Fuel Carbon Dioxide." *Annual Review of Earth and Planetary Sciences* 37:117–134.

Hansen, J., M. Sato, R. Ruedy, P. Kharecha, A. Lacis, R. Miller, L. Nazarenko, et al. 2007. "Dangerous Human-Made Interference with Climate: A GISS ModelE Study." *Atmospheric Chemistry and Physics* 7:2287–2312.

Principles of Climate Change: Feedback

Frank, J. 2 September 2010. "Explaining How the Water Vapor Greenhouse Effect Works." Skeptical Science. http://www.skepticalscience.com/water-vapor-greenhouse-gas.htm (accessed January 14, 2014).

Kolbert, E. 2006. Field Notes from a Catastrophe: Man, Nature, and Climate Change. New York: Bloomsbury Press.

Mackey, B., I. C. Prentice, W. Steffen, J. I. House, D. Lindenmayer, H. Keith, and S. Berry. 2013. "Untangling the Confusion around Land Carbon Science and Climate Change Mitigation Policy." *Nature Climate Change* 3:552–557.

Schmidt, G. 2005, April 6. "Water Vapour: Feedback or Forcing?" RealClimate. http://www.realclimate.org/ index.php/archives/2005/04/water-vapour-feedback-or-forcing/ (accessed January 14, 2014).

PRINCIPLES OF CLIMATE CHANGE: TIPPING POINTS

Hansen, J. 2008. "Tipping Point: Perspective of a Climatologist," in *State of the Wild 2008–2009: A Global Portrait of Wildlife, Wildlands, and Oceans*, ed. E. Fearn. Washington, DC: Island Press, 7–15.

Lemonick, M. D. 2008, September. "Global Warming: Beyond the Tipping Point." *Scientific American* 18(4):60–67.

Lenton, T. M., H. Held, E. Kriegler, J. W. Hall, W. Lucht, S. Rahmstorf, and H. J. Schellnhuber. 2008. "Tipping Elements in the Earth's Climate System." *Proceedings of the National Academy of Sciences USA* 105:1786–1793.

Pitman, N. C. A., J. W. Terborgh, M. R. Silman, P. Nunez, D. A. Neill, C. E. Ceron, W. A. Palacios, and M. Aulestia. 2002. "A Comparison of Tree Species Diversity in Two Upper Amazonian Forests." *Ecology* 83:3210–3224.

Schmidt, G. 2006, July 5. "Runaway Tipping Points of No Return." RealClimate. http://www.realclimate.org/ index.php/archives/2006/07/runaway-tipping-points-of-no-return/ (*accessed October 9, 2014).

Shepherd, A., E. R. Ivins, A. Geruo, V. R. Barletta, M. J. Bentley, S. Bettadpur, K. H. Briggs, D. H. Bromwich, et al. 2012. "A Reconciled Estimate of Ice-Sheet Mass Balance." *Science* 338:1183–1189.

Wight, J. 2012, August 2. "Is Greenland Close to a Climate Tipping Point?" Skeptical Science. http://www.skepticalscience.com/is-greenland-close-to-a-climate-tipping-point.html (accessed January 14, 2014).

CONCLUSIONS: A PROBLEM AT THE INTERSECTION OF SCIENCE, POLITICS, ECONOMICS, AND ETHICS

Gardiner, S. M. 2011. *A Perfect Moral Storm: The Ethical Tragedy of Climate Change*. New York: Oxford University Press.

APPENDIX 2. HOW DOES EVOLUTION WORK?

Bloom, G., and P. W. Sherman. 2005. "Dairying Barriers Affect the Distribution of Lactose Malabsorption." *Evolution and Human Behavior* 26:301–312.

Cook, L. M., B. S. Grant, I. J. Saccheri, and J. Mallet. 2012. "Selective Bird Predation on the Peppered Moth: The Last Experiment of Michael Majerus." *Biology Letters* 8:609–612.

Dobzhansky, T. 1973. "Nothing in Biology Makes Sense Except in the Light of Evolution." *American Biology Teacher* 35:125–129.

Kettlewell, H. B. D. 1959. "Darwin's Missing Evidence." *Scientific American* 200(3):48–53.

Majerus, M. E. N. 2009. "Industrial Melanism in the Peppered Moth, *Biston betularia*: An Excellent Teaching Example of Darwinian Evolution in Action." *Evolution: Education and Outreach* 2:63–74.

Rudge, D. W. 2005. "The Beauty of Kettlewell's Classic Experimental Demonstration of Natural Selection." *Bioscience* 55:369–375.

APPENDIX 3. SENSORY WORLDS OF HUMANS AND OTHER ANIMALS

Von Frisch, K. 1971. *Bees: Their Vision, Chemical Senses, and Language Revised*. Ithaca, NY: Cornell University Press.

APPENDIX 4. GLOBAL CLIMATE CHANGE—HOW CAN AMATEURS COMPREHEND COMPLEX MODELS?

Houghton, J. T., et al. 2001. *Climate Change 2001: The Scientific Basis. Contribution of Working Group I to the Third Assessment Report of the Intergovernmental Panel on Climate Change*. Cambridge: Cambridge University Press.

Pope, V. 2007, February 2. "Models 'Key to Climate Forecasts.'" BBC News. http://news.bbc.co.uk/2/hi/science/nature/6320515.stm (accessed December 6, 2013).

Schmidt, G. 2009, October. "Wrong but Useful." *Physics World* 22(10):33–35.

APPENDIX 5. THE MYSTERY OF MISSING HERITABILITY

Flint, J., and M. Munafò. 2013. "Herit-Ability." *Science* 340:1416–1417.

Turkheimer, E. 2000. "Three Laws of Behavior Genetics and What They Mean." *Current Directions in Psychological Science* 9:160–164.

Visscher, P. M., W. G. Hill, and N. R. Wray. 2008. "Heritability in the Genomics Era—Concepts and Misconceptions." *Nature Reviews Genetics* 9:255–266.

Visscher, P. M., J. Yang, and M. E. Goddard. 2010. "A Commentary on 'Common SNPs Explain a Large Proportion of the Heritability for Human Height' by Yang et al. (2010)." *Twin Research and Human Genetics* 13:517–524.

Weedon, M. N., and T. M. Frayling. 2008. "Reaching New Heights: Insights into the Genetics of Human Stature." *Trends in Genetics* 24:595–603.

Yang, J., B. Benyamin, B. P. McEvoy, S. Gordon, A. K. Henders, D. R. Nyholt, P. A. Madden, et al. 2010. "Common SNPs Explain a Large Proportion of the Heritability for Human Height." *Nature Genetics* 42:565–569.

Zuk, O., E. Hechter, S. R. Sunyaev, and E. S. Lander. 2012. "The Mystery of Missing Heritability: Genetic Interactions Create Phantom Heritability." *Proceedings of the National Academy of Sciences USA* 109:1193–1198.

APPENDIX 6. MAPPING ARGUMENTS TO AID CRITICAL THINKING ABOUT THE WEIGHT OF EVIDENCE

Fallows, J. 2007, June. "What Was I Thinking?" *Atlantic*, 131–133.

Van Gelder, T. 2005. "Teaching Critical Thinking: Some Lessons from Cognitive Science." *College Teaching* 53:41–46.

APPENDIX 7. WHAT ARE THE BENEFITS OF ORGANIC FARMING? THE WEIGHT OF EVIDENCE

Barański, M., D. Średnicka-Tober, N. Volakakis, C. Seal, R. Sanderson, G. B. Stewart, C. Benbrook, et al. 2014. "Higher Antioxidant and Lower Cadmium Concentrations and Lower Incidence of Pesticide Residues in Organically Grown Crops: A Systematic Literature Review and Meta-Analyses." *British Journal of Nutrition* 112:794–811.

Benbrook, C. 2012, September 4. "The Devil in the Details." http://csanr.wsu.edu/devil-in-the-details/ (accessed January 8, 2014).

Bittman, M. 2012, October 2. "That Flawed Stanford Study." *New York Times*. http://opinionator.blogs. nytimes.com/2012/10/02/that-flawed-stanford-study/?_r=0.

Brandt, K., C. Leifert, R. Sanderson, and C. J. Seal. 2011. "Agroecosystem Management and Nutritional Quality of Plant Foods: The Case of Organic Fruits and Vegetables." *Critical Reviews in Plant Sciences* 30:177–197.

Chang, K. 2012, October 15. "Parsing of Data Led to Mixed Messages on Organic Food's Value." *New York Times*. http://www.nytimes.com/2012/10/16/science/stanford-organic-food-study-and-vagaries-of-met a-analyses.html.

Chang, K. 2013, December 9. "More Helpful Fatty Acids Found in Organic Milk." *New York Times*. http:// www.nytimes.com/2013/12/10/health/organic-milk-high-in-helpful-fatty-acids-study-finds.html?_r=0.

Jenkins, S. H. 2004. *How Science Works: Evaluating Evidence in Biology and Medicine*. New York: Oxford University Press.

Philpott, T. 2012, September 5. "5 Ways the Stanford Study Sells Organics Short." *Mother Jones*. http:// www.motherjones.com/tom-philpott/2012/09/five-ways-stanford-study-underestimates-organic-food (accessed January 8, 2014).

Smith-Spangler, C., M. L. Brandeau, G. E. Hunter, C. Bavinger, M. Pearson, P. J. Eschbach, V. Sundaram, et al. 2012. "Are Organic Foods Safer or Healthier than Conventional Alternatives?" *Annals of Internal Medicine* 157:348–366.

Andrews, P. K. 2013. "Are Organic Foods Safer or Healthier?" *Annals of Internal Medicine* 158:295–296.

Benbrook, C. 2013. "Are Organic Foods Safer or Healthier?" *Annals of Internal Medicine* 158:296–297.

Brandt, K. 2013. "Are Organic Foods Safer or Healthier?" *Annals of Internal Medicine* 158:295.

Davis, D. R. 2013. "Are Organic Foods Safer or Healthier?" *Annals of Internal Medicine* 158:297.

Davison, S. L. 2013. "Are Organic Foods Safer or Healthier?" *Annals of Internal Medicine* 158:296.

Smith-Spangler, C., M. L. Brandeau, I. Olkin, and D. M. Bravata. 2013. "Are Organic Foods Safer or Healthier than Conventional Alternatives?" *Annals of Internal Medicine* 158:297–300.

Thomas, K. 2013, June 29. "Breaking the Seal on Drug Research." *New York Times*.

APPENDIX 8. GLOBAL CLIMATE CHANGE—EVALUATING EXPERT OPINION

Anderegg, W. R. L. 2010. "Moving Beyond Scientific Agreement: An Editorial Comment on 'Climate Change: A Profile of US Climate Scientists' Perspectives.'" *Climatic Change* 101:331–337.

Anderegg, W. R. L., J. W. Prall, J. Harold, and S. H. Schneider. 2010. "Expert Credibility in Climate Change." *Proceedings of the National Academy of Sciences USA* 107:12107–12109.

Anderson, E. 2011. "Democracy, Public Policy, and Lay Assessments of Scientific Testimony." *Episteme* 8:144–164.

Craven, G. 2009. *What's the Worst That Could Happen? A Rational Response to the Climate Change Debate.* New York: Penguin Group.

Lefsrud, L. M., and R. E. Meyer. 2012. "Science or Science Fiction? Professionals' Discursive Construction of Climate Change." *Organization Studies* 33:1477–1506.

Rosenberg, S., A. Vedlitz, D. F. Cowman, and S. Zahran. 2009. "Climate Change: A Profile of US Climate Scientists' Perspectives." *Climatic Change* 101:311–329.

Credits

Figure 3.7   Data from "Mammalian Metabolic Allometry: Do Intraspecific Variation, Phylogeny, and Regression Models Matter?" by A. E. Sieg et al., 2009, *The American Naturalist* 174:720–733.

Figure 4.1   Modified from Figure 3 of "Cannabis in Painful HIV-Associated Sensory Neuropathy: A Randomized Placebo-Controlled Trial" by D. I. Abrams et al., *Neurology* 68:515–521, copyright © 2007 by Lippincott Williams & Wilkins.

Figures 4.2 and 4.3   Modified from Figure 2 of "Dose-Dependent Effects of Smoked Cannabis on Capsaicin-Induced Pain and Hyperalgesia in Healthy Volunteers" by M. Wallace et al., *Anesthesiology* 107:785–96, copyright © 2007 by Lippincott Williams & Wilkins.

Figure 4.4   Modified with permission from a figure drawn by Tanya Wolfson, University of California, San Diego.

Figure 4.5   Reprinted from the National Human Genome Research Institute of the US National Institutes of Health (http://commons.wikimedia.org/wiki/File:NHGRI_human_male_karyotype.png, accessed January 17, 2014).

Figure 4.7   Modified from Figure 1 of "Birth Sex Ratios Relate to Mare Condition at Conception in Kaimanawa Horses by E. Z. Cameron et al., *Behavioral Ecology* 10:472–475, copyright ©1999 by Oxford University Press.

Figure 4.8   Modified from Figure 1 of "Experimental Alteration of Litter Sex Ratios in a Mammal" by E. Z. Cameron et al., *Proceedings of the Royal Society of London* B 275:323–327, copyright © 2008 by Royal Society Publishing.

Figure 5.1   Data from "A New Parameter for Sex Education" by H. Sies, *Nature* 332:495, copyright © 1988 by the Nature Publishing Group.

Figure 5.2   Licensed by Randall Munroe under a Creative Commons Attribution-NonCommercial 2.5 License (http://xkcd.com/552/).

Figure 5.3   Modified from "A New Parameter for Sex Education" by H. Sies, *Nature* 332:495, copyright © 1988 by the Nature Publishing Group.

Figure 5.4   Modified from Figure 1 of "Association between Number of Cell Phone Contracts and Brain Tumor Incidence in Nineteen U.S. States" by S. Lehrer et al., *Journal of Neurooncology* 101:505–507, copyright © 2011 by Springer Publishing.

Figures 5.5–5.7   Data from "Association between Number of Cell Phone Contracts and Brain Tumor Incidence in Nineteen U.S. States" by S. Lehrer et al., *Journal of Neurooncology* 101:505–507, copyright © 2011 by Springer Publishing.

Figure 5.8   Data from Table 2 of "Brain Tumour Risk in Relation to Mobile Telephone Use: Results of the INTERPHONE International Case-Control Study" by the INTERPHONE Study Group, *International Journal of Epidemiology* 39:675–694, 2010.

Figure 5.9   Modified from Figure 1 of "Understanding International Crime Trends: The Legacy of Preschool Lead Exposure" by R. Nevin, *Environmental Research* 104:315–336, copyright © 2007 by Elsevier, with additional data on crime in recent years from Uniform Crime Reports of the US Federal Bureau of Investigation: "Crime in the

United States 2012" (http://www.fbi.gov/about-us/cjis/ucr/crime-in-the-u.s/2012/crime-in-the-u.s.-2012).

Figure 5.10 Modified from Figure 12 of "How Lead Exposure Relates to Temporal Changes in IQ, Violent Crime, and Unwed Pregnancy" by R. Nevin, *Environmental Research* 83:1–22, copyright © 2000 by Elsevier.

Figure 5.11 Modified from Figure 2 of "New Evidence for the Theory of the Stork" by T. Höfer et al., *Paediatric and Perinatal Epidemiology* 18:88–92, copyright © 2004 by John Wiley & Sons.

Figure 6.3 Reproduced with permission from the James D. Watson Collection, Cold Spring Harbor Laboratory Archives.

Figure 6.5 Modified with permission from "Tragic Choices: Autism, Measles, and the MMR Vaccine" by M. P. Rowe, copyright © 2011 by the National Center for Case Study Teaching in Science (http://sciencecases.lib.buffalo.edu/cs/). Data on measles from Public Health England (http://www.hpa.org.uk/web/HPAweb&HPAwebStandard/HPAweb_C/1195733833790, accessed December 7, 2013).

Figure 6.8 Photograph of vedalia beetles on cottony cushion scale provided by Florida Department of Agriculture and Consumer Services, Division of Plant Industry and used with their permission. History of cottony cushion scale in California modified from Figure 1.8 in *Biological Control by Natural Enemies* by P. Debach and M. Rosen, copyright © 1974 by Cambridge University Press, which in turn was redrawn from "The Integrated Control Concept" by V. M. Stern et al., *Hilgardia* 29:81–101, 1959.

Figures 7.1 and 7.2 Data provided by Judy Silberg, Director of the Mid-Atlantic Twin Registry, and used with her permission.

Figure 7.3 Modified from Figure 1 of "Environmental Transmission of a Personality Trait: Foster Parent Exploration Behaviour Predicts Offspring Exploration Behaviour in Zebra Finches" by W. Schuett et al., *Biology Letters* 9: 20130120, copyright © 2013 by Royal Society Publishing.

Figure 7.4 Data from "Experimental Studies on the Nature of Species. III. Environmental Responses of Climatic Races of *Achillea*" by J. Clausen, D. D. Keck, and W. M. Hiesey, 1948, Carnegie Institution of Washington Publication No. 581, Washington, DC.

Figure 7.6 Modified from Figure 2 of "The Mysterious Trend in American Heights in the 20th Century" by J. Komlos and B. E. Lauderdale, *Annals of Human Biology* 34: 206–215, copyright © 2007 by Informa Healthcare.

Figure 8.3 Modified from Keystone (architecture) page of Wikipedia (http://en.wikipedia.org/wiki/Keystone_%28architecture%29, accessed December 23, 2013), licensed under a Creative Commons Attribution-Share Alike license.

Figure 8.4 Data from Tables 2 and 3 of "Sea Otters and Kelp Forests in Alaska: Generality and Variation in a Community Ecological Paradigm" by J. A. Estes and D. O. Duggins, *Ecological Monographs* 65:75–100, copyright © 1995 by the Ecological Society of America.

Figure 8.5 Modified from Figure 3 of "Sea Otters, Sea Urchins, and Kelp Beds: Some Questions of Scale" by J. A. Estes and C. Harrold, pages 116–150 in *The Community*

Ecology of Sea Otters, edited by G. R. VanBlaricom and J. A. Estes, copyright © 1988 by Springer-Verlag.

Figure 8.6   Reprinted with permission from "Killer Appetites: Assessing the Role of Predators in Ecological Communities" by T. M. Williams et al., *Ecology* 85:3373–3384, copyright © 2004 by the Ecological Society of America.

Figure 8.7   Modified from Figure 2 of "Killer Whale Predation on Sea Otters Linking Oceanic and Nearshore Ecosystems" by J. A. Estes et al., *Science* 282:473–476, copyright © 1998 by the American Association for the Advancement of Science.

Figure 8.8   In addition to the data used for Figure 8.4, this uses data from Figure 1 of "Killer Whale Predation on Sea Otters Linking Oceanic and Nearshore Ecosystems" by J. A. Estes et al., *Science* 282:473–476, copyright © 1998 by the American Association for the Advancement of Science.

Figure 8.9   Modified from Figure 2 of "Sequential Megafaunal Collapse in the North Pacific Ocean: An Ongoing Legacy of Industrial Whaling?" by A. M. Springer et al., *Proceedings of the National Academy of Sciences* 100:12223–12228, copyright © 2003 by the National Academy of Sciences, USA.

Figure 10.1   Modified with permission from "2012 Updates to Model-Observation Comparisons" by G. Schmidt, RealClimate, February 7, 2013 (http://www.realclimate. org/index.php/archives/2013/02/2012-updates-to-model-observation-comparions/, accessed January 22, 2014).

Figure 10.2   Data from the National Snow and Ice Data Center as reported by Schmidt (see Figure 10.1 credit).

Figure 10.3   Modified with permission from Figure 1 of "The Human Fingerprint in Global Warming" by J. Cook, *Skeptical Science*, March 29, 2010 (http://www.skepticalscience.com/human-fingerprint-in-global-warming.html, accessed January 22, 2014). Ice core data from Figure SPM-10a of Intergovernmental Panel on Climate Change Third Assessment Report, Summary for Policymakers; Mauna Loa data and emissions data from Carbon Dioxide Information Analysis Center of the US Department of Energy.

Figure 10.5   Reprinted with permission from "Climate Impact of Increasing Atmospheric Carbon Dioxide" by J. Hansen et al., *Science* 213:957–966, copyright © 1981 by the American Association for the Advancement of Science.

Figure 10.6   Modified from Figure 9 of "Dangerous Human-Made Interference with Climate: A GISS ModelE Study" by J. Hansen et al., *Atmospheric Chemistry and Physics* 7:2287–2312, licensed under a Creative Commons Attribution-Share Alike license.

Figure 10.7   Data from Figure 3 of "Is Greenland Close to a Climate Tipping Point?" by J. Wight, *Skeptical Science*, August 2, 2012 (http://www.skepticalscience.com/ is-greenland-close-to-a-climate-tipping-point.html, accessed January 22, 2014).

Figure A2.1   Reprinted from *Ecological Genetics* by E. B. Ford, copyright © 1964 by Methuen & Co.

Plate 13   Photo a used with permission of Dave Cowles (http://rosario.wallawalla.edu/inverts); photos b to d licensed under Creative Commons Attribution-Share Alike licenses.

Plate 14   Photo a by the National Park Service, photo b by Chuck Kopczak, Curator of Ecology, California Science Center, and used with his permission.

Plate 15   Reprinted with permission from "Trophic Cascades in Yellowstone: The First 15 Years after Wolf Reintroduction" by W. J. Ripple and R. L. Beschta, *Biological Conservation* 145:205–213, copyright © 2012 by Elsevier.

Plate 16   Photos by R. Haley of the National Park Service (a), M. Lavin of Montana State University (c), and N. Dochtermann of North Dakota State University (d); diagram in b reprinted from *The Wise One* by F. Conibear and J. L. Blundell, copyright © 1949 by Frank Conibear and J. L. Blundell.

Plates 17 and 18   Reprinted with permission from Figure 1b and Figure 4, respectively, of *Climate Change 2001: The Scientific Basis. Contribution of Working Group I to the Third Assessment Report of the Intergovernmental Panel on Climate Change*, Cambridge University Press.

QUOTATIONS

Quotation on page 6 reprinted from "Discovering the Monarch's Mexican Haven" by Fred A. Urquhart, *National Geographic* 150(2), copyright © 1976 by Fred A. Urquhart.

Quotation on page 40 reprinted with permissions from "The Power of Stories over Statistics" by T. B. Newman, *British Medical Journal* 327:1424–1427, copyright © 2003 by the British Medical Journal.

Quotation on page 221 reprinted with permission from "The Chronic Fatigue Virus: De-Discovered?" by Carl Zimmer, *Phenomena: The Loom*, May 31, 2011 (http://phenomena.nationalgeographic.com/2011/05/31/the-chronic-fatigue-virus-de-discovered/).

Quotations on pages 227, 228, and 240 reprinted with permission from "A Field Guide to Critical Thinking" by James Lett, *Skeptical Inquirer* 14(4):153–160, copyright © 1990 by the Skeptical Inquirer.

Quotation on page 230 reprinted from "Paradox" by Clarence R. Wylie Jr., *Scientific Monthly* 67(1):63, 1948.

Index

DATE DUE

			PRINTED IN U.S.A.